Complementarity:
Applications, Algorithms and Extensions

Applied Optimization

Volume 50

Series Editors:

Panos M. Pardalos
University of Florida, U.S.A.

Donald Hearn
University of Florida, U.S.A.

The titles published in this series are listed at the end of this volume.

Complementarity: Applications, Algorithms and Extensions

Edited by

Michael C. Ferris

Computer Sciences Department,
University of Wisconsin,
Madison, Wisconsin, U.S.A.

Olvi L. Mangasarian

Computer Sciences Department,
University of Wisconsin,
Madison, Wisconsin, U.S.A.

and

Jong-Shi Pang

Department of Mathematical Sciences,
The Johns Hopkins University,
Baltimore, Maryland, U.S.A.

KLUWER ACADEMIC PUBLISHERS
DORDRECHT / BOSTON / LONDON

A C.I.P. Catalogue record for this book is available from the Library of Congress.

ISBN 0-7923-6816-9

Published by Kluwer Academic Publishers,
P.O. Box 17, 3300 AA Dordrecht, The Netherlands.

Sold and distributed in North, Central and South America
by Kluwer Academic Publishers,
101 Philip Drive, Norwell, MA 02061, U.S.A.

In all other countries, sold and distributed
by Kluwer Academic Publishers,
P.O. Box 322, 3300 AH Dordrecht, The Netherlands.

Printed on acid-free paper

Printed in the Netherlands.

Contents

Preface

This volume presents state-of-the-art complementarity applications, algorithms, extensions and theory in the form of eighteen papers. These invited papers were presented at the International Conference on Complementarity 99 (ICCP99) held in Madison, Wisconsin during June 9-12, 1999 with support from the National Science Foundation under Grant DMS-9970102.

Complementarity is becoming more widely used in a variety of application areas. In this volume, there are papers studying the impact of complementarity in such diverse fields as deregulation of electricity markets, engineering mechanics, optimal control and asset pricing. Furthermore, application of complementarity and optimization ideas to related problems in the burgeoning fields of machine learning and data mining are also covered in a series of three articles.

In order to effectively process the complementarity problems that arise in such applications, various algorithmic, theoretical and computational extensions are covered in this volume. Nonsmooth analysis has an important role to play in this area as can be seen from articles using these tools to develop Newton and path following methods for constrained nonlinear systems and complementarity problems. Convergence issues are covered in the context of active set methods, global algorithms for pseudomonotone variational inequalities, successive convex relaxation and proximal point algorithms.

Theoretical contributions to the connectedness of solution sets and constraint qualifications in the growing area of mathematical programs with equilibrium constraints are also presented. A relaxation approach is given for solving such problems. Finally, computational issues related to preprocessing mixed complementarity problems are addressed.

The use of complementarity within other problem types, such as bilevel optimization and data mining is considered. Positive semidefinite relaxations are a new and growing area of research that has significant ramifications for discrete optimization. These and other application specific extensions of the field are also covered in this volume.

It is hoped that the results presented here will inspire further contributions to the field of complementarity, especially in the form of novel applications and computational techniques with significant underlying theory.

MICHAEL FERRIS, OLVI MANGASARIAN AND JONG-SHI PANG

Chapter 1

APPROXIMATING MAXIMUM STABLE SET AND MINIMUM GRAPH COLORING PROBLEMS WITH THE POSITIVE SEMIDEFINITE RELAXATION

S. J. Benson

Division of Mathematics and Computer Science, Argonne National Laboratory, Argonne, IL 60439

benson@mcs.anl.gov

Y. Ye

Department of Management Sciences, The University of Iowa, Iowa City, Iowa 52242

yyye@yuan.biz.uiowa.edu

Abstract We compute approximate solutions to the maximum stable set problem and the minimum graph coloring problem using a positive semidefinite relaxation. The positive semidefinite programs are solved using an implementation of the dual scaling algorithm that takes advantage of the sparsity inherent in most graphs and the structure inherent in the problem formulation. From the solution to the relaxation, we apply a randomized algorithm to find approximate maximum stable sets and a modification of a popular heuristic to find graph colorings. We obtained high quality answers for graphs with over 1000 vertices and over 6000 edges.

Keywords: Stable Set, Independent Set, Maximum Clique, Graph Coloring, Positive Semidefinite Relaxation.

M.C. Ferris et al. (eds.), Complementarity: Applications, Algorithms and Extensions, 1–17.

1 INTRODUCTION

Given an undirected graph $G = (V, E)$, a *stable set* of vertices (or *vertex packing* or *independent set*) is a subset of V such that no two vertices are adjacent. The *Maximum Stable Set Problem* (MSS) asks for the stable set with the maximum cardinality. A *clique* of graph G is a subset set of vertices such that every pair of vertices is adjacent. A *vertex cover* is a subset of vertices that are incident to each edge in the graph. Denoting $\bar{G}$ as the graph complement of G, the following statements concerning any $S \subset V$ are known to be equivalent:

1. S is a stable set of G,
2. S is a clique of $\bar{G}$,
3. $V \setminus S$ is vertex cover of G

Accordingly, the problems of finding a maximum stable set of G, a maximum clique in $\bar{G}$, and a minimum vertex cover in G are equivalent.

A *vertex coloring* of a graph is an assignment of colors to the vertices V such that no two adjacent vertices receive the same color. Equivalently, the problem looks to partition the vertices into independent sets. The smallest number of colors needed for this coloring is called the *chromatic number* of G. A graph is *k-colorable* if it can be colored with k colors or less. Obviously, the cardinality of any clique in G is a lower bound on the chromatic number of G. When a graph, and every node induced subgraph, have a chromatic number that equals the cardinality of the largest clique, it is known as a *perfect graph.* For this special class of graphs, the MSS problem can be solved to optimality using a polynomial algorithm.

These problems are classical problems in combinatorial optimization and are well known to be NP-complete[19]. The MSS problem can be solved using polynomial time algorithms for special classes of graphs such as perfect graphs and t-perfect graphs, circle graphs and their complements, circular arc graphs and their complements, claw-free graphs, and graphs with long odd cycles[27], but the existence of a polynomial time algorithm for arbitrary graphs seems unlikely.

Various exact solution methods have been developed for these combinatorial optimization problems. An implicit enumeration technique of Carrahan and Pardalos[12], integer programming with branch and bound by Babel and Tinhofer[3][4], Balas, Xue, and Yu[6][7], Mannino and Sassano [27], and Nemhauser[30], integer programming with cutting planes by Balas [5], Nemhauser[31], and Nemhauser and Sigismondi [30], and a tabu search by Friden[17] have all been applied to the maximum stable set problem. There effectiveness, however, has usually been limited to graphs with less than 500 vertices. For the minimum graph coloring problem, implicit enumeration and branch and bound based methods of Kubale[25] [26] have been limited to very small instances, and a column generation approach based upon the stable set formulation by Mehrotra and Trick[28] has been applied to graphs with up to 600 vertices. Of course all of these algorithms have exponential complexity, so for larger graphs, the only option available is heuristic methods [34] [21] [22] [29], which have the cost of regularly suboptimal solutions.

Aside from its theoretical interest, the MSS problem arises in applications in information retrieval, experimental design, signal transmission, and computer vision[7]. Graph coloring arises when using finite differences to approximate sparse Hessian matrices, and well as applications in computer register allocation[11][14][13], timetable scheduling[9][15][43], and electronic bandwidth allocation[18]. In many of these applications, it suffices to find an approximately optimal solution. This fact and the difficulty of finding exact solutions, have encouraged considerable effort on finding good approximation algorithms.

2 POSITIVE SEMIDEFINITE RELAXATIONS

The standard form of a positive semidefinite program is:

$$\text{(SDP)} \qquad \begin{array}{ll} \text{Minimize} & C \bullet X \\ \text{Subject to} & A_i \bullet X = b_i, \quad i = 1, \ldots, m, \\ & X \in K \end{array}$$

where $K = K_1 \oplus K_2 \oplus \cdots \oplus K_r$ and K_l is the cone of $n_l \times n_l$ symmetric positive semidefinite matrices, $C, A_i \in \boldsymbol{R}^{n \times n}$ are symmetric, and $A \bullet C = \text{tr}\,(A^T C)$.

The dual of (SDP) can be written as:

$$\text{(DSP)} \qquad \begin{array}{ll} \text{Maximize} & b^T y \\ \text{Subject to} & \displaystyle\sum_{i=1}^{m} y_i A_i + S = C, \quad S \in K, \end{array}$$

where $y \in \boldsymbol{R}^m$.

There are some very strong connections between positive semidefinite programming and combinatorial optimization. The famous Lovász number, which provides an upper bound to the maximum stable set of a graph and a lower bound to its chromatic number, is the solution to a positive semidefinite program. Many more combinatorial problems have can be relaxed into a positive semidefinite program, and some of these relaxations offer rounding techniques that are guaranteed to be within a specified fraction of optimality.

Most linear programming relaxations do not offer a performance guarantee, but Geomans and Williamson[20], in a now classic result, applied the solution of a maximum cut positive semidefinite relaxation to a randomized algorithm and proved that the answers it generates have an expectation greater than 0.878 of optimality. Although the stable set problem cannot be approximated within a constant fraction in polynomial time unless $P = NP$, provably good approximation algorithms using a positive semidefinite relaxation have been found for MAX-SAT, MAX-2-SAT, MAX-3-SAT, MAX-4-SAT, MAX k-CUT[36], MAX-3-CSP, minimum bandwidth, graph bisection, bound constrained quadratic programming[32][45], graph coloring[23], and some scheduling problems.

Much like the formulation of Kleinberg and Goemans[24] the SDP relaxation of the MSS problem will assign each vertex an integer value of -1 or $+1$. One of the two sets will be a stable set. Given a graph G with $n-1$ vertices, our formulation, adds an artificial vertex v_n with no edges connecting it to other vertices. Since the artificial vertex is obviously a member of the maximal stable set of the new graph, it will used to identify the stable set and enforce the constraints of the problem. The MSS problem can be stated as:

$$\text{(MSS)}\qquad \begin{array}{ll} \text{Maximize} & \frac{1}{2}\left(\sum_{i=1}^{n-1} v_i^2 + v_n v_i\right) \\ \text{Subject to} & v \in \{-1, 1\}^n, \\ & |v_i + v_j + v_n| = 1 \text{ if } (v_i, v_j) \in \text{E} \end{array}$$

Denoting $e_{i,j,n} \in \boldsymbol{R}^n$ as the vector with zeros at all indices except i, j, and n, whose elements equal 1, the positive semidefinite relaxation of

MSS is

$$
\text{(MSSSDP)} \quad
\begin{array}{ll}
\text{Maximize} & \begin{pmatrix} .5 & & & .25 \\ & \ddots & & \vdots \\ & & .5 & .25 \\ .25 & \cdots & .25 & 0 \end{pmatrix} \bullet X \\
\text{Subject to} & diag(X) = e, \\
& (e_{i,j,n} e_{i,j,n}^T) \bullet X = 1 \qquad \forall (i,j) \in \text{E} \\
& X \succeq 0
\end{array}
$$

Imposing the additional constraint upon (MSSSDP) that the matrix X have rank one would make it equivalent to (MSS). Relaxing this constraint to include all symmetric positive semidefinite matrices makes the feasible region convex, and the solution to this problem provides an upper bound to the integer program (MSS). A randomized algorithm uses a solution of the relaxed problem, X^*, to identify stable sets. The randomized algorithm goes as follows:

1. Given a solution X^* to (MSSSDP), find a $V \in \boldsymbol{R}^{n \times n}$ such that $X^* = V^T V$.
2. Select a unit vector $u \in \boldsymbol{R}^n$ from the unit sphere and let $v = sign(V^T u)$.
3. For each $(i, j) \in \text{E}$, if $|v_i + v_j + v_n| \neq 1$, change the sign of either v_i or v_j.

The stable set will be the set of vertices with the same sign as v_n. For arbitrary graphs, the constraints corresponding to the edges of the graph will be satisfied with a frequency greater than 91% [10]. The third step of the randomized algorithm ensures that no edge connects vertices in the set by selectively removing vertices from the set. The choice of whether to switch vertices v_i or v_j may be arbitrary, but a better choice may be made by switching the vertex whose value is farthest from v_n: if $|v_i - v_n| > |v_j - v_n|$, change the sign of v_i, otherwise change the sign of v_j. This randomized algorithm can be applied multiple times to calculate multiple stable sets.

In the linear programming relaxation of the maximal stable set problem, utilizing larger cliques is crucial for a tight approximation to the convex hull of the integer program. These cliques can also improve the positive semidefinite relaxation. Given cliques $C^1, \ldots, C^d$, such that C^k

has n_k vertices, stable sets $v \in \{-1,1\}^n$ must satisfy

$$|(n_k - 1)v_n + \sum_{v_i \in C^k} v_i| = 1$$

for $k = 1, \ldots, d$. This formulation has a positive semidefinite relaxation that more closely approximates the convex hull of the integer program. This formulation has fewer constraints which can significantly reduce the time required to solve the positive semidefinite program.

To favor the inclusion of selected vertices into the stable set, the *weighted maximal stable set* problem has a similar formulation. Given a weights w_i on the vertices, this problem seeks to maximize

$$\frac{1}{2}\sum_{i=1}^{n-1} w_i \left(v_i^2 + v_n v_i\right)$$

subject to the same constraints as (MSS). These problems can also be addressed using the positive semidefinite relaxation.

For the graph coloring problem, instead of assigning colors or integers to the vertices of the graph, a unit vector $v_i \in \boldsymbol{R}^n$ is assigned to the each of the n vertices i in V. To capture the property of coloring, the vectors of adjacent vertices should different in a natural way. Using the definition of [23], the vector k- coloring of G is an assignment of unit vectors $v_i \in \boldsymbol{R}^n$ to each vertex i in V such that for any two adjacent vertices i and j, the dot product of the vectors satisfies the inequality $v_i^T v_j \leq -\frac{1}{k-1}$. In other words, the angle between the vectors of adjacent vertices must be sufficiently large. Define the matrix V such that column i is given by v_i and let $X = V^T V$. The matrix X is positive semidefinite and satisfies the inequalities $X_{ij} = X_{ji} \leq -\frac{1}{k-1}$ for each pair of adjacent edges (i,j). Obviously, any matrix is n-colorable, so the graph coloring problem can be posed as:

$$\text{(COLOR)} \quad \begin{array}{ll} \text{Minimize} & \text{rank}(X) \\ \text{Subject to} & diag(X) = e, \\ & X_{ij} \leq -\frac{1}{n-1} \text{ for } (i,j) \in \text{E} \\ & X \succeq 0 \end{array} \tag{1.1}$$

Ignoring the objective function, the problem is now a positive semidefinite program which seeks to find a feasible point. Heuristic algorithms can then be applied to the solution to color the graph.

Let $a_{ij} \in \boldsymbol{R}^n$ be a vector of zeros except indices i and j, whose elements equal one. A positive semidefinite relaxation of the graph k

coloring problem can be rewritten as:

$$
\text{(COLORSDP)} \quad \begin{array}{ll} \text{Minimize} & 0 \bullet X \\ \text{Subject to} & diag(X) = e, \\ & (a_{ij}a_{ij}^T) \bullet X \leq 2 - \frac{2}{k-1} \text{ if } (i,j) \in \text{E} \end{array} \tag{1.2}
$$

A solution X^* with rank less than or equal to k, identifies a legal k-coloring. The problem can be solved exactly. More generally, Karger, Motwani, Sudan propose a randomized algorithm that produces a k-semicoloring, an assignment of colors with relatively few adjacent vertices with the same color. We propose a heuristic procedure for to obtain a legal coloring, albeit with more than k colors if necessary.

Coloring Algorithm For $k = 1, \ldots$,

1. Let U^k be the uncolored vertices. If U^k is empty, terminate the algorithm.

2. Sort the vertices of U^k in decreasing order of degree in $G[U^k]$, the graph induced by the uncolored vertices, and let i be the vertex with highest degree.

3. Build a vertex set W^k by examining vertices $j \in U^k$ in the decreasing order of X_{ij}. Add j to W^k if it is not adjacent to any of the vertices in W^k.

4. Assign the vertices in W^k color k.

This algorithm is a modification of the algorithm proposed by [35]. In their algorithm, only step 3 is different. Instead of using the solution to the a positive semidefinite program, they examine the vertices in decreasing order of degree in $G[U^k]$. This algorithm remains one of the simplest and most popular, although other heuristics have been proposed and can be modified to include information inherent in the positive semidefinite program.

3 POSITIVE SEMIDEFINITE PROGRAMMING ALGORITHMS

There are actually several polynomial algorithms that can solve positive semidefinite programs. One is the primal-scaling algorithm (Nesterov and Nemirovskii [33], Alizadeh [1], Vandenberghe and Boyd [42], and Ye [44]), which is the analogue of the primal potential reduction algorithm for linear programming. This algorithm uses X to generate the iterate direction. Another is the dual-scaling algorithm (Vandenberghe and Boyd [42], Anstreicher and Fampa [2], and Ye [44]), which is the analogue of the dual-scaling algorithm for linear programming. The dual-scaling algorithm uses only S to generate the iterate direction. The third is the primal-dual scaling algorithm which uses both X and S to generate iterate directions, including Alizadeh-Haeberly-Overton, Helmberg-Rendl-Vanderbei-Wolkowicz/ Kojima-Shida-Hara/ Monteiro, Nesterov-Todd, Gu, and Toh directions, as well as directions called the MTW and Half directions (see Todd [39] and references therein). All these algorithms possess $O(\sqrt{n}\log(1/\epsilon))$ iteration complexity to yield accuracy ϵ.

The features of the positive semidefinite program should determine which algorithm and which implementation of the algorithm is most appropriate. In contrast to applications of SDP in control theory and truss topology design, positive semidefinite programs arising in combinatorial optimization typically have many variables, contain sparse low rank constraint matrices, and require relatively low precision solutions. Although rank one matrices reduce the complexity of interior point algorithms for positive semidefinite programming by a factor of n, not all implementations utilize this structure to reduce the complexity. Our implementation of the dual scaling algorithm explicitly accounts for these features[8]. Furthermore, the dual matrix S has a sparsity pattern like that of the graph's adjacency matrix. This sparsity offers the potential for savings in computation time and memory requirements, which the dual scaling algorithm can exploit better than primal dual algorithms. Although the rate of convergence of the dual algorithms is only linear, the relatively low precision required by combinatorial problems lessens the disadvantage of slower convergence.

One assumption for the convergence of the dual scaling algorithm is that the feasible primal region has a relative interior.

Theorem 1.1 *The positive semidefinite relaxation (MSSSDP) has a relative interior.*

Proof Define the vectors $v^1, v^2, \ldots, v^n$ by

$$v_i^j = \begin{cases} -1 & \text{if } i = j \text{ or } i = n+1 \\ 1 & \text{otherwise} \end{cases}$$

and

$$v_i^{n+1} = \begin{cases} -1 & \text{if } 1 \le i \le n \\ 1 & \text{if } i = n+1 \end{cases}$$

These vectors satisfy the constraints of (MSS) and the matrices $X^i = v^i(v^i)^T$ satisfy the positive semidefinite relaxation (MSSSDP). Let

$$X = \frac{1}{n+1}\sum_{i=1}^{n+1} X^i.$$

This matrix is a strict convex combination of symmetric rank one matrices and is therefore positive semidefinite. To prove it is positive definite, it suffices to show that $\{v^i : i \in \{1, 2, \ldots, n+1\}\}$ is linearly independent. Linear independence can be shown by evaluating the determinant of V^n, whose columns are the vectors v^i. Since $\det(V_n) = -(2)^n \neq 0$, the convex hull of the feasible solutions of the nonconvex optimization problems in $\boldsymbol{R}^n$ contains $n+1$ linearly independent vectors, so the SDP relaxation has a feasible solution that is positive definite. □

Theorem 1.2 *The feasible region of the n-coloring problem relaxation (COLORSDP) contains a positive definite matrix.*

Proof Let

$$X_{ij} = \begin{cases} 1 & \text{if } i = j \\ -\frac{1}{n+1} & \text{otherwise} \end{cases}$$

Then $X = -\frac{1}{n+1}ee^T + \frac{n+2}{n+1}I$. The matrix has one eigenvalue of $\frac{2}{n+1}$ and $n-1$ eigenvalue equal to $\frac{n+2}{n+1}$, which implies it is positive definite. □

Since the primal and dual problems of these SDP relaxations always have a feasible solution whose S and X part is positive definite, it follows that the primal and dual optimal values are attained and equal[37]. (Quite recently, Tuncel extended these theorems to the SDP relaxations of rather general nonconvex sets [41].)

4 COMPUTATIONAL RESULTS

In our computational experiments, we used a variety of previously tested graphs drawn from a large number of sources. For each of these graphs, we formulated the positive semidefinite relaxation of the integer combinatorial problem and solved the relaxation until a relative duality gap of 10^{-3} has been achieved.

For the maximum stable set problems, most of the graphs are taken from the 2nd DIMACS Challenge [16]. These graphs were contributed as test problems for solving the maximum clique problem. For these graphs, we took the complement of these graphs and applied our maximum stable set algorithm. The results are supplied in Table 1.1. A second set of test problems are examples of Mycielski graphs[40]. These graphs are interesting because they contain no cliques of size larger than 2. For these graphs, we expect our relaxation to be very tight. The results are also included in Table 1.1. A third set of graphs are line graphs created from other randomly generated graphs. Three line graphs were created from a graph with 100 vertices and 248 nonzero edges. Another three line graphs were created from graphs with 200 vertices and 597 nonzero edges. These line graphs are interesting because the SDP relaxation methods and successive relaxation methods perform very poorly for the maximum stable set problem in the worst case[38]. For these line graphs, lower bounds for the maximum stable set was calculated using the program "dfmax.c", also available from the DIMACS web site[16].

For each graph, we solved the positive semidefinite relaxation, without using cliques of size 3 or larger, and applied the randomized procedure for finding stable sets. Since the time required by the randomization procedure is very small relative to the time spent on solving the positive semidefinite program, we applied the randomized procedure n times on each problem. The data in Table 1.1 includes the number of vertices ($|V|$) and edges ($|E|$) in each graph, the upper bound provided by the semidefinite relaxation (SDP), the size of the maximum stable set (Optimal), and the size the the largest stable set found using our implementation of the algorithm (DSDP).

Of the 24 graphs, we solved (MSS) exactly 14 times. In 13 of those 14 instances, the positive semidefinite relaxation was extremely tight. These 13 instances include the five Mycielski graphs, which have no large cliques. This evidence demonstrates the importance of using large cliques when such knowledge is available. The ten instances in which DSDP did not find the optimal answer included all of the line graphs. Even in the line graphs, however, the SDP relaxation proved to be within about 10% of the optimal answer. The worst results were from the problem sanr200_0.7, whose SDP bound is 23.9, maximum stable set size is 18, and DSDP answer is 11. In most cases, however, the SDP relaxation was strong and our answers were good.

For the graph coloring problem, we used examples collected by Trick and Mehrotra [40]. For these problems, we formulated and solved the relaxed of the $n-$coloring problem (COLORSDP). From this solution, we applied the graph coloring heuristic to obtain one graph coloring. Table 1.2 shows the minimal number of colors used, the number of colors we used in DSDP, and the number of colors used by the heuristic [35].

Of these test problems, the optimal coloring is known for 34 of them. In 24 of these 34 problems, we correctly identified an optimal coloring of the graph. Although the heuristic also found an optimal coloring

Table 1.1 Maximum Stable Set Problems

Graph	$\|V\|$	$\|E\|$	SDP	Optimal	DSDP
hamming10-2	1024	5120	512.1	512	512
hamming6-2	64	192	32.0	32	32
hamming6-4	64	1312	5.35	4	4
hamming8-2	256	1024	128.0	128	128
johnson16-2-4	120	1600	8.0	8	8
johnson8-2-4	28	168	4.0	4	4
brock200_1	200	5066	27.5	21	14
brock200_3	200	7852	18.8	15	9
brock200_4	200	6811	21.3	17	9
keller4	172	5100	14.0	11	7
san200_0.9_1	200	1990	70.0	70	70
san200_0.9_2	200	1990	60.0	60	60
san200_0.9_3	200	1990	44.1	44	44
sanr200_0.7	200	6032	23.9	18	11
sanr200_0.9	200	2037	49.3	40	34
myciel3	11	20	5.0	5	5
myciel4	23	71	11.0	11	11
myciel5	47	236	23.0	23	23
myciel6	95	755	47.0	47	47
myciel7	191	2360	95.0	95	95
line1	248	1202	50.0	≥ 47	39
line2	248	1220	49.5	≥ 47	40
line3	248	1212	49.5	≥ 47	42
line4	597	3414	100.0	≥ 89	79
line5	597	3481	100.0	≥ 85	76
line6	597	3635	100.0	≥ 85	82

in many of these graphs, problem queen5.5, utilized the solution to the positive semidefinite program to find an optimal coloring which the heuristic could not do. In a total of four problems, the coloring obtained using the SDP relaxation was better than the coloring obtained by the heuristic, but in the three DSJC125 graphs, the coloring was actually worse.

For seven of the graphs in which we definitely did not compute the optimal coloring, we formulated a tighter formulation. Instead of using the n-color formulation, we used the k-color formulation where k is the minimal graph coloring. We solved these tighter relaxations and applied our heuristic to these solutions, hoping to identify a better coloring. The results are in Table 1.3. The number of colors required when using the tighter formulation is in the last column (DSDP2). In only one of the seven instances did the tighter formulation actually improve the coloring. On the other hand, there was one instance where the tighter formulation actually worsened the coloring of the graph. Hence, it seems sufficient to pose the n-coloring relaxation.

The time required to solve these problems ranged from less than a second for queen5.5 to over twelve hours to find the maximum stable set of brock200_1. The heuristic can find answers very quickly, but the positive semidefinite relaxation may offer improved answers. For other combinatorial problems, performance guarantees for algorithms using the positive semidefinite relaxation exist. The cost of these guarantees, however, is the significant additional cost in computation time and memory requirements. This contributes the growing mountain of evidence demonstrating the high quality of solutions that can be obtained from the semidefinite relaxation.

Acknowledgments

This work was supported by the Mathematical, Information, and Computational Sciences Division subprogram of the Office of Advanced Scientific Computing Research, U.S. Department of Energy, under Contract W-31-109-Eng-38.

References

[1] F. Alizadeh. *Combinatorial optimization with interior point methods and semidefinite matrices.* PhD thesis, University of Minnesota, Minneapolis, MN, 1991.

[2] K. M. Anstreicher and M. Fampa. A long-step path following algorithm for semidefinite programming problems. Working Paper, Department of Management Science, The University of Iowa, Iowa City, IA, 1996.

Table 1.2 Graph Coloring Problems

Graph	$\|V\|$	$\|E\|$	Optimal	DSDP	Heuristic
anna	138	493	11	11	11
david	87	406	11	11	11
homer	561	1629	13	13	13
huck	74	301	11	11	11
jean	80	254	10	10	10
games120	120	638	9	9	9
miles250	128	387	8	8	8
miles500	128	1170	20	20	20
miles750	128	2113	31	32	32
miles1000	128	3216	42	42	42
miles1500	128	5198	73	73	73
queen5.5	25	160	5	5	7
queen6.6	36	290	7	9	9
queen7.7	49	476	7	11	11
queen8.8	64	728	9	11	12
queen9.9	81	1055	10	13	13
queen10.10	100	1470	?	14	14
queen11.11	121	1980	11	15	15
queen12.12	144	2596	?	17	17
queen13.13	169	3328	13	18	18
queen14.14	196	4186	?	19	19
myciel3	11	20	4	4	4
myciel4	23	71	5	5	5
myciel5	47	236	6	6	6
myciel6	95	755	7	7	7
myciel7	191	2360	8	8	8
zeroin.i.1	211	4100	49	49	49
zeroin.i.2	211	3541	30	30	30
zeroin.i.3	206	3540	30	30	30
mulsol.i.1	197	3925	49	49	49
mulsol.i.2	188	3885	31	31	31
mulsol.i.3	184	3916	31	31	31
mulsol.i.4	185	3946	31	31	31
mulsol.i.5	186	3973	31	31	31
DSJC125.1	125	736	?	6	7
DSJC125.5	125	3891	?	21	22
DSJC125.9	125	6961	?	49	50
DSJC250.1	250	3218	?	11	11
DSJR500.1	500	3555	?	13	13

Table 1.3 Graph Coloring Problems with a Tighter Relaxation

Graph	$\|V\|$	$\|E\|$	Optimal	DSDP	DSDP2
miles750	128	2113	31	32	32
queen6.6	36	290	7	9	9
queen7.7	49	476	7	11	10
queen8.8	64	728	9	11	12
queen9.9	81	1055	10	13	13
queen11.11	121	1980	11	15	15
queen13.13	169	3328	13	18	18

[3] L. Babel. Finding maximum cliques in arbitrary and in special graphs. *Computing*, 46:321–341, 1991.

[4] L. Babel and G. Tinhofer. A branch and bound algorithm for the maximum clique problem. *J. of Global Optimization*, 4, 1994.

[5] Egon Balas and H. Samuelsson. A node covering algorithm. *Naval Research Logistics Quarterly*, 24(2):213–233, 1977.

[6] Egon Balas and Jue Xue. Minimum weighted coloring of triangulated graphs, with application to maximum weight vertex packing and clique finding in arbitrary graphs. *SIAM Journal on Computing*, 20(2):209–221, April 1991.

[7] Egon Balas and Chang Sung Yu. Finding a maximum clique in an arbitrary graph. *SIAM Journal on Computing*, 15(4):1054–1068, November 1986.

[8] S. Benson, Y. Ye, and X. Zhang. Solving large–scale sparse semidefinite programs for combinatorial optimization. Technical report, Department of Management Science, University of Iowa, Iowa City, IA 52242, USA, September 1997. To appear in *SIAM J. of Optimization.*

[9] C. Berge. *Graphs and Hypergraphs.* North–Holland, Amsterdam, 1973.

[10] D. Bertsimas and Y. Ye. Semidefinite relaxations, multivariate normal distributions, and order statistics. In D.-Z. Du and P.M. Pardalos, editors, *Handbook of Combinatorial Optimization*, volume 3, pages 1–19. Kluwer Academic Publishers, 1998.

[11] P. Briggs, K. Cooper, K. Kennedy, and L. Torczon. Coloring heuristics for register allocation. In *ASCM Conference on Program Lan-*

guage Design and Implementation, pages 275–284. The Association for Computing Machinery, 1998.

[12] R. Carrahan and P. M. Pardalos. An exact algorithm for the maximum clique problem. *Operations Research Letters*, 9:375–382, 1990.

[13] G.J. Chaitin, M. Auslander, A.K. Chandra, J. Cocke, M.E. Hopkins, and P. Markstein. Register allocation via coloring. *Computer Languages*, 6:47–57, 1981.

[14] Gregory J. Chaitin. Register allocation and spilling via graph coloring. *SIGPLAN Notices (Proceedings of the SIGPLAN '82 Symposium on Compiler Construction, Boston, Mass.)*, 17(6):98–101, June 1982.

[15] D. De Werra. An introduction to timetabling. *European Journal of Operations Research*, 19:151–162, 1985.

[16] DIMACS Center Web Page. The Second DIMACS Implementation Challenge: 1992-1993. ftp://dimacs.rutgers.edu/pub/challenge/graph/benchmarks/clique/.

[17] C. Friden, A. Hertz, and D. de Werra. An exact algorithm based on tabu search for finding a maximum independent set in a graph. *Computers Operations Research*, 17(5):375–382, 1990.

[18] Andreas Gamst. Some lower bounds for a class of frequency assignment problems. *IEEE Transactions of Vehicular Technology*, 35(1):8–14, 1986.

[19] Michael R. Garey and David S. Johnson. *Computers and Intractability: A Guide to the Theory of NP–Completeness.* W.H Freeman, San Francisco, CA, 1979.

[20] M. X. Goemans and D. P. Williamson. .878-approximation for MAX CUT and MAX 2SAT. In *Proc.* 26th *ACM Symp. Theor. Computing*, pages 422–431, 1994.

[21] David S. Johnson. Approximation algorithms for combinatorial problems. *Journal of Computer and System Sciences*, 9:256–278, 1974.

[22] David S. Johnson. Worst case behavior of graph coloring algorithms. In *Proceedings of 5th Southeastern Conference on Combinatorics, Graph Theory, and Computing*, pages 513–527. Utilitas Mathematica, Winnipeg, Canada, 1974.

[23] David Karger, Rajeev Motwani, and Madhu Sudan. Approximate graph coloring by semidefinite programming. Technical report, MIT, Cambridge, MA 52242, USA, 1994.

[24] Jon Kleinberg and Michel X. Goemans. The Lovász theta function and a semidefinite programming relaxation of vertex cover. *SIAM Journal on Discrete Mathematics*, 11(2):196–204, May 1998.

[25] M. Kubale and B. Jackowski. A generalized implicit enumeration algorithm for graph coloring. *Communications of the ACM*, 28:412–418, 1985.

[26] M. Kubale and E. Kusz. Computational experience with implicit enumeration algorithms for graph coloring. In *Proceedings of the WG'83 International Workshop on Graphtheoretic Concepts in Computer Science*, pages 167–176, Linz, 1983. Trauner Verlag.

[27] Carlo Mannino and Antonio Sassano. An exact algorithm for the maximum cardinality stable set problem. *Networks*, page (submitted), 1993. ftp://dimacs.rutgers.edu/pub/challenge/graph/ contributed.

[28] Anuj Mehrotra and Michael A. Trick. A column generation approach for graph coloring. http://mat.gsia.cmu.edu/trick.html, April 1995.

[29] Craig A. Morgenstern and Harry D. Shapiro. Coloration neighborhood structures for general graph coloring. In *Proceedings of the First Annual ACM-SIAM Symposium on Discrete Algorithms, San Francisco, Jan, 1990.* Society for Industrial and Applied Mathematics, Philadelphia, 1990.

[30] George L. Nemhauser and G. L. Sigismondi. A strong cutting plane / branch and bound algorithm for node packing. *Journal of the Operational Research Society*, 43(5), 1992.

[31] George. L. Nemhauser and Les. E. Trotter, Jr. Vertex packings: Structural properties and algorithms. *Mathematical Programming*, 8(2):232–248, 1975.

[32] Yu. E. Nesterov. Semidefinite relaxation and nonconvex quadratic optimization. *Optimization Methods and Software*, 9:141–160, 1998.

[33] Yu. E. Nesterov and A. S. Nemirovskii. *Interior Point Polynomial Methods in Convex Programming: Theory and Algorithms.* SIAM Publications, SIAM, Philadelphia, 1993.

[34] B. Pittel. On the probable behaviour of some algorithms for finding the stability number of a graph. *Mathematical Proceedings of the Cambridge Philosophical Society*, 92:511–526, 1982.

[35] M. J. D. Powell and P. L. Toint. On the estimation of sparse Hessian matrices. *SIAM Journal on Numerical Analysis*, 16:1060–1074, 1979.

[36] Proc. 4th IPCO Conference. *Improved approximation algorithms for max k-cut and max bisection*, 1995.

[37] M. V. Ramana, L. Tunçel, and H. Wolkowicz. Strong duality for semidefinite programming. *SIAM Journal on Optimization*, 7:641–662, 1997.

[38] T. Stephen and L. Tuncel. On a representation of the matching polytope via semidefinite liftings. *Mathematics of Operations Research*, 24(1):1–7, 1999.

[39] M. J. Todd. On search directions in interior-point methods for semidefinite programming. Technical Report 1205, School of Operations Research and Industrial Engineering, Cornell University, Itheca, NY 14853–3801, October 1997.

[40] M. Trick. Graph coloring instances. http://mat.gsia.cmu.edu/COLOR/instances.html.

[41] L. Tuncel. On the slater condition for the sdp relaxations of nonconvex sets. Technical Report CORR2000-13, Department of Combinatorica and Optimization, University of Waterloo, Waterloo, Ontario N2L3G1, Canada, February 2000.

[42] L. Vandenberghe and S. Boyd. Semidefinite programming. *SIAM Review*, 38(1):49–95, 1996.

[43] D. C. Wood. A technique for coloring a graph applicable to large scale time-tabling problems. *The Computer Journal*, 3:317–319, 1969.

[44] Y. Ye. *Interior Point Algorithms : Theory and Analysis*. Wiley-Interscience Series in Discrete Mathematics and Optimization. John Wiley & Sons, New York, 1997.

[45] Y. Ye. Approximating quadratic programming with bound and quadratic constraints. *Mathematical Programming*, 84:219–226, 1999.

Chapter 2

NONMONOTONE PATH FOLLOWING METHODS FOR NONSMOOTH EQUATIONS AND COMPLEMENTARITY PROBLEMS

Stephen C. Billups*
Department of Mathematics, University of Colorado, Denver, Colorado 80217-3364
sbillups@carbon.cudenver.edu

Adam L. Speight
Department of Mathematics, University of Colorado, Denver, Colorado 80217-3364
aspeight@math.cudenver.edu

Layne T. Watson
Departments of Computer Science and Mathematics, Virginia Polytechnic Institute and State University, Blacksburg, VA 24061-0106
ltw@cayuga.cs.vt.edu

Abstract We present a homotopy path following method for nonsmooth equations, which is effective at solving highly nonlinear equations for which the norm of the residual has nonglobal local minima. The method is based on a class of homotopy mappings that are smooth in the interior of their domain. Sufficient conditions are presented that guarantee the existence of well-behaved zero curves of these homotopy mappings, which can be followed to a solution. These zero curves need not be monotonic in the homotopy parameter. The method is specialized to solve complementarity problems through the use of MCP functions and associated smoothers.

Keywords: Nonsmooth equations, complementarity problems, homotopy methods, smoothing, path following.

*Research partially supported through NSF grant DMS-9973321.

M.C. Ferris et al. (eds.), Complementarity: Applications, Algorithms and Extensions, 19–41.

1 INTRODUCTION

Nonsmooth systems of equations, where the functions involved are not continuously differentiable, provide a unifying framework for the study of many important classes of problems. Among these are constrained optimization, finite-dimensional variational inequalities, complementarity problems, equilibrium problems, generalized equations, partial differential equations, and fixed point problems. In recent years, significant progress has been made in extending locally convergent methods for smooth equations to the realm of nonsmooth equations. Principal among these are the Newton-based methods, which have been extended to solve nonsmooth equations [22]. Techniques developed in this vein are attractive because they maintain the fast local convergence properties that are characteristic of Newton-based methods for smooth equations. To improve the domain of convergence of these methods, a globalizing strategy, such as a trust region or line search technique, is usually employed, which ensures that the iterates converge, at least, to a stationary point of an associated merit function (such as the norm of the residual). Such methods perform with great efficacy when applied to nonsmooth problems that are nearly linear in the sense that their merit functions do not contain local minima that are not solutions.

For problems that are *highly nonlinear*, (that is, where the merit function contains nonglobal, local minima) these Newton-based schemes tend to fail without a carefully chosen starting point. The reason, of course, is that unless started close to a solution, the iterates may converge only to a local minimum of the merit function.

An alternative approach for solving systems of equations is a path following method, which works by defining a family of functions parameterized by a scalar μ. Under certain conditions, the zeros of these functions, over all values of μ, form a path, which can be followed to reach a solution of the original system of equations. Some path following methods receiving considerable attention recently are interior point methods [31], smoothing methods [8, 9, 16, 32], and very recently, noninterior point methods [18, 7]. These methods rely on monotonic decrease of the parameter μ as the path is followed. This poses a problem for highly nonlinear problems because the actual path may not be monotonic in μ.

For smooth equations, a highly developed path following technique for solving highly nonlinear systems of equations is the probability-one homotopy method discussed in [10, 27, 29, 30]. For problems of this class, homotopy methods often prove more robust than Newton-based methods or the monotonic path following methods mentioned above. The reason is that the homotopy methods view the extra parameter

μ as simply another variable, and therefore do not insist that μ vary monotonically along the path.

To date, only a small amount of progress has been made in extending probability-one homotopy methods to the domain of nonsmooth systems of equations. The principal difficulty lies in the fact that the essential convergence theory requires that the system of equations be twice continuously differentiable. In the realm of complementarity problems, a few papers have appeared. Watson [28] proposed such a method to solve the nonlinear complementarity problem. Watson's method involved reformulating the nonlinear complementarity problem as a system of smooth (C^2) equations and applying a homotopy method to solve this system. Another approach was developed by Sellami and Robinson [23, 25, 24] based on the theoretical framework for piecewise smooth continuation methods presented in [1, 2, 3]. This approach was complicated by the fact that a special procedure was needed to make the transition from one smooth segment of the homotopy zero curve to another. Another approach was presented in [4]. The central idea of this method was to use a homotopy method on a smooth approximation of the original system as a globalizing strategy for Newton's method.

This paper proposes a more direct approach, constructing a single homotopy mapping that is twice continuously differentiable on the interior of its domain. This mapping is constructed using a family of smoothing approximations of the nonsmooth function F, parameterized by some parameter μ, which is a function of the homotopy parameter. In essence the smoothing parameter and the homotopy parameter are combined together into a single parameter.

Section 2 defines notation and gives background material on nonsmooth equations, complementarity problems, probability-one homotopy convergence theory, and smoothing operators. Section 3 describes a probability-one homotopy algorithm for nonsmooth equations, and contrasts it with an earlier hybrid homotopy/damped Newton scheme of Billups [4]. This general algorithm for nonsmooth systems of equations is then specialized for the mixed complementarity problem in Section 4. Section 5 addresses implementation details and computational results, and Section 6 concludes.

2 BACKGROUND

2.1 NOTATION

When discussing vectors and vector-valued functions, subscripts are used to indicate components, whereas superscripts are used to indicate the iteration number or some other label. In contrast, for scalars or scalar-valued functions, subscripts refer to labels so that superscripts can be reserved for exponentiation. The vector of all ones is represented by e.

Unless otherwise specified, $\|\cdot\|$ denotes the Euclidean norm. The notation $(\cdot)^+$, $(\cdot)^-$, and $|\cdot|$ represent the componentwise plus, minus, and absolute value operators, respectively, for vectors. That is, $x^+ := (\max(x_1, 0); \ldots ; \max(x_n, 0))$, $x^- := (\max(-x_1, 0); \ldots ; \max(-x_n, 0))$, and $|x| := (|x_1|; \ldots ; |x_n|)$.

The symbols $\mathbb{R}_+$ and $\mathbb{R}_{++}$ refer to the nonnegative real numbers and the positive real numbers respectively. The extended real numbers are denoted by $\overline{\mathbb{R}} := \mathbb{R} \bigcup \{-\infty, +\infty\}$.

Real-valued functions are denoted with lower-case letters like f or ϕ whereas vector-valued functions are represented by upper-case letters like F or Φ. For a function $F : C \subset \mathbb{R}^n \to \mathbb{R}^m$, define $\nabla_i F_j(x) := \partial F_j(x)/\partial x_i$. $\nabla F(x)$ is the $n \times m$ matrix whose ijth element is $\nabla_i F_j(x)$. Let $D \subset \mathbb{R}^m$. Then $F^{-1}(D)$ is the set-valued inverse defined by $F^{-1}(D) := \{x \mid F(x) \in D\}$.

For a set $C \subset \mathbb{R}^n$, denote its closure and boundary by $\overline{C}$ and ∂C respectively. $\pi_C(x)$ represents the orthogonal projection (with respect to the Euclidean norm) of x onto the set C.

2.2 NONSMOOTH EQUATIONS

By Rademacher's theorem, if $F : \mathbb{R}^n \to \mathbb{R}^m$ is locally Lipschitzian, it is differentiable almost everywhere. Let D_F be the set where F is differentiable. We define the B-subdifferential by

$$\partial_B F(x) := \left\{ V \,\middle|\, \exists \{x^k\} \to x, x^k \in D_F, \quad \text{with} \quad V = \lim_{k \to \infty} \nabla F(x_k) \right\}.$$

The Clarke subdifferential $\partial F(x)$ is the convex hull of $\partial_B F(x)$.

Definition 2.1 *We say that F is* semismooth *at x if*

$$\lim_{\substack{V \in \partial F(x + th') \\ h' \to h, t \downarrow 0}} \{Vh'\}$$

exists for any $h \in \mathbb{R}^n$.

Definition 2.2 *We say that a semismooth function $F : \mathbb{R}^n \to \mathbb{R}^n$ is* BD-regular *at x if all elements in $\partial_B F(x)$ are nonsingular, and F is* strongly regular *at x if all elements in $\partial F(x)$ are nonsingular.*

This section closes with an implicit function theorem for nonsmooth functions. Let $H : \mathbb{R}^m \times \mathbb{R}^k \to \mathbb{R}^k$ be a Lipschitz continuous function. Define $\pi_z \partial H(\hat{y}, \hat{z})$ to be the set of $k \times k$ matrices M such that $[\ N \quad M\ ] \in \partial H(\hat{y}, \hat{z})$.

Theorem 2.3 *[11][Corollary, Page 256] Let* $H : \mathbb{R}^m \times \mathbb{R}^k \to \mathbb{R}^k$ *be Lipschitz continuous in a neighborhood of* $(\hat{y}, \hat{z})$*, and suppose that all matrices in* $\pi_z \partial H(\hat{y}, \hat{z})$ *are nonsingular. Then there exists a neighborhood* Y *of* $\hat{y}$*, a neighborhood* Z *of* $\hat{z}$*, and a Lipschitz function* $\zeta : Y \to \mathbb{R}^k$ *such that* $\zeta(\hat{y}) = \hat{z}$*, and for every* $y \in Y$*,*

$$z = \zeta(y) \quad \textit{if and only if } H(y, z) = 0 \quad \textit{and} \quad z \in Z.$$

Note: The "if" part of the above theorem is not explicitly stated in [11]. However, it can be proved easily by a simple modification to the given proof.

2.3 COMPLEMENTARITY PROBLEMS

Given a continuously differentiable function $F : \mathbb{R}^n \to \mathbb{R}^n$, the nonlinear complementarity problem NCP(F) is to find some $x \in \mathbb{R}^n$ so that

$$0 \leq x \perp F(x) \geq 0, \tag{2.1}$$

where $x \perp F(x)$ means that $x^T F(x) = 0$.

It is well known that NCP(F) can be reformulated as a system of equations. This was first shown by Mangasarian [19]. An excellent review of reformulations of NCP can be found in [21]. A reformulation that has been very popular recently is based on the Fischer-Burmeister function [15] $\phi : \mathbb{R}^2 \to \mathbb{R}$, defined by

$$\phi(a, b) := a + b - \sqrt{a^2 + b^2}. \tag{2.2}$$

It is easily seen that $\phi(a, b) = 0$ if and only if $0 \leq a \perp b \geq 0$. Thus, by defining the function $H : \mathbb{R}^n \to \mathbb{R}^n$ by

$$H_i(x) := \phi(x_i, F_i(x)), \tag{2.3}$$

it is clear that $x \in \mathbb{R}^n$ solves NCP(F) if and only if $H(x) = 0$.

While ϕ is not differentiable at the origin, ϕ^2 is continuously differentiable everywhere. This property, together with the fact that ϕ is semismooth, makes this reformulation well suited for use in globalization strategies for nonsmooth Newton-based methods (see, for example, [12]).

Given a rectangular region $\mathcal{B}_{l,u} := \prod_{i=1}^{n} [l_i, u_i] \subset \overline{\mathbb{R}}^n$ defined by two vectors, l and u in $\mathbb{R}^n$ where $-\infty \leq l < u \leq \infty$, and a function $F : \mathbb{R}^n \to \mathbb{R}^n$, the mixed complementarity problem MCP($F, \mathcal{B}_{l,u}$) is to find an $x \in$

$\mathcal{B}_{l,u}$ such that for each $i \in \{1, \ldots, n\}$, either 1) $x_i = l_i$ and $F_i(x) \geq 0$, 2) $F_i(x) = 0$, or 3) $x_i = u_i$ and $F_i(x) \leq 0$. This is equivalent to the condition that $\text{mid}(x - l, x - u, F(x)) = 0$, where mid represents the componentwise median function. When these conditions are satisfied, write $F(x) \perp x$ and say that x is complementary to $F(x)$. Typically, one assumes that the function F above is continuously differentiable.

A reformulation of the MCP as a system of nonsmooth equations can be constructed based on the following function $\psi : \mathbb{R}^4 \to \mathbb{R}$ [4]:

$$\psi(l, u, x, f) := \phi(x - l, -\phi(u - x, -f)). \tag{2.4}$$

Constructing a function $H : \mathbb{R}^n \to \mathbb{R}^n$ by

$$H_i(x) := \psi(l_i, u_i, x_i, F_i(x)), \tag{2.5}$$

yields a reformulation of the MCP$(F, \mathcal{B}_{l,u})$. $H(x) = 0$ if and only if x is a solution to MCP$(\mathcal{B}_{l,u}, F)$ [4].

2.4 HOMOTOPY METHODS

A homotopy is a topological construction representing a function that continuously deforms an element of a function space into another element of that space. Consider the function $F : \mathbb{R}^n \to \mathbb{R}^n$. An example of a simple homotopy is

$$\rho(a, \lambda, x) := \lambda F(x) + (1 - \lambda)G_a(x),$$

where G_a is some trivial function on $\mathbb{R}^n$, (for example, $G_a(x) = x - a$), $\lambda \in [0, 1]$ is called the *homotopy parameter*, and a is some *parameter vector* in $\mathbb{R}^n$. It is clear that ρ continuously deforms G_a to F as λ goes from 0 to 1.

2.4.1 Theory and Definitions.

Much of the theory of probability-one homotopy methods can be derived from generalizations of a result in differential geometry called Sard's Theorem. This class of theorems uses the notion of a *regular value.*

Definition 2.4 *Let $U \subset \mathbb{R}^n$ be open and $F : \mathbb{R}^n \to \mathbb{R}^m$ be a smooth function. We say that $y \in \mathbb{R}^m$ is a* regular value *of F if*

$$\text{range } \nabla F(x) = \mathbb{R}^m \text{ for all } x \in F^{-1}(\{y\}).$$

Theorem 2.5 (Parameterized Sard's Theorem) *[10] Let $V \subset \mathbb{R}^q$, and $U \subset \mathbb{R}^m$ be open sets and let $\Upsilon : V \times U \to \mathbb{R}^p$ be of class C^r, where $r > \max\{0, m - p\}$. If $0 \in \mathbb{R}^p$ is a regular value of Υ, then for* *almost every $a \in V$, 0 is a regular value of the function Υ_a defined by $\Upsilon_a(\cdot) := \Upsilon(a, \cdot)$.*

Corollary 2.6 *[10] Assume the conditions of the previous theorem. If $m = p + 1$, then each component of $\Upsilon_a^{-1}(\{0\})$ is a smooth curve for almost every $a \in V$ with respect to q-dimensional Lebesgue measure.*

The following proposition gives conditions under which a well-behaved zero curve exists. It is similar to results presented in [26] and [10, Theorem 2.4]; however, it does not assume F itself to be differentiable. The path γ_a defined in the proposition "reaches a zero of F" in the sense that it contains a sequence $\{(\lambda_k, x^k)\}$ that converges to $(1, \bar{x})$, where $\bar{x}$ is a zero of F.

Proposition 2.7 *Let $F : \mathbb{R}^n \to \mathbb{R}^n$ be a Lipschitz continuous function and suppose there is a C^2 map*

$$\rho : \mathbb{R}^m \times [0, 1) \times \mathbb{R}^n \to \mathbb{R}^n$$

such that

1. *$\nabla\rho(a, \lambda, x)$ has rank n on the set $\rho^{-1}(\{0\})$,*
2. *the equation $\rho_a(0, x) = 0$, where $\rho_a(\lambda, x) := \rho(a, \lambda, x)$, has a unique solution $x^a \in \mathbb{R}^n$ for every fixed $a \in \mathbb{R}^m$,*
3. *$\nabla_x\rho_a(0, x^a)$ has rank n for every $a \in \mathbb{R}^m$,*
4. *ρ is continuously extendible (in the sense of Buck [6]) to the domain $\mathbb{R}^m \times [0, 1] \times \mathbb{R}^n$, and $\rho_a(1, x) = F(x)$ for all $x \in \mathbb{R}^n$ and $a \in \mathbb{R}^m$, and*
5. *γ_a, the connected component of $\rho_a^{-1}(\{0\})$ containing $(0, x^a)$, is bounded for almost every $a \in \mathbb{R}^m$.*

Then for almost every $a \in \mathbb{R}^m$ there is a zero curve γ_a of ρ_a, along which $\nabla\rho_a$ has rank n, emanating from $(0, x^a)$ and reaching a zero $\bar{x}$ of F at $\lambda = 1$. Further, γ_a does not intersect itself and is disjoint from any other zeros of ρ_a. Also, if γ_a reaches a point $(1, \bar{x})$ and F is strongly regular at $\bar{x}$. then γ_a has finite arc length.

Proof Assumption 1 and the Parameterized Sard's Theorem give that 0 is a regular value of ρ_a for almost every $a \in \mathbb{R}^m$. By Corollary 2.6 each component of $\left(\rho_a|_{(0,1)\times\mathbb{R}^n}\right)^{-1}(\{0\})$ is a smooth curve and, as such, either diffeomorphic to a circle or an interval [10]. Because $\nabla_x\rho_a(0, x^a)$ has rank n, the Implicit Function Theorem gives x in terms of λ in a neighborhood of $(0, x^a)$. This implies that γ_a is not diffeomorphic to a circle. Since ρ_a is of class C^2, all limit points of γ_a must lie in $\rho_a^{-1}(\{0\})$ (with ρ_a's extended domain $\mathbb{R}^n \times [0,1] \times \mathbb{R}^m$), so the only limit point of γ_a in $\{0\} \times \mathbb{R}^n$ is $(0, x^a)$. Furthermore, $\gamma_a \cap ((0,1) \times \mathbb{R}^n)$ is diffeomorphic to an interval and $(0, x^a)$ corresponds to one end of it. Since γ_a is bounded, there is some bounded closed set $B \subset \mathbb{R}^n$ such that $\gamma_a \subset [0,1] \times B$. By the Implicit Function Theorem, γ_a cannot terminate (have an end point) in $(0,1) \times \mathbb{R}^n$, and by compactness γ_a has finite arc length in every compact subset of $[0,1) \times B$. (For each point $z \in \gamma_a$, there is a neighborhood of z in which γ_a has finite arc length by the Implicit Function Theorem and the fact that 0 is a regular value of ρ_a. For $z \notin \gamma_a$, there is a neighborhood of z disjoint from γ_a since $\mathbb{R}^n$ is normal. These neighborhoods form an open covering of $[0,1) \times B$, for which the finite subcovering of a compact subset $[0, 1-\epsilon] \times B$ yields finite arc length for the portion of γ_a in $[0, 1-\epsilon] \times B$.) Therefore γ_a must leave every set $[0, 1-\epsilon] \times B$, $\epsilon > 0$, and have a limit point in $\{1\} \times B$. By continuity of ρ_a if $(1, x^*)$ is a limit point of γ_a, then x^* is a zero of F.

That γ_a does not intersect itself or any other component of $\rho_a^{-1}(\{0\})$ follows from the Implicit Function Theorem and the full rank of $\nabla \rho_a(\lambda, x)$ on $\rho_a^{-1}(\{0\})$: for each $z \in \gamma_a$ there is a neighborhood $B(z)$ such that $B(z) \cap \gamma_a$ is diffeomorphic to an open interval.

If $(1, \bar{x})$ is a limit point of γ_a and F is strongly regular at $\bar{x}$, then every element of $\pi_x \partial \rho_a(\lambda, x)$ is nonsingular. By Theorem 2.3 γ_a can be described as a Lipschitz continuous function of λ in a neighborhood of $(1, \bar{x})$, and therefore has finite arc length in this neighborhood. Further, the zero curve has finite arc length outside this neighborhood by the above argument, and hence finite arc length everywhere. □

Because γ_a is a smooth curve, it can be parameterized by its arc length away from $(0, x^a)$. This yields a function $\gamma_a(s)$, the point on γ_a of arc length s away from $(0, x^a)$.

The construction of a globally convergent probability-one homotopy algorithm entails: (1) constructing a map ρ according to Proposition 2.7, (2) choosing $a \in \mathbb{R}^m$, (3) finding x^a solving $\rho_a(0, x) = 0$, and (4) tracking γ_a starting from $(0, x^a)$ until $\lambda = 1$. Assuming an appropriate ρ exists, the theory guarantees that for almost all a (in the sense of Lebesgue measure), γ_a exists and leads to a solution, hence the term "probability-one".

A simple (and occasionally useful in practice) homotopy mapping is $\rho : \mathbb{R}^n \times [0,1) \times \mathbb{R}^n \rightarrow \mathbb{R}^n$ given by

$$\rho(a, \lambda, x) := \lambda F(x) + (1 - \lambda)(x - a). \tag{2.6}$$

If F is C^2 then ρ trivially satisfies properties (1), (2), (3), and (4) but not necessarily (5) of Proposition 2.7. The following theorem gives conditions on F under which the fifth condition is satisfied. This result will be generalized to nonsmooth functions in Theorem 2.11.

Theorem 2.8 *[28] Let $F : \mathbb{R}^n \to \mathbb{R}^n$ be a C^2 function such that for some $\tilde{x} \in \mathbb{R}^n$ and $r > 0$,*

$$(x - \tilde{x})^T F(x) \geq 0 \text{ whenever } \|x - \tilde{x}\| = r. \tag{2.7}$$

Then F has a zero in a closed ball of radius r about $\tilde{x}$, and for almost every a in the interior of this ball there is a zero curve γ_a of

$$\rho_a(\lambda, x) := \lambda F(x) + (1 - \lambda)(x - a),$$

along which $\nabla \rho_a(\lambda, x)$ has full rank, emanating from $(0, a)$ and reaching a zero $\bar{x}$ of F at $\lambda = 1$. Further, γ_a has finite arc length if $\nabla F(\bar{x})$ is nonsingular.

The actual statement of the theorem in [28] fixes $\tilde{x} = 0$. However, the proof can be modified trivially to yield the more general statement above. Since the required modifications are included in the proof of Theorem 2.11, we omit the details here. It is interesting to note that in many applications, (2.7) holds for all r sufficiently large (not just for some fixed r). This makes the choice of $\tilde{x}$ irrelevant. Furthermore, in such cases, we are free to choose a arbitrarily, (instead of from some neighborhood of $\tilde{x}$), thus making the method truly globally convergent.

(2.7) will be referred to as the *global monotonicity* property. If a C^2 function F possesses this property, these theoretical results have some profound implications: the guaranteed existence of a path of finite arc length between almost any starting point and a solution to $F(x) = 0$. In theory, to find a solution, one must simply follow the path to a point of γ_a where $\lambda = 1$. In practice, however, the task of constructing a ρ for which γ_a is short and smooth is very difficult, although this has been done for large classes of problems.

Several packages exist to solve root finding problems using homotopy-based techniques [30]. This work will use the routine STEPNX from the HOMPACK90 suite of software [29] [30, Section 3], which tracks the zero curve of a homotopy mapping specified by the user.

2.5 SMOOTHING OPERATORS

Consider the system $F(x) = 0$ where F is a nonsmooth function, and suppose there exists a family of functions F^μ parameterized by a *smoothing parameter* μ so that $\lim_{\mu \downarrow 0} F^\mu = F$ in some sense. Under suitable conditions, the solutions to the systems $F^\mu(x) = 0$ converge to a solution to $F(x) = 0$ along a smooth trajectory [8].

Definition 2.9 *Given a nonsmooth continuous function* $\varphi : \mathbb{R}^p \to \mathbb{R}$, *a* smoother *for* φ *is a continuous function* $\tilde{\varphi} : \mathbb{R}^p \times \mathbb{R}_+ \to \mathbb{R}$ *such that*

1. $\tilde{\varphi}(x, 0) = \varphi(x)$, *and*

2. $\tilde{\varphi}$ *is twice continuously differentiable on the set* $\mathbb{R}^p \times \mathbb{R}_{++}$.

For convenience, let $\varphi_\mu(x)$ denote the smoother $\tilde{\varphi}(x, \mu)$. To define smoothers for functions $F : \mathbb{R}^n \to \mathbb{R}^n$, say that $F^\mu : \mathbb{R}^n \times \mathbb{R}_+ \to \mathbb{R}^n$ is a smoother for F if for each $i \in \{1 \dots n\}$ F_i^μ is a smoother for F_i.

2.6 COMPLEMENTARITY SMOOTHERS

The reformulations (2.3) and (2.5) for the NCP and MCP, respectively, are both based on the Fischer-Burmeister function (2.2), which is not smooth at the origin. Smoothers for these functions can be constructed based on Kanzow's smoother for the Fischer-Burmeister function [17]:

$$\phi_\mu(a, b) := \tilde{\phi}(a, b, \mu) := a + b - \sqrt{a^2 + b^2 + 2\mu}, \tag{2.8}$$

and the associated MCP smooth

$$\psi_\mu(l, u, x, f) := \tilde{\psi}(l, u, x, f, \mu) := \phi_\mu(x - l, -\phi_\mu(u - x, -f)). \tag{2.9}$$

Smoothers for (2.3) and (2.5) are then given, respectively, by

$$H_i^\mu(x) := \phi_\mu(x_i, F_i(x)), \quad \text{and} \tag{2.10}$$

$$H_i^\mu(x) := \psi_\mu(l_i, u_i, x_i, F_i(x)). \tag{2.11}$$

3 THE ALGORITHM

This section presents a new probability-one homotopy algorithm for solving nonsmooth systems of equations. It contrasts with another hybrid Newton-homotopy method presented by Billups [4]. Billups' method begins by using a nonsmooth version of a damped-Newton's method to solve the root finding problem $F(x) = 0$. If the Newton algorithm stalls, a standard homotopy method is invoked to solve a particular smoothed version of the original problem, $F^\mu(x) = 0$, where μ is fixed. The smoothing parameter μ is chosen based on the level of a merit function on F at the last point $\hat{x}$ generated by the Newton method. Starting from $\hat{x}$, a homotopy method is carried out until it produces a point that yields a better merit value than the previous Newton iterate. The Newton method is then started again and the process repeats until a point is produced that is close enough to a solution or the homotopy method fails. One key feature of that hybrid method is that each time the Newton method stalls, a different homotopy map is constructed. The smoothing parameter μ is chosen based on the level of the merit function when the Newton method stalls, so the homotopy that is then used is

$$\rho_a^\mu(\lambda, x) := \lambda F^\mu(x) + (1-\lambda)(x-a).$$

Computational experience with that algorithm indicates that starting the homotopy algorithm from a point at which Newton's method failed often results in zero curves that are very difficult to track. In fact, picking a random starting point from which to start a homotopy algorithm usually produces better results. The logical alternative, proposed here, is to adopt a pure probability-one homotopy algorithm: fix the homotopy map, and track a single homotopy zero curve into the Newton domain of convergence around a solution. Essentially, the idea is to use a standard probability-one homotopy algorithm, but with a specially designed "end game" near a solution.

3.1 THE HOMOTOPY MAP

Given a function F and an associated smoother F^μ consistent with Definition 2.9, construct a single homotopy mapping with F^μ whereby the smoothing parameter μ is a function of the homotopy parameter λ so that $\mu \downarrow 0$ as $\lambda \uparrow 1$. If this homotopy satisfies the conditions in Proposition 2.7, a well behaved path exists from almost any starting point to a solution, and standard curve tracking techniques can reliably solve the equation $F(x) = 0$.

Throughout this section, assume that F is a function on $\mathbb{R}^n$ and that F^μ is a smoother for F. Take $\mu(\lambda)$ to be a nonincreasing differentiable function such that $\lim_{\lambda \uparrow 1} \mu(\lambda) = 0$, for example,

$$\mu(\lambda) := \alpha(1-\lambda)^2 \tag{2.12}$$

for some parameter $\alpha > 0$. Define the homotopy map, nonlinear in λ, by

$$\varrho_a(\lambda, x) := \lambda\, F^{\mu(\lambda)}(x) + (1-\lambda)(x-a) \tag{2.13}$$

and let γ_a be the connected component of the set $\varrho_a^{-1}(\{0\})$ that contains $(0, a)$. Notice that this mapping is a generalization of (2.6), since if F is C^2, we can choose $F^\mu := F$.

Since the smoothing parameter $\mu(\lambda)$ converges to zero as $\lambda \uparrow 1$, the function $F^{\mu(\lambda)}$ may be nearly nonsmooth near $\lambda = 1$. This means the second derivatives of some components of $F^{\mu(\lambda)}$ may be very large. This behavior may result in the zero curve γ_a having large curvature as it approaches a point where $\lambda = 1$, making curve tracking difficult. In addition to potential curve tracking difficulty near $\lambda = 1$, technically the smoothers are not defined for $\mu < 0$. A standard homotopy curve tracking algorithm typically tracks the zero curve to a point where $\lambda > 1$. While this poses no difficulty for the smoothing parameter in (2.12), using a function like $\mu(\lambda) := \alpha(1 - \lambda)$ for a smoothing parameter would yield disastrous results if a point were ever evaluated such that $\lambda > 1$.

This discussion suggests that the best strategy is to use an algorithm similar to the general phase of the routine STEPNX to track the zero curve γ_a to a point somewhere short of $\lambda = 1$, and then switch to a nonsmooth Newton algorithm for the end game to converge to the solution. What is required are conditions under which a point (λ, x) on the zero curve with λ sufficiently close to one will have x arbitrarily close to a solution. This criterion then becomes the basis for switching from tracking γ_a to the nonsmooth Newton iteration end game.

3.2 PROPERTIES OF ϱ_A

The fundamental issue is to state conditions on F and its smoother so that Proposition 2.7 can be invoked. The following weak assumption on the smoother will be useful in the theory that follows.

Assumption 2.10 *There is a nondecreasing function $\eta : \mathbb{R}_+ \to \mathbb{R}_+$ satisfying $\lim_{\nu \downarrow 0} \eta(\nu) = 0$ such that for all x in $\mathbb{R}^n$ and all ν in $\mathbb{R}_+$*

$$\|F^\nu(x) - F(x)\|_\infty \leq \eta(\nu).$$

It is important to note [4, Proposition 2.14] that the Kanzow MCP smoother (2.9) satisfies Assumption 2.10 with

$$\eta(\nu) := 3\sqrt{2\nu}. \tag{2.14}$$

The following theorem is a generalization of Theorem 2.8.

Theorem 2.11 *Let $F : \mathbb{R}^n \to \mathbb{R}^n$ be a Lipschitz continuous function such that for some fixed $r > 0$ and $\tilde{x} \in \mathbb{R}^n$,*

$$(x - \tilde{x})^T F(x) \geq 0 \text{ whenever } \|x - \tilde{x}\| = r,$$

and let F^{μ} be a smoother for F satisfying Assumption 2.10. Further, suppose that the smoothing parameter $\mu(\lambda)$ is such that

$$\eta(\mu(\lambda)) < \frac{1-\lambda}{\lambda} M \text{ for } 0 < \lambda \leq 1 \tag{2.15}$$

for some $M \in (0, r)$. Then γ_a is bounded for almost every $a \in \mathbb{R}^n$ such that $\|a - \tilde{x}\| < \tilde{r} := r - M$.

Proof Consider any point (λ, x) with $0 < \lambda < 1$, $\|x - \tilde{x}\| = r$, and let $\|a - \tilde{x}\| < \tilde{r}$. Starting with

$$\varrho_a(\lambda, x) = \lambda F^{\mu(\lambda)}(x) + (1-\lambda)(x - a),$$

multiplying by $x - \tilde{x}$ and dividing by $1 - \lambda$ results in

$$\frac{\varrho_a(\lambda, x)^T (x - \tilde{x})}{1-\lambda} = \frac{\lambda}{1-\lambda} F^{\mu(\lambda)}(x)^T (x - \tilde{x}) + (x - a)^T (x - \tilde{x}). \tag{2.16}$$

By assumption

$$\begin{aligned} F^{\mu(\lambda)}(x)^T (x - \tilde{x}) &= (F^{\mu(\lambda)}(x) - F(x))^T (x - \tilde{x}) + F(x)^T (x - \tilde{x}) \\ &\geq -\left\| F^{\mu(\lambda)}(x) - F(x) \right\| \|x - \tilde{x}\| + 0 \\ &\geq -\eta(\mu(\lambda)) \|x - \tilde{x}\| \\ &\geq -\frac{1-\lambda}{\lambda} Mr. \end{aligned} \tag{2.17}$$

Combining (2.16) and (2.17) gives

$$\begin{aligned} \frac{\varrho_a(\lambda, x)^T (x - \tilde{x})}{1-\lambda} &\geq -\frac{\lambda}{1-\lambda} \frac{1-\lambda}{\lambda} Mr + (x - a)^T (x - \tilde{x}) \\ &\geq -Mr + ((x - \tilde{x}) + (\tilde{x} - a))(x - \tilde{x}) \\ &> -Mr + r^2 - \tilde{r}r = 0. \end{aligned}$$

Therefore ϱ_a is not zero on the set

$$B := \left\{ (\lambda, x) \mid \|x - \tilde{x}\| = r, \ \lambda \in [0, 1) \right\}.$$

Since $(0, a) \in \gamma_a$ and $\|a - \tilde{x}\| < \tilde{r} < r$, γ_a is bounded (being contained in the convex hull of B). □

A direct application of Proposition 2.7 gives the main convergence theorem.

Theorem 2.12 *Under the assumptions of Theorem 2.11, F has a zero in a closed ball of radius r about $\tilde{x}$, and for almost every a in the interior of a ball of radius $\tilde{r}$ about $\tilde{x}$, there is a zero curve γ_a of*

$$\varrho(a, \lambda, x) := \varrho_a(\lambda, x) := \lambda F^{\mu(\lambda)}(x) + (1 - \lambda)(x - a),$$

along which $\nabla \varrho_a(\lambda, x)$ has full rank, emanating from $(0, a)$ and reaching a zero $\bar{x}$ of F at $\lambda = 1$. Further, γ_a has finite arc length if F is strongly regular at $\bar{x}$.

Observe that in applications, the r in Theorem 2.11 can be arbitrarily large, hence so can $\tilde{r} = r - M$, and thus $\|a - \tilde{x}\| < \tilde{r}$ is really no restriction at all.

The fact that a zero curve may be followed arbitrarily close to a solution follows easily from the continuity of ϱ_a and the boundedness of γ_a:

Theorem 2.13 *Suppose F and $F^{\mu(\lambda)}$ satisfy the conditions in Theorem 2.11. Then for each $\epsilon > 0$ there exists $\delta > 0$ such that $(\lambda, x) \in \gamma_a$ and $\lambda \in (1 - \delta, 1)$ imply*

$$\|x - x^*\| < \epsilon$$

for some zero x^* of F.

3.3 THE PROBABILITY-ONE HOMOTOPY ALGORITHM

The modification to a standard probability-one homotopy algorithm being proposed here is to replace the standard end game (a robust blend of secant iterations with Newton corrections), begun after a point (λ, x) with $\lambda > 1$ is found, with a nonsmooth version of Newton's method, described in [4, Figure 1], begun while $\lambda < 1$ still. In order to solve the system $F(x) = 0$, the nonsmooth Newton's method requires that F be semismooth. If, in addition, F is BD-regular at a solution x^*, Newton's method will converge superlinearly in some neighborhood about x^*. To use the homotopy approach, F should satisfy the global monotonicity property and be strongly regular at every solution. This guarantees that the homotopy's zero curve crosses the hyperplane $\lambda = 1$ transversally rather than tangentially, and ensures that the zero curve will have finite arc length.

Denote the nonsmooth Newton's method by $A(x, \epsilon)$, which stops when it either has produced a point $\tilde{x}$ such that $\theta(\tilde{x}) < \epsilon$, has converged to a local minimum of θ that is not a solution, or has stagnated without progress in decreasing θ. Figure 2.1 describes a probability-one homotopy algorithm, which will produce a solution provided that the assump-

Step 1 [Initialization] Pick parameters σ and δ_0 in $(0,1)$ and a tolerance $\epsilon > 0$. Let $k = 0$, $\lambda_0 = 0$ and choose some $x^0 \in \mathbb{R}^n$.

Step 2 [Homotopy] Starting from the point (λ_k, x^k), use a curve tracking algorithm (STEPNX from HOMPACK90) to find a point (λ_{k+1}, x^{k+1}) on γ_a such that $\lambda_{k+1} \geq 1 - \delta_k$.

Step 3 [Local Convergence] Run $A(x^{k+1}, \epsilon)$. If a point $\tilde{x}$ is produced such that $\theta(\tilde{x}) < \epsilon$ report $\tilde{x}$ as a solution and terminate. Otherwise proceed to Step 4.

Step 4 [Adjustment] Let $\delta_k := \sigma \min\{\delta_k, 1 - \lambda_{k+1}\}$, increment k by one and go to Step 2.

Figure 2.1 Probability-one homotopy algorithm

tions in Theorem 2.12 are satisfied and the curve tracking is done accurately enough. If the algorithm fails numerically, it will do so in the homotopy step. Possible remedies are to change the homotopy map ϱ_a (the best!), or reduce the curve tracking tolerances after refining the point (λ_{k+1}, x^{k+1}) on γ_a. One notable feature of this algorithm is that if the Newton method in Step 3 fails, the homotopy Step 2 resumes from the last point found on γ_a. This contrasts with the algorithm described in [4], which follows the zero curve of a different homotopy mapping at each failure of the Newton method.

4 SOLVING MIXED COMPLEMENTARITY PROBLEMS

The algorithm described above can be used to solve mixed complementarity problems by constructing the nonsmooth operator H given in (2.5) and its associated smoother H^μ defined by (2.11). When all bounds of the MCP are finite, these functions are particularly well suited for use with the algorithm since Billups [4] showed that H satisfies the global monotonicity property required in Theorem 2.11. Thus, if H is strongly regular at all solutions then the homotopy map ϱ_a (see (2.13)) induced by H and H^μ yields a well-behaved zero curve leading to a solution. Also, the smoother H^μ satisfies Assumption 2.10 as shown in (2.14). Since (2.14) and (2.12) imply (2.15), Theorem 2.11 applies. By Theorem 2.13 with $F^{\mu(\lambda)} = H^{\mu(\lambda)}$, the zero curve γ_a reaches the domain of convergence of Newton's method, which converges superlinearly to a solution because H is semismooth and strongly regular at all solutions.

5 SOLVER IMPLEMENTATION & TESTING

To implement the algorithm presented in Figure 2.1, we use in Step 3 the nonsmooth Newton's method described in Figure 1 of [4]. To implement Step 2, we use a routine similar to HOMPACK90's FIXPNF to track the zero curve of the homotopy map. This routine is a MATLAB function that uses a MEX interface to the STEPNX routine from HOMPACK90 to generate iterates along the zero curve. Its implementation differs from FIXPNF's in that it omits the end game strategy. If the STEPNX routine produces an iterate (λ, x) with $\lambda > 1$, the algorithm simply throws the point out and calls STEPNX again with the maximum allowable stepsize reduced by some prescribed factor (default is 0.5). If the routine generates 1000 points on the zero curve without producing a point with $\lambda \in (1 - \delta_k, 1)$, then the curve tracking phase is terminated. Finally, an iteration limit of 30 invocations of Step 2 is enforced.

The algorithm was tested using the MATLAB interface to the MCPLIB suite of complementarity problems given in [14]. Because our prototype implementation uses dense linear algebra, testing was limited to problems having fewer than 110 variables. The algorithm was tested in two ways. In the first configuration, δ_0 was chosen to be 1.0. This causes the curve tracking in Step 2 to be skipped initially so that the Newton method in Step 3 is attempted first. If the Newton method solves the problem, the homotopy curve tracking method is never invoked. The second configuration used $\delta_0 = 0.01$. This causes the homotopy curve tracking method to do most of the work, using Newton's method only for the 'end game' phase. Both configurations used $\sigma = 0.5$ and $\epsilon = 10^{-10}$. Additionally, the calls to STEPNX used curve tracking tolerances `RELERR`=`ABSERR`=0.01 and default values for the `SSPAR` parameters.

The results of the testing are shown in Table 2.1. For each choice of δ, the table displays how many times the curve tracking method in Step 2 was called, as well as the total number of Jacobian evaluations required to compute the solution (or the word "fail" if the algorithm failed). Note that a "failure" usually means that the curve tracking was abandoned because the path was long, not that the path would not have eventually reached a solution. Execution times are not included because our prototype implementation was written largely in MATLAB, without regard to computational efficiency.

Implicit in the results is an indication of the performance of Newton's method on these problems. Observe that in numerous cases the algorithm solved the problem with no Step 2 calls for $\delta_0 = 1$. This indicates that Newton's method solved the problem without any calls to the curve tracking method. In contrast, a positive number of Step 2 calls indicates that Newton's method failed.

The computational results indicate that the algorithm is reasonably robust, solving the majority of the test cases, and that it clearly outperforms Newton's method in terms of numbers of problems solved. Interestingly, the more robust version of the algorithm uses $\delta = 1.0$, so that Newton's method is attempted first. In fact, there are several problems that were solved by Newton's method alone, which were not solved using $\delta = 0.01$. This is not surprising since the homotopy map used here is the "default" trivial one, which can result in long and ill-conditioned paths. More sophisticated maps can ameliorate these difficulties. A "failure" with $\delta = 0.01$ often means the path was very long, not that the path would not eventually reach a solution. Simply scaling the solutions to lie inside the unit ball would remove such failures.

It should also be noted that this algorithm does not seem to perform as well on the test suite as other existing complementarity solvers such as the PATH solver [13], which are able to solve nearly all of the problems in the test suite [5]. However, it is notable that the algorithm solves several problems that have so far been considered very difficult. Among these are the billups, pgvon105, and simple-ex problems, which are among the few problems that the PATH solver does not solve [20].

6 CONCLUSIONS

This paper introduced a class of homotopy path following methods for solving nonsmooth systems of equations. These methods are an extension to nonsmooth equations of the probability-one homotopy methods described in [10, 27, 29, 30]. These homotopy methods are complementary to Newton based methods in that they are designed to solve problems that are often difficult for Newton based methods. This is justified both theoretically and computationally. Theoretically, we proved sufficient conditions guaranteeing the existence of a well-behaved zero curve leading to a solution. This result shows that the method will, in principal, be effective at solving highly nonlinear problems where the norm of the residual has nonglobal local minima, provided that the problems have the global structure specified in Theorem 2.11. On the computational side, we implemented a prototype algorithm in MATLAB and

Table 2.1 MCPLIB Test Problems

			$\delta = 1$		$\delta = 0.01$	
Problem Name	st. pt.	size	Step 2 calls	Jac evals	Step 2 calls	Jac evals
badfree	1	5	5	23	1	18
bertsekas	1	15	9	431	3	656
bertsekas	2	15	9	552	3	526
bertsekas	3	15	12	fail	3	627
billups	1	1	1	13	1	18
box	1	44	1	fail	1	fail
colvdual	1	20	1	49	1	83
colvdual	2	20	1	60	1	66
colvnlp	1	15	1	47	1	83
colvnlp	2	15	0	10	1	48
colvtemp	1	20	1	49	1	83
cycle	1	1	0	4	1	18
degen	1	2	0	5	1	12
ehl_k40	1	41	1	106	1	128
ehl_k60	1	61	1	fail	1	580
ehl_k80	1	81	1	416	1	fail
ehl_kost	1	101	1	fail	1	fail
explcp	1	16	0	5	1	24
freebert	1	15	9	616	3	494
freebert	2	15	0	10	1	fail
freebert	3	15	9	293	3	295
freebert	4	15	9	577	3	577
freebert	5	15	0	13	2	fail
freebert	6	15	9	427	3	433
gafni	1	5	0	13	1	44
gafni	2	5	0	11	1	128
gafni	3	5	0	12	1	fail
games	1	16	0	32	4	fail
hanskoop	1	14	1	25	1	41
hanskoop	2	14	1	24	1	39
hanskoop	3	14	1	23	1	32
hanskoop	4	14	1	23	1	34
hanskoop	5	14	2	85	1	176
hydroc06	1	29	0	5	1	fail

Table 2.2 MCPLIB Test Problems (continued)

			$\delta = 1$		$\delta = 0.01$	
Problem Name	st. pt.	size	Step 2 calls	Jac evals	Step 2 calls	Jac evals
jel	1	6	0	9	1	120
josephy	1	4	0	8	1	23
josephy	2	4	0	7	1	14
josephy	3	4	4	333	1	208
josephy	4	4	0	5	1	30
josephy	5	4	0	5	1	18
josephy	6	4	0	8	1	21
kojshin	1	4	0	8	1	25
kojshin	2	4	0	8	1	22
kojshin	3	4	0	8	1	210
kojshin	4	4	0	5	1	26
kojshin	5	4	0	6	1	21
kojshin	6	4	0	6	1	23
mathinum	1	3	0	6	1	19
mathinum	2	3	0	6	1	22
mathinum	3	3	0	11	1	19
mathinum	4	3	0	8	1	21
mathisum	1	4	0	5	1	22
mathisum	2	4	0	7	1	14
mathisum	3	4	0	11	1	29
mathisum	4	4	0	6	1	19
methan08	1	31	0	4	1	fail
multi-v	1	48	1	fail	1	fail
nash	1	10	0	8	1	32
nash	2	10	0	11	1	115
ne-hard	1	3	0	25	5	276
pgvon105	1	105	2	203	3	339
pgvon105	2	105	1	110	1	86
pgvon105	3	105	1	101	1	110
pgvon106	1	106	12	fail	5	fail
pies	1	42	0	15	1	251

Table 2.3 MCPLIB Test Problems (continued)

			$\delta = 1$		$\delta = 0.01$	
Problem Name	st. pt.	size	Step 2 calls	Jac evals	Step 2 calls	Jac evals
powell	1	16	0	9	2	99
powell	2	16	0	12	1	fail
powell	3	16	1	241	1	265
powell	4	16	0	12	1	62
powell_mcp	1	8	0	6	1	24
powell_mcp	2	8	0	7	1	26
powell_mcp	3	8	0	9	1	42
powell_mcp	4	8	0	8	1	29
qp	1	4	0	2	1	16
scarbsum	1	40	4	fail	2	fail
scarfanum	1	13	1	39	1	37
scarfanum	2	13	0	13	1	39
scarfanum	3	13	0	12	3	69
scarfasum	1	14	0	9	1	fail
scarfasum	2	14	1	94	1	fail
scarfasum	3	14	2	86	1	fail
scarfbnum	1	39	12	589	5	224
scarfbnum	2	39	11	504	5	259
scarfbsum	1	40	1	68	2	235
scarfbsum	2	40	9	346	3	183
shubik	1	42	20	fail	11	fail
simple-ex	1	17	1	42	1	39
simple-red	1	13	0	11	1	30
sppe	1	27	1	69	1	114
sppe	2	27	1	46	3	87
tobin	1	42	0	7	1	82
tobin	2	42	0	9	1	79

showed that it is successful in solving some complementarity problems that are not solved by the current state-of-the-art solvers. More sophisticated homotopy maps and solution scaling will be the topic of future work.

References

[1] J. C. Alexander. The topological theory of an embedding method. In H. Wacker, editor, *Continuation Methods*, pages 37–68. Academic Press, New York, 1978.

[2] J. C. Alexander, R. B. Kellogg, T.-Y. Li, and J. A. Yorke. Piecewise smooth continuation. manuscript, 1979.

[3] J. C. Alexander, T.-Y. Li, and J .A. Yorke. Piecewise smooth homotopies. In B. C. Eaves, F. J. Gould, H.-O. Peitgen, and M. J. Todd, editors, *Homotopy Methods and Global Convergence*, pages 1–14. Plenum Press, New York, 1983.

[4] S. C. Billups. A homotopy based algorithm for mixed complementarity problems. UCD/CCM Report No. 124, Department of Mathematics, University of Colorado at Denver, Denver, Colorado, 1998.

[5] S. C. Billups, S. P. Dirkse, and M. C. Ferris. A comparison of large scale mixed complementarity problem solvers. *Computational Optimization and Applications*, 7:3–25, 1997.

[6] C. Buck. *Advanced Calculus*. McGraw–Hill, New York, NY, 3rd edition, 1978.

[7] J. V. Burke and S. Xu. The global linear convergence of a non-interior path-following algorithm for linear complementarity problems. *Mathematics of Operations Research*, 23:719–735, 1998.

[8] B. Chen and X. Chen. A global and local superlinear continuation-smoothing method for $P_0 + R_0$ NCP or monotone NCP. *SIAM Journal on Optimization*, 9:624–645, 1999.

[9] C. Chen and O. L. Mangasarian. A class of smoothing functions for nonlinear and mixed complementarity problems. *Computational Optimization and Applications*, 5:97–138, 1996.

[10] S.-N. Chow, J. Mallet-Paret, and J. A. Yorke. Finding zeros of maps: homotopy methods that are constructive with probability one. *Mathematics of Computation*, 32:887–899, 1978.

[11] F. H. Clarke. *Optimization and Nonsmooth Analysis*. John Wiley & Sons, New York, 1983.

[12] T. De Luca, F. Facchinei, and C. Kanzow. A semismooth equation approach to the solution of nonlinear complementarity problems. *Mathematical Programming*, 75:407–439, 1996.

[13] S. P. Dirkse and M. C. Ferris. The PATH solver: A non-monotone stabilization scheme for mixed complementarity problems. *Optimization Methods and Software*, 5:123–156, 1995.

[14] M. C. Ferris and T. F. Rutherford. Accessing realistic complementarity problems within Matlab. In G. Di Pillo and F. Giannessi, editors, *Nonlinear Optimization and Applications*, pages 141–153. Plenum Press, New York, 1996.

[15] A. Fischer. A special Newton–type optimization method. *Optimization*, 24:269–284, 1992.

[16] S. A. Gabriel and J. J. Moré. Smoothing of mixed complementarity problems. In M. C. Ferris and J. S. Pang, editors, *Complementarity and Variational Problems: State of the Art*, Philadelphia, Pennsylvania, 1997. SIAM Publications.

[17] C. Kanzow. Some equation–based methods for the nonlinear complementarity problem. *Optimization Methods and Software*, 3:327–340, 1994.

[18] C. Kanzow. Some noninterior continuation methods for linear complementarity problems. *SIAM Journal on Matrix Analysis and Applications*, 17:851–868, 1996.

[19] O. L. Mangasarian. Equivalence of the complementarity problem to a system of nonlinear equations. *SIAM Journal on Applied Mathematics*, 31:89–92, 1976.

[20] T. S. Munson. Private communication, 1999.

[21] L. Qi. Regular pseudo-smooth NCP and BVIP functions and globally and quadratically convergent generalized Newton methods for complementarity and variational inequality problems. *Mathematics of Operations Research*, 24(2):440–471, 1999.

[22] L. Qi and J. Sun. A nonsmooth version of Newton's method. *Mathematical Programming*, 58:353–368, 1993.

[23] H. Sellami. A homotopy continuation method for normal maps. *Mathematical Programming*, pages 317–337, 1998.

[24] H. Sellami and S. M. Robinson. Homotopies based on nonsmooth equations for solving nonlinear variational inequalities. In G. Di Pillo and F. Giannessi, editors, *Nonlinear Optimization and Applications*, pages 327–343. Plenum Press, New York, 1996.

[25] H. Sellami and S. M. Robinson. Implementation of a continuation method for normal maps. *Mathematical Programming*, pages 563–578, 1997.

[26] L. T. Watson. An algorithm that is globally convergent with probability one for a class of nonlinear two-point boundary value problems. *SIAM Journal on Numerical Analysis*, 16:394–401, 1979.

[27] L. T. Watson. A globally convergent algorithm for computing fixed points of C^2 maps. *Applied Mathematics and Computation*, 5:297–311, 1979.

[28] L. T. Watson. Solving the nonlinear complementarity problem by a homotopy method. *SIAM Journal on Control and Optimization*, 17:36–46, 1979.

[29] L. T. Watson, S. C. Billups, and A. P. Morgan. Algorithm 652: HOMPACK: A suite of codes for globally convergent homotopy algorithms. *ACM Transactions on Mathematical Software*, 13:281–310, 1987.

[30] L. T. Watson, R. C. Melville, A. P. Morgan, and H. F. Walker. Algorithm 777: HOMPACK90: A suite of FORTRAN 90 codes for globally convergent homotopy algorithms. *ACM Transactions on Mathematical Software*, 23:514–549, 1997.

[31] S. J. Wright. *Primal–Dual Interior–Point Methods*. SIAM, Philadelphia, Pennsylvania, 1997.

[32] S. Xu and J. V. Burke. A polynomial time interior-point path-following algorithm for LCP based on Chen-Harker-Kanzow smoothing techniques. *Mathematical Programming*, 86:91–104, 1999.

Chapter 3

SCALABLE PROBABILISTIC CLUSTERING

P. S. Bradley

Microsoft Research, Microsoft Corporation, One Microsoft Way, Redmond, WA 98052

bradley@microsoft.com

U. M. Fayyad

digiMine.com, 11250 Kirkland Way, Kirkland, WA 98033

usama@digiMine.com

C. A. Reina

Microsoft Research, Microsoft Corporation, One Microsoft Way, Redmond, WA 98052

coryr@microsoft.com

Abstract The Expectation-Maximization (EM) algorithm is a popular approach to probabilistic database clustering. A database of observations is clustered by identifying k sub-populations and summarizing each sub- population with a model or probability density function. The EM algorithm is an approach that iteratively estimates the memberships of the observations in each cluster and the parameters of the k density functions for each cluster. Typical EM implementations require a full database scan at each iteration and the number of iterations required to converge is arbitrary. For large databases, these scans become prohibitively expensive. We present a scalable implementation of the EM algorithm based upon identifying regions of the data that are compressible and regions that must be maintained in memory. The approach operates within the confines of a limited main memory buffer. Data resolution is preserved to the extent possible based upon the size of the memory buffer and the fit of the current clustering model to the data. We extend the framework to update multiple cluster models simultaneously. Computational tests indicate that this scalable scheme outperforms sampling-based and

M.C. Ferris et al. (eds.), Complementarity: Applications, Algorithms and Extensions, 43–65.

incremental approaches – the straightforward alternatives to "scaling" existing traditional in-memory implementations to large databases.

1 INTRODUCTION

Data clustering is important in many fields, including data mining [17], statistical data analysis [19, 3], compression [31], and vector quantization. Applications include data analysis and modeling [15, 16], image segmentation, marketing, fraud detection, predictive modeling, data summarization, and general data reporting tasks [6]. Clustering is a crucial data mining step and performing this task over large databases is essential.

A general view of clustering places it in the framework of density estimation [28, 26, 3]. Clustering can be viewed as identifying the dense regions of the data source or, equivalently, local maxima of the probability density function governing the data source. An effective representation of the probability density function is the *mixture model* which asserts that the underlying density is a convex combination of k component density functions, corresponding to k clusters. The problem can be succinctly stated as follows. Given a database of observations, identify k sub-populations and provide a model (density) summarizing the observations in each sub-population.

The Expectation-Maximization (EM) algorithm [12, 10] is an effective technique for estimating the mixture model parameters over a given database. The EM algorithm takes as input initial mixture model parameters, which may be determined randomly or by some other process, and iteratively refines them to maximize the log-likelihood of the data given the mixture model. The EM algorithm terminates at local maximum or, rarely, a saddle point [5].

The EM algorithm is similar to the K-Mean [2, 14, 18] and K-Median [9] algorithms in that it alternates between a *cluster assignment* step and a *cluster update* step at each iteration. The cluster assignment step in EM differs from that of K-Mean or K-Median in that each data point is assigned to each cluster with a probability of membership, whereas K-Mean and K-Median assign each data point to one cluster only. Given initial estimates of the k distributions summarizing each cluster, each data point x is assigned to cluster h based upon the probability that data point x was generated by cluster h [5]. Once the cluster memberships have been determined, parameters of the density functions for each cluster are updated. This process is repeated until the parameters of the density functions cease to change or the underlying objective function value (log-likelihood) remains unchanged.

In estimating the parameters of the mixture model over large databases, a single database scan is an expensive proposition. Typical EM implementations require a full database scan at each iteration in the cluster assignment step. The number of iterations required for EM to converge is arbitrary and hundreds to thousands of iterations have been observed over large databases. Hence application of the standard EM implementation to large databases is unacceptable. We present a scalable version of EM, which satisfies the following requirements.

1. The database is scanned at most once, with early termination when appropriate.
2. The algorithm operates within the confines of a limited main memory (RAM) buffer.
3. The database is scanned with a forward-only cursor.

The single forward-only database scan addresses the fact that the individual data points provided to the clustering algorithm may be the result of an expensive query over a potentially distributed data warehouse, requiring much processing to bring together all dimensions that constitute a point. Operating within the confines of a limited main memory buffer not only makes the approach applicable to wide array of hardware, but has the added benefit of constraining the EM process from overtaking the resources of the database server.

The problem of computing the mixture model parameters is decomposed resulting in the ability to cluster arbitrarily large databases while utilizing a small amount of main memory. The fundamental observation is that all data is not of equal importance when estimating the mixture model parameters. Data points may be classified into one of three categories while maintaining integrity of the mixture model: points that can be safely discarded, points that can be compressed, and points that must be maintained in memory.

The utility of mixture models estimated via EM has been demonstrated in approximating typical database count queries over a specified region [27] (e.g. return the number of records in the database with salary between 25,000 and 40,000 and age between 25 and 35). Mixture models estimated via EM have also been used to speed up nearest neighbor queries over large databases [4]. These methods require statistical semantics and theory, and clustering methods that do not derive proper statistical models are at a substantial disadvantage.

Prior work most relevant to the proposed method is the analysis of incremental versions of the general EM algorithm presented in [21]. These EM variants require multiple data scans to converge, although

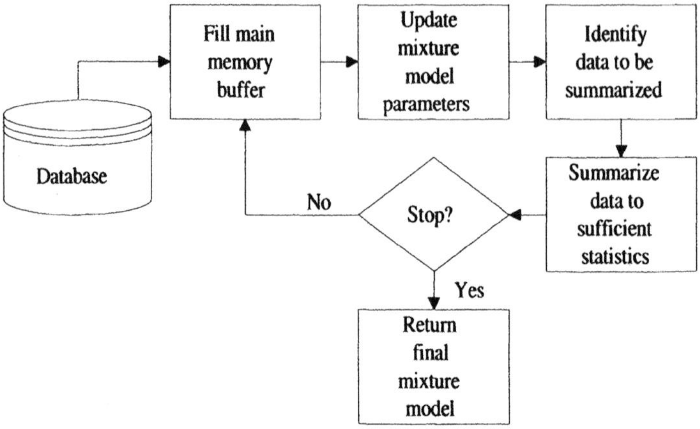

Figure 3.1 Overview of the Scalable EM process.

the method proposed here is compared with one of the incremental algorithms proposed in [21] with the added computational constraint of a single database scan.

Recently, the problem of mixture model estimation has been formulated as a mathematical program with equilibrium constraints (MPEC) in [23] and applied to a multispectral minefield classification application.

Previous scalable clustering work has focused on K-Mean-type approaches [31, 7] and region growing [22, 24, 1]. These techniques, while effective, do not derive statistical models of the data (i.e. they are based on notions of distance metrics, etc.) and do not allow for cluster overlap (a data record may belong to different clusters with different membership probabilities). The mixture model requires the specification of a probability density function instead of a distance metric. Hence clustering databases with continuous and discrete attributes (e.g. eye color) can be accomplished by defining the appropriate probability density functions over these different attribute types. Due to the probabilistic nature of the mixture model, a data point may belong to multiple clusters with different membership probabilities.

Before formally describing mixture models and the scalable EM approach, we discuss the architecture of the scalable EM algorithm at a high level.

Scalable EM Architecture

In making the EM algorithm scalable to large databases, we begin with the assumption that an interface to the database providing a uniform random data sample of observations is available, noting that obtaining a random sample from a database may be a non-trivial task. The scalable EM approach requires the specification of initial mixture model parameters (as does standard EM) along with main a memory (RAM) buffer size, in addition to a stopping tolerance, etc. We emphasize that the scalable EM process does not require additional memory to that specified in the buffer size, although best results are obtained by specifying a buffer size sufficient so that each cluster contains a representative sample.

The scalable EM algorithm operates as follows (see Figure 3.1).

1. The algorithm begins by loading a uniform random sample into the allocated main memory buffer.

2. The standard EM algorithm is then applied to the contents of the main memory buffer, updating the mixture model parameters over the data in the buffer. Recall that the standard EM algorithm requires multiple data passes to converge. Since these occur only over the contents of the memory buffer, this operation is inexpensive. Standard EM is run until it converges over the data in the memory buffer.

3. After the mixture model parameters are updated in step 2, we identify data that may "safely" be summarized. The goal of this step is to summarize data via sufficient statistics, while minimizing the resulting error in updating the mixture model. *Sufficient statistics* are functions of the data that provide just the summary needed to compute or perform inference on a parameter of interest (e.g. the mean and covariance matrix for a cluster). The sufficient statistics derived from a set of data require less memory than the set. Hence, after summarization, memory is available to process more data on the next scalable EM iteration. See Figure 3.2.

4. After identifying and summarizing data via sufficient statistics, we check to see if the mixture model satisfies the stopping criteria. We explicitly stop the scalable EM algorithm when the mixture model has not changed or if the database has been fully scanned. Otherwise, the process repeats.

In summary, EM is applied to the database in "chunks". After the mixture model is updated over a chunk, portions of the chunk are summarized. Future updates to the mixture model occur over new data chunks coupled with the summary of past chunks processed.

Identifying data to summarize (Step 3, above) is implemented in two phases. Figure 3.2 is an example of a mixture model with 3 Gaussian distributions summarizing 3 clusters in 2 dimensional real-space. Figure

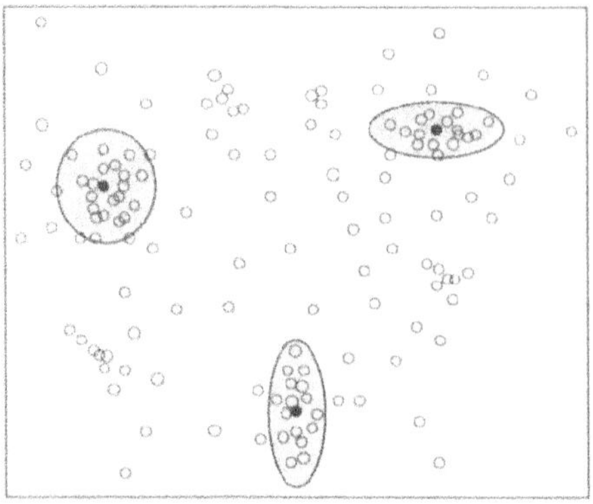

(a) Primary data summarization

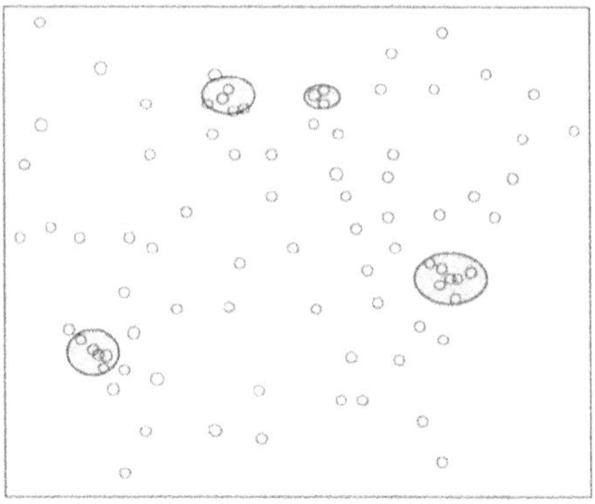

(b) Secondary data summarization

Figure 3.2 An example of data summarization for 3 Gaussian distributions in 2 dimensions. Data points are open circles, shaded areas indicate data regions having points to be summarized.

3.2(a) depicts the first phase of data summarization. In the Gaussian case, the first phase of data summarization occurs near the means of the $k = 3$ Gaussian distributions. The assumption is made that the 3 Gaussians are stable in the sense that the location of their means will not change drastically as the result of processing more data. Note that this assumption is valid as long as the random samples obtained from the database are large enough so that each cluster contains a sufficient number of points (i.e. a confidence interval on the Gaussian mean is small when a sufficient number of points are used to estimate it).

Figure 3.2(b) depicts the second phase of data summarization, which occurs in small, "dense" regions of the data not near the Gaussian means. The motivation for identifying and summarizing these portions of the data space is that each "dense" region updates each cluster as a "group". Updates to the mixture model can be made by approximating each group with a single data point (the centroid of the group) weighted by the number of data points in the group.

The paper is organized as follows. Section 2 formally describes mixture models and estimation of mixture model parameters via EM. Section 3 describes the scalable EM variant in detail. We make some observations on extending scalable EM to update multiple mixture models simultaneously in Section 4. Numerical results are presented in Section 5 and Section 6 concludes the paper.

2 PRELIMINARIES

The mixture model form of the probability density function is a convex combination of component density functions modeling each cluster $h = 1, \ldots, k$. Let the database D consist of m records, each having n attributes. Then the probability of observing a database record x is:

$$p(x) = \sum_{h=1}^{k} p_h \cdot f_h(x|\phi_h). \tag{3.1}$$

The weights in the convex combination (3.1), denoted as p_h, represent the fraction of the database belonging to cluster h. The weight p_h equivalently represents the probability of observing a data point from the h-th cluster and the set of weights form a probability distribution over the k clusters: $\sum_{h=1}^{k} p_h = 1$, $p_h \geq 0$. The functions $f_h(x|\phi_h)$, $h = 1, \ldots, k$ are the cluster or component density functions modeling the records with each cluster, and ϕ_h specifies the parameter values for the h-th component density function (e.g. for a Gaussian distribution $f_h(x|\phi_h)$, ϕ_h is the mean and covariance matrix). The component density functions are normalized so that $\int f_h(x|\phi_h)dx = 1$, $f_h(x|\phi_h) \geq 0$, and can be regarded as class-conditional densities [5].

To generate a data record from the mixture model density (3.1), one of the components h is selected with probability p_h. Then a record is generated from the component density function $f_h(x|\phi_h)$. The mixture model can approximate any continuous density function given enough components and properly chosen mixture model parameters [5].

The mixture model also allows for "overlapping" clusters in the sense that data records may belong to all k clusters with different membership probabilities. This assignment generalizes the "hard" assignment employed by K-Mean [2, 14, 18] and K-Median [9] type algorithms where data records are members of one and only one cluster. The probability of membership or weight of data record x in cluster h is:

$$w_h(x) = \frac{p_h \cdot f_h(x|\phi_h)}{p(x)}. \tag{3.2}$$

The denominator is given by (3.1).

By making the assumption that the attributes of a database record are independent within a cluster, the component density function can be decomposed as a product of density functions over each of the attributes x_j of the data point x:

$$f_h(x|\phi_h) = \prod_{j=1}^{n} f_h(x_j|\phi_h). \tag{3.3}$$

Let Φ denote the vector of mixture model parameters (parameters for each of the component density functions, ϕ_h, and the weights p_h in (3.1)). The EM algorithm iteratively refines an initial estimate of Φ and terminates at stationary point of the following maximum likelihood problem [12, 5]:

$$\max_{\Phi} \sum_{x \in D} \log(p(x)). \tag{3.4}$$

We focus on the application of the EM algorithm to estimate mixture model parameters over a database D in which all n attributes take continuous values and data records can be regarded as points in n-dimensional real-space R^n. Additionally, we choose multivariate Gaussian distributions as the component density functions modeling each cluster.

Throughout this paper, all vectors will be column vectors unless transposed to a row vector by a prime superscript $'$. For an $n \times n$ matrix $A \in R^{n \times n}$, $|A|$ will denote the determinant of A and A^{-1} indicates its matrix inverse.

The multivariate Gaussian characterizing cluster h is parameterized by a mean vector $\mu(h) \in R^n$ and covariance matrix $\Sigma(h) \in R^{n \times n}$:

$$f_h(x|\mu(h), \Sigma(h)) = \frac{1}{\sqrt{(2\pi)^n |\Sigma(h)|}} \exp\left(-\frac{1}{2}(x - \mu(h))'(\Sigma(h))^{-1}(x - \mu(h))\right). \tag{3.5}$$

The Gaussian mixture model parameters consist of the mixture weights, the means and covariances for each component Gaussian:

$$\Phi = \{p_h \in [0,1],\ \mu(h) \in R^n,\ \Sigma(h) \in R^{n \times n},\ h = 1, \ldots, k\}. \tag{3.6}$$

For a specific value of the parameter vector Φ, the log-likelihood objective (3.4) is denoted by $L(\Phi)$:

$$L(\Phi) = \sum_{x \in D} \log \left(\sum_{h=1}^{k} p_h \cdot f_h(x|\mu(h), \Sigma(h)) \right). \tag{3.7}$$

We next outline the standard EM algorithm. The method begins with initial mixture model parameters and iteratively updates them so that $L(\Phi)$ is monotonically non-decreasing at each iteration [12, 5].

Algorithm 3.1 Standard Expectation-Maximization (EM) *Given a database D of m records in R^n, a stopping tolerance $\epsilon > 0$ and mixture model parameters $\Phi^t = \{p_h^t, \mu^t(h), \Sigma^t(h), h = 1, \ldots, k\}$ at iteration t, compute Φ^{t+1} at iteration $t+1$ in the following 2 steps:*

1. *Compute probabilistic assignments (3.2):*

$$w_h^t(x) = \frac{p_h^t \cdot f_h(x|\mu^t(h), \Sigma^t(h))}{\sum_{\hat{h}=1}^{k} p_{\hat{h}}^t \cdot f_{\hat{h}}(x|\mu^t(\hat{h}), \Sigma^t(\hat{h}))}, \quad h = 1, \ldots, k, \ \forall x \in D. \tag{3.8}$$

2. *Compute updated parameters Φ^{t+1}: for $h = 1, \ldots, k$ set*

$$\mu^{t+1}(h) = \frac{\sum_{x \in D} w_h^t(x) \cdot x}{\sum_{x \in D} w_h^t(x)}, \tag{3.9}$$

$$\Sigma^{t+1}(h) = \frac{\sum_{x \in D} w_h^t(x) \cdot (x - \mu^{t+1}(h))(x - \mu^{t+1}(h))'}{\sum_{x \in D} w_h^t(x)}, \tag{3.10}$$

$$p_h^{t+1} = \frac{\sum_{x \in D} w_h^t(x)}{m}. \tag{3.11}$$

If $|L(\Phi^{t+1}) - L(\Phi^t)| \leq \epsilon$ then terminate, otherwise increment t by 1 and go to 1.

Note that a full data scan is required at each iteration of the standard EM algorithm and the number of iterations required to converge is arbitrary. We next discuss a scalable variant of EM requiring at most one full data scan.

3 MODEL ESTIMATION OVER LARGE DATABASES

Motivation for this work centers on the observation that EM can be sequentially applied to samples from the database D. After a data sample has been processed, it is possible to effectively continue to update the mixture model parameters by selectively summarizing regions of the data sample. The approach assumes an interface to the database allowing the algorithm to load sequential random data samples. The allocated main memory buffer is initially filled with a random sample. The initial mixture model parameters Φ^0 are updated over the contents of the buffer via the standard EM Algorithm 3.1. Then each data point in the buffer is classified as belonging to one of three sets, depending on the degree to which the data point is summarized by the current mixture model: (1) the retained set $\mathcal{R}$ (points that will remain in the main memory buffer), (2) the discard set $\mathcal{D}$ (subsets which can be safely discarded and modeled effectively by the sufficient statistics associated with one of the clusters), and (3) the summarized set $\mathcal{C}$ (subsets that update each cluster as a unit and can be modeled by sufficient statistics). Storing the sufficient statistics derived from data subsets requires less memory, hence freeing space in the main memory buffer allowing the next sample of data to be loaded from the database.

Prior to formalizing the scalable EM (SEM) algorithm, we further define a few useful terms. For a set S, let $card(S)$ denote its cardinality. The average log-likelihood for a mixture model with parameters Φ over a database with m records is:

$$\bar{L}(\Phi, m) = \frac{1}{m} L(\Phi). \tag{3.12}$$

And $L(\Phi)$ is given in (3.7).

We next specify the scalable EM algorithm, noting that specific components are described in detail in the following sections.

Algorithm 3.2 Scalable EM (SEM) *Initialize the summary sets* $\mathcal{R}^0 = \emptyset$, $\mathcal{C}^0 = \emptyset$, $\mathcal{D}^0 = \emptyset$. *Initialize the number of data points processed* $m^0 = 0$. *Given a database* D *with* m *records in* R^n, *initial mixture model parameters* Φ^0, *an interface to* D *that supplies sample subsets* $S^t \subset D$, *a specified main memory buffer size and a stopping tolerance* ϵ, *set iteration counter* $t = 0$ *and update* Φ^t *to* Φ^{t+1} *as follows.*

1. *Set* $\mathcal{R}^{t+1} = \mathcal{R}^t \cup S^t$ *so that the memory requirement for* $\mathcal{R}^{t+1} \cup \mathcal{C}^t \cup \mathcal{D}^t$ *equals the specified main memory buffer size. Set* $m^{t+1} = m^t + card(S^t)$.

2. *Apply extended EM to data* $\mathcal{R}^{t+1} \cup \mathcal{C}^t \cup \mathcal{D}^t$, *updating* Φ^t *to* Φ^{t+1} *(see Model Updates over Sample + Sufficient Statistics, below).*

3. *Partition the singleton data points in* $\mathcal{R}^{t+1}$:

i. *Identify data subsets updating a single cluster (see Primary Data Summarization, below). Augment $\mathcal{D}^{t+1}$ by incorporating the sufficient statistics derived from these subsets (see Data Summarization and Storage, below). Remove data from the subsets identified from $\mathcal{R}^{t+1}$.*

ii. *Identify data subsets updating clusters as a single unit (see Secondary Data Summarization, below). Augment $\mathcal{C}^{t+1}$ by incorporating the sufficient statistics derived from these subsets (see Data Summarization and Storage, below). Remove data from the subsets identified from $\mathcal{R}^{t+1}$.*

4. *If $t > 1$ and $|\bar{L}(\Phi^t, m^t) - \bar{L}(\Phi^{t+1}, m^{t+1})| \leq \epsilon$, terminate. If $m^{t+1} = m$, terminate (i.e. terminate if the database has been fully scanned). Otherwise, increment t by 1 and go to 1.*

We elaborate on Algorithm 3.2 first by describing sufficient statistics for the Gaussian mixture model. Note that sufficient statistics are tailored to the specific choice of the cluster or component densities.

Data Summarization and Storage

Step 3 of Algorithm 3.2 places data to be summarized into one of two sets $\mathcal{D}$ or $\mathcal{C}$. This summarized data then updates the mixture model parameters through sufficient statistics in future iterations. Let $T = \{x^1, x^2, \ldots, x^N\} \subset D$ be a subset of data to be summarized by the sufficient statistics triple (θ, Γ, N) : $\theta = \sum_{i=1}^{N} x^i \in R^n$, $\Gamma = \sum_{i=1}^{N} (x^i)(x^i)' \in R^{n \times n}$.

$$\begin{aligned} \mu_T &= \tfrac{1}{N} \cdot \theta, \\ \Sigma_T &= \tfrac{1}{N} \left(\Gamma - \tfrac{1}{N}\theta\theta'\right). \end{aligned} \tag{3.13}$$

The sufficiency of the triple (θ, Γ, N) for the mean μ_T follows from the definition of the empirical mean of a set of values. The sufficiency of the triple for the covariance matrix Σ_T is as follows. Note that element u, v of the covariance matrix Σ_T is the covariace between dimension u and dimension v over T. Hence:

$$\begin{aligned} (\Sigma_T)_{u,v} &= Cov(x_u, x_v) \text{ over } T, \\ &= \tfrac{1}{N} \textstyle\sum_{i=1}^{N} (x_u^i) \cdot (x_v^i) - ((\mu_T)_u) \cdot ((\mu_T)_v), \\ &= \tfrac{1}{N} \left[\textstyle\sum_{i=1}^{N} (x_u^i) \cdot (x_v^i) - \tfrac{1}{N} \left(\textstyle\sum_{i=1}^{N} x_u^i \right) \cdot \left(\textstyle\sum_{i=1}^{N} x_v^i \right) \right], \\ &= \tfrac{1}{N} \left[\Gamma_{u,v} - \tfrac{1}{N} \theta_u \cdot \theta_v \right] \end{aligned} \tag{3.14}$$

Line 2 above follows from taking the definition of covariance, expanding terms and taking expectation of each term separately [13].

The sufficient statistics resulting from merging two sets T^1 and T^2, represented by $(\theta^1, \Gamma^1, N^1)$ and $(\theta^2, \Gamma^2, N^2)$, respectively, are obtained for $T^1 \cup T^2$ by $(\theta^1 + \theta^2, \Gamma^1 + \Gamma^2, N^1 + N^2)$.

The stopping criteria in Algorithm 3.2 is estimated over the contents of the buffer $(\mathcal{R}^{t+1} \cup \mathcal{C}^t \cup \mathcal{D}^t)$ to avoid doing a full data scan to compute the log-likelihood value (3.7):

$$\bar{L}(\Phi, m) \approx \frac{1}{m}\left[\sum_{x} \log(p(x)) + \sum_{(\theta,\Gamma,N)} N \cdot \log(p(\theta/N))\right]. \tag{3.15}$$

The value of p is given in (3.1).

Model Updates over Sample + Sufficient Statistics

Step 2 of Algorithm 3.2 consists of performing EM iterations over $\mathcal{R}^{t+1} \cup \mathcal{C}^t \cup \mathcal{D}^t$ consisting of singleton data points $x \in \mathcal{R}^{t+1}$ and sets of sufficient statistics $(\theta, \Gamma, N) \in \mathcal{C}^t \cup \mathcal{D}^t$. The Extended EM (ExEM) Algorithm is a slight augmentation of the standard EM Algorithm 3.1 which performs this operation. ExEM updates the mixture model parameters exactly as standard EM over singleton data points in $\mathcal{R}^{t+1}$. Updates over sufficient statistics are accomplished by treating the data subset represented by $(\theta, \Gamma, N) \in \mathcal{C}^t \cup \mathcal{D}^t$ as a single data point (the mean of the subset) weighted by N. A detailed description of ExEM is given in [8].

We apply a post-processing step after the ExEM algorithm has converged which resets the parameters of "empty" clusters – those with p_h^{t+1} less than a user-specified tolerance. Termination with "empty" clusters is a problem that plagues both standard EM and K-Mean, especially when applied to high-dimensional data. Empty clusters are viewed as a problem since we make the assumption that the value of k supplied by the user is "correct". We also take the view that empty clusters are poor local maxima of the log-likelihood objective and a higher likelihood value is conceivable by resetting an empty cluster so that it summarizes at least one data point.

We address this problem by resetting the means and covariances of a component Gaussian distribution that has gone "empty". Suppose cluster $h_{\mathbf{empty}}$ is "empty". Its mean is reset to the mean of a subset in $\mathcal{C}^t \cup \mathcal{D}^t$ which is least likely under the current model with parameters $\Phi^{t+1} = \{p_h^{t+1}, \mu^{t+1}(h), \Sigma^{t+1}(h), h = 1, \dots, k\}$:

$$\mu^{t+1}(h_{\text{empty}}) \in \arg\min_{\tilde{\mu}} \left\{ \log\left(\sum_{h=1}^{k} p_h^{t+1} \cdot f_h(\tilde{\mu} | \mu^{t+1}(h), \Sigma^{t+1}(h)) \right) \middle| \begin{array}{r} \tilde{\mu} = \frac{\theta}{N}, \\ (\theta, \Gamma, N) \in \mathcal{C}^t \cup \mathcal{D}^t. \end{array} \right\} \quad (3.16)$$

Cluster h_{empty} is then reset with the mean determined in (3.16) along with the corresponding covariance matrix. By resetting an empty cluster to mean and covariance of this subset, we essentially gain the greatest increase in log-likelihood over candidates in $\mathcal{C}^t \cup \mathcal{D}^t$.

After the "empty" cluster has been reset, the ExEM Algorithm is applied to the parameters Φ^{t+1} with the new values for $\mu(h_{\text{empty}})$ and $\Sigma(h_{\text{empty}})$.

Data Summarization

Data summarization involves two phases. *Primary summarization* occurs near the means of the k component Gaussians and produces the set of sufficient statistics $\mathcal{D}$. *Secondary summarization* occurs in dense regions of data not near the k Gaussians means and produces the set of sufficient statistics $\mathcal{C}$. Updating the mixture model parameters over sets of sufficient statistics is no longer exact (in the sense that updates differ from those in the standard EM Algorithm 3.1). However, by accurately determining data regions to summarize, we aim to minimize mixture model update error when constrained to the single data scan and limited main memory buffer. We note that a subset of data is summarized only if the resulting sufficient statistics require less memory than the subset generating them.

Primary Data Summarization

If the current mixture model is accurate, areas near the component Gaussian means $\mu(h)$ will have greatest data density. The full mixture model can be sufficiently updated by locally modeling these regions with sufficient statistics. The basic intuition is to discard data points that are not likely to change degree of membership in the clusters. If the mixture model is stable, the assumption is made that data near the component Gaussian means will not change cluster membership. The assumption is made that a data sample of sufficient size is available so that the confidence interval on the Gaussian means are sufficiently small. Prior to performing primary data summarization near the mean of cluster h, it is prudent to check the confidence interval.

The first primary data summarization approach is a simple approximation that works well in practice. If the Gaussian clusters are separated, then data near the means of the k component Gaussians is unlikely to change cluster assignment and only affects updates to the nearest cluster. This data may be summarized. To identify data sufficiently close to the Gaussian means, we threshold the Mahalanobis distance [14] between a data point $x \in \mathcal{R}$ and the nearest cluster with mean μ and covariance matrix Σ:

$$D_M(x, \mu, \Sigma) = \sqrt{(x - \mu)'\Sigma^{-1}(x - \mu)}. \qquad (3.17)$$

The set of data points to be summarized near the mean of cluster h is then given by the set $\mathcal{D}(h) = \{x \in \mathcal{R} \,|\, D_M(x, \mu(h), \Sigma(h)) \leq r^h\}$ for a specified radius r^h. One approach is to determine the radii r^h, $h = 1, \ldots, k$ so that $p\%$ of the data in $\mathcal{R}$ is summarized.

Once the sets $\mathcal{D}(h)$ have been determined, the sufficient statistics are merged with sufficient statistics determined in step 3(i) of Algorithm 3.2 in previous iterations. The set $\mathcal{D}$ is then given by $\mathcal{D} = \{\mathcal{D}(1), \ldots, \mathcal{D}(k)\}$. The data points from $\mathcal{R}$ generating the sufficient statistics contributing to the set $\mathcal{D}$ are then removed from $\mathcal{R}$.

Another primary data summarization approach performs sensitivity analysis computations over regions bounding the location of mixture model parameters. See [8] for details.

Figure 3.2(a) illustrates an example with 3 Gaussians in 2 dimensions. The shaded area indicates primary data summarization occurring near the 3 Gaussian means.

Secondary Data summarization

Secondary data summarization identifies sub-clusters of points that contribute to the component Gaussian updates as a single unit. Secondary data summarization has two parts: (1) locate candidate subsets of $\mathcal{R}$ which have not been summarized in the primary summarization phase (see above), and (2) determine if these candidate subsets pass a criterion to be treated as a unit. We employ the approximation that a candidate subset with a sufficiently small covariance updates the component Gaussians as a unit. This is valid when the data set lies within the region of the component Gaussian tails. Candidate subsets are determined by applying the standard K-Mean clustering algorithm [2, 14, 18] to the current set of singleton data in $\mathcal{R}$ (after removing the data items summarized in the primary phase). To increase the likelihood of finding "dense" candidate subsets (having small covariances), K-Mean searches for a large number of candidate sub-clusters $k' > k$ and is initialized by randomly selecting k' singleton data records from $\mathcal{R}$. The K-Mean algorithm is used here since "hard" cluster memberships are needed for data summarization – a data point must be an element of a set to be summarized or not.

Once k' sub-clusters have been determined, a criterion requiring the maximum element of the covariance matrix diagonal be bounded by a threshold β is applied. For a subset of data with sufficient statistics (θ, Γ, N), the criterion is satisfied if:

$$\max_{j=1,\dots,n}\left\{\sqrt{\frac{1}{N}\cdot\left[\Gamma_{jj}-\frac{1}{N}\theta_j^2\right]}\right\} \leq \beta. \tag{3.18}$$

Suppose $k'' \leq k'$ sub-clusters satisfy this criterion. The sufficient statistics for these k'' sub-clusters are appended to the set $\mathcal{C}$. Specifically, $\mathcal{C}$ is updated to be $\mathcal{C} \cup \{(\theta^1, \Gamma^1, N^1), \dots, (\theta^{k''}, \Gamma^{k''}, N^{k''})\}$. After $\mathcal{C}$ is updated, its elements are merged via hierarchical agglomerative clustering (HAC) [14]. The nearest two sub-clusters in $\mathcal{C}$ are determined and merged as long as the new merged sub-cluster does not violate (3.18). If the new merged sub-cluster satisfies (3.18), the merged sub-cluster is kept in $\mathcal{C}$ while the smaller sub-clusters generating it are removed. The goal of this merging procedure is so that the resulting set $\mathcal{C}$ has a minimal number of elements satisfying (3.18).

The shaded area of Figure 3.2(b) indicates secondary data summarization occurring in dense regions of the data after the data summarized in the primary phase (Figure 3.2(a)) have been removed.

4 ONE SCAN, MANY SOLUTIONS

The scheme presented involves updating a single model over a database. However, the summarization machinery also admits the possibility of updating multiple mixture models simultaneously within a single data scan. The key insights are: (1) the retained singleton data points in $\mathcal{R}$ and the sufficient statistics in the set $\mathcal{C}$ (see Secondary Data Summarization, above) may be shared among multiple mixture models; and (2) each mixture model M_i, $i = 1, \dots, \nu$ will have its own set $\mathcal{D}_i$ associated with it (see Primary Data Summarization, above). Each model M_i requires its own set of sufficient statistics $\mathcal{D}_i$ summarizing data near the Gaussian means of M_i. Each model does not require its own set $\mathcal{C}$ since this data summarization occurs over regions of the data *not* near the Gaussian means of *any* model.

The Scalable EM Algorithm 3.2 is altered slightly to allow updating and data summarization to occur over multiple models. Recall that the

Extended EM algorithm (see Model Updates over Sample + Sufficient Statistics, above) performs model updates over the singleton data elements in $\mathcal{R}$ and over the sufficient statistics in $\mathcal{C} \cup \mathcal{D}$. When updating model M_i, ExEM is called with the singleton data elements in $\mathcal{R}$ and the sufficient statistics in $\mathcal{C} \cup_{i=1}^{\nu} \mathcal{D}_i$. The primary data summarization step (step 3(i) of Algorithm 3.2), is altered slightly when updating multiple models simultaneously. Each data point $x \in \mathcal{R}$ is evaluated to determine whether it can be summarized into any of the sets $\mathcal{D}_i$ associated with models M_i, $i = 1, \ldots, \nu$. Suppose a data point x satisfies the criteria for summarization in two models M_i and M_j. The point x is summarized in model M_i if M_i has a Gaussian cluster mean which is nearer to x than any of the cluster means in M_j (distance is measured via the Mahalanobis metric (3.17)). A data point cannot be allowed to enter more than one set $\mathcal{D}_i$ as it would be accounted for more than once during model updating.

It is worthy to note that there is an implicit coupling among the models M_i, $i = 1, \ldots, \nu$ since all sufficient statistics $\mathcal{C}$ and $\mathcal{D}_i$, $i = 1, \ldots, \nu$ are used in updating the models in the ExEM algorithm. It is empirically observed that when the Scalable EM Algorithm 3.2 is run with multiple models, more models are converging to better solutions as measured by the log-likelihood objective (3.7) than if the models were computed sequentially.

A weakness of the presented scheme is that data summarization, once performed, cannot be "undone" in the single-scan framework. We conjecture that the error introduced from summarization of data to sufficient statistics is more than accounted for by the fact that the algorithm processes more data (and has "memory" of data already processed) than sampling alternatives. We now focus on empirical evaluations.

5 EMPIRICAL EVALUATION

In the following experiments, the Scalable EM Algorithm 3.2 (SEM) parameters were fixed to the following values. Primary summarization factor p (percentage of data summarized in the primary phase) was set to 0.5. The secondary data summarization tolerance β was set to 0.5 and k' (number of sub-clusters to search for in secondary summarization) was set to $2k$ where k is the number of mixture model components or clusters. Synthetic data was generated so that values in every dimension have similar scale. Real-world databases have been normalized so that the average value along each dimension is zero and standard deviation is one. Sensitivity of SEM to parameter settings is studied in [8]. The

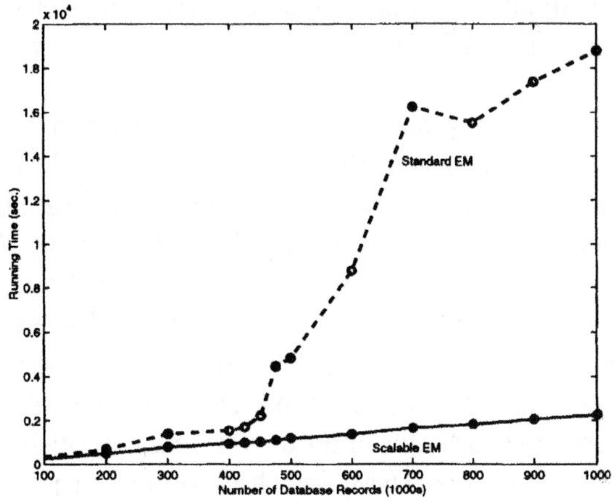

Figure 3.3 Running time vs. database size.

parameter values chosen here were those producing the highest quality models in the study.

Synthetic Data

Evaluations over synthetically generated data target scalability and quality of mixture models computed by the proposed approach.
Scalability: This experiment is intended to illustrate algorithm behavior as the number of database records grows. Synthetic data was generated by a mixture of 10 Gaussian distributions in R^{25}. Scalable EM (SEM, Algorithm 3.2) was applied to databases in which the number of points varied from 100,000 to 1 million. SEM was allowed a main memory buffer size equivalent to storing 5% of the total database. Ten mixture models were simultaneously updated. Initial mixture model means were chosen by randomly selecting 10 data points and initial covariance matrices were the identity. Running times for SEM and standard EM (EM, Algorithm 3.1) are given in Figure 3.3. SEM scales linearly while standard EM begins to utilize virtual memory and page to disk when the database size reaches 450,000 points. The experiment was conducted on a PII-400 Pentium workstation with 128 MB RAM.

Despite the computational overhead of performing data summarization, SEM runs much faster than standard EM even when the database fits into main memory (RAM) and there is no disk I/O overhead to access data records. It is conjectured that since SEM accesses the sufficient statistics of summarized data so frequently, better use is made of the on-CPU memory cache. Updating over the entire database (standard EM) is more likely to cause a cache fault whereas updating over cache memory is significantly faster as it avoids memory bus access off the CPU.

Model Quality: Mixture model quality is quantified by the average log-likelihood of the data given the mixture model (3.12). SEM was compared with a batch variant of an incremental EM (BEM) implementation in which the main memory buffer allocated to the clustering process is filled with a random sample from the database, the mixture model is updated over the contents of the buffer via the standard EM algorithm, then the buffer is purged and filled with another sample. This process is repeated until the entire data set has been scanned. SEM was also compared with an incremental version (or on-line version) of EM [21], which we refer to as 1EM, in which one data point is scanned, the mixture model is updated, another data point is scanned, etc. The method is limited to a single data scan in this case in order to stay within the framework of interest. SEM is also compared with a simple sample-based approach (SampEM) where the main memory buffer is filled with a random sample and standard EM is applied only to this sample. Results are summarized in Table 3.1 for databases synthetically generated by a mixture of 10 Gaussians with 50,000 points in R^{10}, R^{25} and R^{50}. Average log-likelihood of the data given the model is computed for the 10 mixture models determined by SEM and the 10 best models computed by BEM, 1EM and SampEM. Entries in Table 3.1 are averages over the 10 models plus/minus one standard deviation. As dimensionality increases, quality of the mixture models for all methods decreases, but SEM mixture model quality degrades more gracefully than BEM, 1EM and SampEM.

n	SEM	BEM	1EM	SampEM
10	**-10.3**±0.3	-23.0±3.2	-16.0±0.0	-38.4±5.0
25	**-25.6**±2.2	-37.6±2.0	-37.3±0.0	-103.7±13
50	**-49.0**±4.8	-84.1±8.2	-72.6±0.0	-229.7±19

Table 3.1 Cluster quality vs. dimensionality n (best in bold).

For fixed database dimensionality of $n = 25$, the number of data points m was varied: $m = 10,000$, $m = 50,000$, $m = 100,000$. Each database was again generated by a Gaussian mixture model with 10 components. As the number of data points increases, each approach improves as expected, but SEM computes models superior to BEM (average log-likelihood improvement by as much as 32% for $m = 50,000$)

and 1EM (average log-likelihood improvement by as much as 36% for $m = 100,000$).

Real Databases

SEM, BEM, 1EM and SampEM were applied to a subset of the Reuters text classification database [20] consisting of word counts for 302 most frequently occurring words in 12,902 documents. SEM, BEM and SampEM were allocated main memory buffer sizes equivalent to 10% and 5% total database storage. Table 3.2 lists averages over 10 mixture models for SEM and 1EM. Results for BEM are averaged over the 4 best mixture models (all others had average log-likelihood at least an order of magnitude poorer). SampEM performance was, at best, an order of magnitude poorer and hence is not listed. The percentages following the methods in Table 3.2 indicate the size of the main memory buffer used to compute the corresponding mixture models. SEM with a larger buffer size computed the best Gaussian mixture models.

SEM 10%	SEM 5%	BEM 10%	BEM 5%	1EM
-39±40	-216±40	-425±883	-375±167	-396±1

Table 3.2 Reuters cluster quality (best in bold).

We note that BEM with the smaller buffer size (5%), on average, performed better in comparison to BEM with a 10% buffer size. This can be explained as follows. Since the data is partitioned into different sized blocks and updates occur over each block sequentially in BEM, it is conceivable that the local maxima with respect to the particular blocks are better for the 5% block-size versus the 10% block-size. Given the large variances associated with these two measurements, the difference is not significant.

Similar numerical results are observed over a subset of the US Census "Adult" database and the REV Digits Recognition database [8].

6 CONCLUDING REMARKS

The mixture model is an extremely useful probabilistic data mining tool and the approach presented effectively enables the EM algorithm to estimate mixture model parameters over large databases. The memory requirement is simply to hold a small sample of the total number of database records in main memory. All updates occur over the contents of this main memory buffer. We have observed that simultaneously updating multiple models in a single database scan results in more opportunity for summarization and models of higher quality. Results on both synthetic and real world data sets indicate that the proposed scheme produces mixture models which outperform other similarly constrained EM approaches: on-line EM (1EM), batch EM (BEM) and EM simply applied to the sample (SampEM).

The single-scan framework presented here raises many interesting research questions. We list a few as research challenges.

- We have not provided an analytic result linking the solution computed via the scalable EM algorithm and the solution computed with the standard EM algorithm. An interesting research issue involves developing a bound on the difference between the two models. Furthermore, a characterization of the solutions would be beneficial. For instance, is the solution computed via scalable EM "far" from a local maximum of the log-likelihood objective measured over the entire database? We have quantified the quality of the mixture models through empirical studies and, generally, the models computed are satisfactory.

- The data presented to the single-scan scalable EM algorithm is essentially dynamic in nature. When a sub-sample of data is first processed, the algorithm (and hence the underlying optimization problem) is presented with the information as a set of exact data points. After the algorithm has updated the mixture model over this particular sub-sample, the information is then changed and summarized with sufficient statistics which may be merged with other sufficient statistics derived subsequently. A generalized framework for studying optimization problems defined by data which is dynamic in such a way may aid in the analysis of many data mining methods computing approximate solutions to underlying mathematical programming problems.

- An open research issue is *efficiently* estimating the "correct" number of clusters k. Typically, many mixture models are computed with different values of k, then each of these models is "scored" and the model (i.e. value of k) with highest score is taken to be the "correct" number of clusters. In [29] the Monte Carlo Cross-Validation method is proposed which scores each model over a hold-out or test set not used to construct the model. Other approaches to determining the "correct" value of k include: hypothesis testing [30]; full Bayesian approaches incorporating prior beliefs on the value of k, approximated in the AutoClass algorithm [10]; and penalized likelihood methods where the objection function optimized includes not only the log-likelihood of the data given the mixture model, but also a term penalizing the number of parameters in the model [11, 25].

Acknowledgments

We thank Daphne Koller for suggesting the batch variant of incremental EM experiments.

References

[1] R. Agrawal, J. Gehrke, D. Gunopulos, and P. Raghavan. Automatic subspace clustering of high dimensional data for data mining applications. In *Proc. ACM SIGMOD Intl. Conf. on Management of Data (SIGMOD'98), Seattle, WA*, pages 94–105. ACM Press, 1998.

[2] M. R. Anderberg. *Cluster Analysis for Applications.* Academic Press, New York, 1973.

[3] J. Banfield and A. Raftery. Model-based Gaussian and non-Gaussian clustering. *Biometrics*, 49:803–821, 1993.

[4] K. P. Bennett, U. M. Fayyad, and D. Geiger. Density-based indexing for approximate nearest neighbor queries. In *Proc. 5th Intl. Conf. on Knowledge Discovery and Data Mining (KDD99)*, pages 233–243, New York, 1999. ACM Press.

[5] C. M. Bishop. *Neural Networks for Pattern Recognition.* Oxford University Press, New York, 1995.

[6] R. Brachman, T. Khabaza, W. Kloesgen, G. Piatetsky-Shapiro, and E. Simoudis. Industrial applications of data mining and knowledge discovery. *Communications of the ACM*, 39(11):42–48, 1996.

[7] P. S. Bradley, U. M. Fayyad, and C. Reina. Scaling clustering to large databases. In *Proc. 4th Intl. Conf. on Knowledge Discovery and Data Mining, KDD98*, pages 9–15, Menlo Park, CA, 1998. AAAI Press.

[8] P. S. Bradley, U. M. Fayyad, and C. Reina. Scaling EM clustering to large databases. Technical Report MSR-TR-98-35, Microsoft Research, Redmond, WA, 1998.

[9] P. S. Bradley, O. L. Mangasarian, and W. N. Street. Clustering via concave minimization. In M. C. Mozer, M. I. Jordan, and T. Petsche, editors, *Advances in Neural Information Processing Systems -9-*, pages 368–374, Cambridge, MA, 1997. MIT Press. ftp://ftp.cs.wisc.edu/math-prog/tech-reports/96-03.ps.Z.

[10] P. Cheeseman and J. Stutz. Bayesian classification (autoclass): Theory and results. In *Advances in Knowledge Discovery and Data Mining*, pages 153–180, Menlo Park, CA, 1996. AAAI Press.

[11] D. M. Chickering and D. Heckerman. Efficient approximations for the marginal likelihood of incomplete data given a Bayesian network. Technical Report MSR-TR-96-08, Microsoft Research, Redmond, Washington 98052, 1996.

[12] A. P. Dempster, N. M. Laird, and D. B. Rubin. Maximum likelihood from incomplete data via the em algorithm. *Journal of the Royal Statistical Society B*, 39:1–38, 1977.

[13] J. L. Devore. *Probability and Statistics for Engineering and the Sciences.* Wadsworth Publishing Company, Belmont, California, 1995. Fourth Edition.

[14] R. O. Duda and P. E. Hart. *Pattern Classification and Scene Analysis.* John Wiley & Sons, New York, 1973.

[15] U. M. Fayyad, S. G. Djorgovski, and N. Weir. Application of classification and clustering to sky survey cataloging and analysis. In E. Wegmen and S. Azen, editors, *Computing Science and Statistics*, volume 29(2), pages 178–186, Fairfax, VA, USA, 1997. Interface Foundation of North America.

[16] U. M. Fayyad, D. Haussler, and P. Stolorz. KDD for science data analysis: Issues and examples. In *Proceedings, Second International Conference on Knowledge Discovery and Data Mining.* AAAI Press, 1996.

[17] U. M. Fayyad, G. Piatetsky-Shapiro, P. Smyth, and R. Uthurasamy. *Advances in Knowledge Discovery and Data Mining.* MIT Press, Cambridge, MA, 1996.

[18] K. Fukunaga. *Statistical Pattern Recognition.* Academic Press, NY, 1990.

[19] L. Kaufman and P. Rousseeuw. *Finding Groups in Data.* John Wiley and Sons, New York, NY, 1989.

[20] D. Lewis, Reuters Inc., and Carnegie Group. Reuters corpus. http://www.research.att.com/~lewis/reuters21578.html.

[21] R. M. Neal and G. E. Hinton. A view of the em algorithm that justifies incremental, sparse, and other variants. In M. I. Jordan, editor, *Learning in Graphical Models*, pages 355–368, Dordrecht, 1998. Kluwer Academic Publishers.

[22] R. Ng and J. Han. Efficient and effective clustering methods for spatial data mining. In *Proc. 20th Int. Conf. on Very Large Databases*

(VLDB'94), Santiago de Chile, Chile, pages 144–155. Morgan Kaufmann, 1994.

[23] T. Olson, J.-S. Pang, and C. Priebe. A likelihood-mpec approach to target classification. *Mathematical Programming*, submitted, February 2000.

[24] J. Sander, M. Ester, H. Kriegel, and X. Xu. Density-based clustering in spatial databases: The algorithm gdbscan and its applications. *Data Mining and Knowledge Discovery*, 2(2):169–194, 1998.

[25] S. C. Sclove. Aplication of the conditional population mixture model to image segmentation. *IEEE Trans. Patt. Anal. Mach. Intell.*, PAMI-5:428–433, 1983.

[26] D. W. Scott. *Multivariate Density Estimation.* John Wiley and Sons, New York, 1992.

[27] J. Shanmugusundaram, U. M. Fayyad, and P. S. Bradley. Compressed data cubes for olap aggregate query approximation on continuous dimensions. In *Proc. 5th Intl. Conf. on Knowledge Discovery and Data Mining (KDD99)*, pages 223–232, New York, 1999. ACM Press.

[28] B. W. Silverman. *Density Estimation for Statistics and Data Analysis.* Chapman & Hall, London, 1986.

[29] P. Smyth. Clustering using monte carlo cross-validation. In E. Simoudis, J. Han, and U. Fayyad, editors, *Proc. 2nd Intl. Conf. on Knowledge Discovery and Data Mining (KDD96)*, pages 126–133, Portland, OR, August 1996. AAAI Press.

[30] D. M. Tittering, A. F. M. Smith, and U. E. Makov. *Statistical Analysis of Finite Mixture Distributions.* John Wiley and Sons, Chichester, UK, 1985.

[31] T. Zhang, R. Ramakrishnan, and M. Livny. Birch: A new data clustering algorithm and its applications. *Data Mining and Knowledge Discovery*, 1(2):141–182, 1997.

Chapter 4

A COMPLEMENTARITY EIGENPROBLEM IN THE STABILITY ANALYSIS OF FINITE DIMENSIONAL ELASTIC SYSTEMS WITH FRICTIONAL CONTACT

A. Pinto da Costa

Inst. Sup. Técnico, Dep. Eng. Civil and ICIST, Av. Rovisco Pais, 1049-001 Lisboa, Portugal

I. N. Figueiredo

Dep. Matemática, Universidade de Coimbra, Apartado 3008, 3000 Coimbra, Portugal

J. J. Júdice

Dep. Matemática, Universidade de Coimbra, Apartado 3008, 3000 Coimbra, Portugal

J. A. C. Martins

Inst. Sup. Técnico, Dep. Eng. Civil and ICIST, Av. Rovisco Pais, 1049-001 Lisboa, Portugal

Abstract In this paper a mixed complementarity eigenproblem (MCEIP) is formulated and a method is proposed for its numerical solution. This mathematical problem is motivated by the study of divergence instabilities of static equilibrium states of finite dimensional mechanical systems with unilateral frictional contact. The complementarity eigenproblem is solved by transforming it into a non-monotone mixed complementarity problem (MCP), which is then solved by using the algorithm PATH. The proposed method is used to study some small sized examples and some large finite element problems.

Keywords: Complementarity, contact, friction, eigenvalues, instability.

M.C. Ferris et al. (eds.), Complementarity: Applications, Algorithms and Extensions, 67–83.

1 INTRODUCTION

The instability of configurations of static equilibrium of finite dimensional plane linearly elastic systems, in frictional contact with a flat obstacle, and subjected to constant applied forces, was recently discussed by Martins et al. [1]. Some of their results were generalized in [2] for the case of non-linear elastic systems in frictional contact with curved obstacles. In the first of the above papers it is shown that a necessary and sufficient condition for the occurrence of divergence instability along a constant admissible direction, is that an appropriate eigenproblem with the form of an inclusion or a variational inequality is solved, with an appropriate sign of the corresponding real eigenvalue (see Section 2 of the present paper). Still in the paper [1], a necessary condition and a sufficient condition for this type of instability were presented, both of which can be easily checked. However, in many circumstances, those conditions do not yield sufficiently sharp results, in particular the necessary condition frequently gives a poor lower bound estimate for the onset of instability [1]. For this reason we propose in the present paper a procedure for the numerical solution of the governing inclusion or variational eigenproblem. This is achieved by transforming that problem into a complementarity eigenproblem (see Section 3), by means of an appropriate change of the contact related variables (see, for instance, Klarbring [3]). The problem is subsequently transformed into a non-monotone mixed complementarity problem (MCP), in which the unknown eigenvalue is treated as a non-negative variable that is complementary with an additional variable involved in a normalizing constraint that prevents the trivial solution (see Section 4). The algorithm PATH (Dirkse and Ferris [4]) is successfully applied to solve some small sized examples and some finite element problems (see Section 5).

2 FORMULATION

We consider a finite dimensional linearly elastic system with plane motion that may establish frictional contact with a fixed flat obstacle, and that is subjected to constant applied forces. The typical situations in mind and the largest examples studied in the present paper involve finite element discretizations of linearly elastic bodies.

For sufficiently smooth time evolutions of the contact candidate particle p of the system, the normal (N) and the tangential (T) components of the vectors (in $\mathbb{R}^2$) of the displacements ($u_p(t)$), the velocities ($\dot{u}_p(t)$), and the reactions ($r_p(t)$) of the particle p satisfy the (Signorini) unilateral contact conditions

$$u_{Np}(t) \leq 0,\ r_{Np}(t) \leq 0,\ u_{Np}(t)\, r_{Np}(t) = 0, \tag{4.1}$$

and the (Coulomb) friction law

$$|r_{Tp}(t)| + \mu\, r_{Np}(t) \leq 0,\ |\dot{u}_{Tp}(t)|\, r_{Tp}(t) - \mu\, \dot{u}_{Tp}(t)\, r_{Np}(t) = 0, \tag{4.2}$$

where $\mu \geq 0$ is the coefficient of friction.

We wish to study a particular type of *dynamic* instability of an *equilibrium* state of the system. The smooth portions of its dynamic evolution under constant applied forces $f^0 \in \mathbb{R}^N$, are governed by the momentum balance equations

$$M\ddot{u}(t) + Ku(t) = f^0 + r(t), \tag{4.3}$$

where M is the $N \times N$ symmetric positive definite (SPD) mass matrix, K is the $N \times N$ SPD stiffness matrix, $t \geq 0$ is the time variable, $\dot{()}$ denotes the time derivative, and $u(t) \in \mathbb{R}^N$ and $r(t) \in \mathbb{R}^N$ are the vectors of the unknown generalized displacements and reactions at time t, respectively. The system of equations (4.3) holds together with the unilateral frictional contact conditions (4.1), (4.2) at each contact candidate particle p, and together with the absence of reactions along the degrees of freedom that are not subjected to any kinematic constraint: the sub-vector $r_F(t)$ of $r(t)$ satisfies

$$r_F(t) = 0. \tag{4.4}$$

On the other hand, an equilibrium state of the system under the same applied forces f^0 is characterized by a displacement vector $u^0 \in \mathbb{R}^N$ and a reaction vector $r^0 \in \mathbb{R}^N$ that satisfy the equilibrium equations (let $\ddot{u} \equiv 0$ in (4.3)):

$$Ku^0 = f^0 + r^0,$$

together with the following form of (4.4) and of the frictional contact conditions at each contact candidate particle p (let $\dot{u}_p \equiv 0$ in (4.2)):

$$r_F^0 = 0,$$
$$u_{Np}^0 \leq 0,\ r_{Np}^0 \leq 0,\ u_{Np}^0 r_{Np}^0 = 0,\ |r_{Tp}^0| \leq -\mu\, r_{Np}^0.$$

In Martins et al. [1] it is shown that, for t in some right neighborhood of some instant τ ($t \in [\tau, \tau + \Delta\tau[$), *there exist dynamic solutions of the form*

$$u(t) = u^0 + \alpha(t)\, v, \qquad r(t) = r^0 + \beta(t)\, w, \tag{4.5}$$

where v and w define constant directions in the sets of right admissible displacement and reaction rates at the equilibrium state (u^0, r^0), the function of time α is twice continuously differentiable, α and $\dot{\alpha}$ are non-negative and non-decreasing in $[\tau, \tau + \Delta\tau[$, the function β is continuous, non-negative and non-decreasing in the same interval, and the initial values $\alpha(\tau) \geq 0$, $\dot{\alpha}(\tau) \geq 0$ are arbitrarily small, *if and only if* there exists a number $\lambda \geq 0$, and two vectors v and w in $\mathbb{R}^N$, $v \neq 0$, such that

$$\begin{gathered}
(\lambda^2 M + K)v = w, \\
w_f = 0, \\
v_d = 0, \\
v_{Nsp} = 0, \\
v_{Tsp}\,\mathrm{sign}(r^0_{Tsp}) \leq 0, \quad w_{Tsp}\,\mathrm{sign}(r^0_{Tsp}) + \mu\, w_{Nsp} \leq 0, \\
\left[w_{Tsp}\,\mathrm{sign}(r^0_{Tp}) + \mu\, w_{Np}\right]\left[v_{Tsp}\,\mathrm{sign}(r^0_{Tsp})\right] = 0, \\
v_{Nzp} \leq 0, \quad w_{Nzp} \leq 0, \quad v_{Nzp} w_{Nzp} = 0, \\
|w_{Tzp}| + \mu\, w_{Nzp} \leq 0, \quad |v_{Tzp}|\, w_{Tzp} - \mu\, v_{Tzp} w_{Nzp} = 0,
\end{gathered} \tag{4.6}$$

where the following notations were introduced for the equilibrium state (u^0, r^0):

f: degrees of freedom not subjected to any kinematic constraint, including those of the contact candidate particles that currently are not in contact (*free*);

z: particles in contact with *zero* reaction;

d: particles in contact with reaction strictly inside the friction cone and consequent vanishing (right) *displacement* rate;

s: particles in contact with non-vanishing reaction on the friction cone and consequent possible *slip* in the near future.

When the conditions indicated above are satisfied, the equilibrium state corresponding to u^0 and r^0 is dynamically unstable: a divergence instability.

The problem (4.6) can be equivalently written as an inclusion or a variational inequality eigenproblem (see [1]). In order to solve it numerically, we shall write it now as a complementarity eigenproblem.

3 THE COMPLEMENTARITY EIGENPROBLEMS

For simplicity of the presentation, we shall restrict ourselves to the transformation of (4.6) into a complementarity eigenproblem in the particular case in which the set of particles in contact with zero reaction (the particles z) is empty. This transformation starts with the elimination of both degrees of freedom of the particles in contact with reaction strictly inside the friction cone (the stick particles d), the elimination of the normal degrees of freedom of the particles in contact with non-vanishing reaction on the friction cone (the impending slip particles s), and the following change of variables (see also [3]):

$$x = \begin{bmatrix} x_f \\ x_{Ts} \end{bmatrix} = \begin{bmatrix} I & 0 \\ 0 & -S \end{bmatrix} \begin{bmatrix} v_f \\ v_{Ts} \end{bmatrix} = \begin{bmatrix} v_f \\ -Sv_{Ts} \end{bmatrix} \in \mathbb{R}^{N^*}, \tag{4.7}$$

$$y = \begin{bmatrix} y_f \\ y_{Ts} \end{bmatrix} = \begin{bmatrix} 0 \\ -(Sw_{Ts} + \mu w_{Ns}) \end{bmatrix} \in \mathbb{R}^{N^*}, \tag{4.8}$$

where I is an identity (sub-)matrix, $S = \text{diag}(sign(r^0_{Tp}))$ and $N^*(\leq N)$ is the number of degrees of freedom of the system that may be (right) active at the equilibrium state (the free f plus the impending slip Ts degrees of freedom). In this manner, the following *mixed complementarity eigenproblem* in λ^2 (MCEIP-λ^2) is obtained: *find* $\lambda^2 \geq 0$ *and* $(x, y) \in \mathbb{R}^{N^*} \times \mathbb{R}^{N^*}$, *with* $x \neq 0$, *such that*

$$(\lambda^2 M^* + K^*)x = y, \tag{4.9}$$

$$y_f = 0, \tag{4.10}$$

$$0 \leq x_{Ts} \perp y_{Ts} \geq 0, \tag{4.11}$$

where $\perp$ denotes orthogonality between the vectors x_{Ts} and y_{Ts} in the euclidean inner product. The matrices M^* and K^* are linear pencils of matrices in μ,

$$M^* = M^*(\mu) = M_0 + \mu M_1, \qquad K^* = K^*(\mu) = K_0 + \mu K_1,$$

with the structure exemplified below for K^*:

$$K_0 = \begin{bmatrix} K_{f,f} & -K_{f,Ts}S \\ -SK_{Ts,f} & SK_{Ts,Ts}S \end{bmatrix}, \quad K_1 = \begin{bmatrix} 0 & 0 \\ -K_{Ns,f} & K_{Ns,Ts}S \end{bmatrix}.$$

The matrices M_0 and K_0 in the linear pencils M^* and K^* are SPD matrices. In fact M_0 and K_0 are similar to the SPD principal submatrices of M and K corresponding to the f plus the Ts degrees of freedom, because the diagonal transformation of variables in (4.7) is orthogonal. By continuity in μ, it follows that, for sufficiently small μ, the matrices M^* and K^* are positive definitive (PD). Note also that in the case of a diagonal mass matrix M, the matrix M^* equals the matrix M_0, so that it is also diagonal and PD.

It is now quite simple to recover the sufficient condition and the necessary condition that were used in [1] to study this problem. *A sufficient* *condition* for the occurrence of a divergence instability of the form (4.5) is that there is $\lambda \geq 0$ and $x \in \mathbb{R}^{N^*}$, with $x \neq 0$, such that

$$(\lambda^2 M^* + K^*)x = 0, \tag{4.12}$$

$$x_{Ts} \geq 0. \tag{4.13}$$

Note that this particular case of (4.9) - (4.11) results from *a priori* assuming that the solution satisfies $y_{Ts} = 0$, i.e. the reactions of the contact particles in impending slip (the s particles) remain on the boundary of the friction cone, which means that those particles remain in impending slip or do initiate sliding. Note also that the simplicity of (4.12), (4.13) relatively to (4.9) - (4.11) is that now (4.12) is a linear eigenproblem, and the additional inequalities in (4.13) can be checked *a posteriori*. On the other hand, doing the inner product of (4.9) with x, it is immediately seen that a *necessary condition* for the occurrence of a divergence instability of the form (4.5) is that

the matrix $M^*(\mu)$ is not PSD or the matrix $K^*(\mu)$ is not PD. (4.14)

Since, as mentioned above, the matrices $M^*(\mu)$ and $K^*(\mu)$ are PD for sufficiently small μ, we can immediately conclude that *no divergence instability of the type (4.5) can occur for sufficiently small coefficients of friction.* Furthermore, in the particular case of a *diagonal mass matrix*, the *necessary condition* (4.14) reduces to:

the matrix $K^*(\mu)$ is not PD, (4.15)

because, as observed above, M^* is then diagonal and PD. A related result in a continuum framework can be found in Chateau and Nguyen [5]. Continuing to consider, for simplicity, this particular case of diagonal M and M^*, the minimum eigenvalue of the symmetric part of K^* (which is relevant for (4.15)) is necessarily smaller or equal [1] to the real part of all eigenvalues of K^* (which are evaluated in (4.12)); this is one of the main reasons for the necessary condition (4.15) to be satisfied (much) earlier than the sufficient condition (4.12).

A related problem that deserves special attention consists of computing the values of the friction coefficient μ and the associated mode shapes that correspond to the transition between stability and instability of a given equilibrium state. This is expressed by the condition $\lambda = 0$ in (4.9), leading thus to the formulation of a *mixed complementarity eigenproblem in μ* (MCEIP-μ): *find $\mu \geq 0$ and $(x, y) \in \mathbb{R}^{N^*} \times \mathbb{R}^{N^*}$, with*

$x \neq 0$, *such that*

$$(K_0 + \mu K_1)x = y, \tag{4.16}$$

$$y_f = 0, \tag{4.17}$$

$$0 \leq x_{Ts} \perp y_{Ts} \geq 0. \tag{4.18}$$

Before turning to the numerical solution of the complementarity eigenproblems introduced in this section, we wish to mention an enumerative procedure that yields all the solutions of the mixed complementarity eigenproblems (4.9) - (4.11) or (4.16) - (4.18). For instance, the problem MCEIP-λ^2 (4.9) - (4.11) can be solved by computing the solutions of a set of 2^{n_s} linear eigenproblems, each of them followed by checking some appropriate inequalities; n_s is the number of particles in impending slip at the equilibrium state. Each of those linear eigenproblems is obtained by *a priori* assuming a specific combination of admissible near future evolutions for the s particles. Some of those particles are assumed to become *stick* (a subvector x_{Ts}^{stick} of x_{Ts} is assumed to be null), while the contact reactions of the other ones (the *slip* or impending slip ones) are assumed to remain on the boundary of the friction cone (a subvector y_{Ts}^{slip} of y_{Ts} is assumed to be null). After elimination of the assumed stick variables ($x_{Ts}^{stick} = 0$) for each specific combination of near future evolutions, and denoting

$$\bar{x} = \begin{bmatrix} x_f \\ x_{Ts}^{slip} \end{bmatrix},$$

the problem (4.9) - (4.11) can be simplified to a linear eigenproblem of the form

$$(\lambda^2 \bar{M} + \bar{K})\bar{x} = 0, \tag{4.19}$$

followed by the verification of the inequalities

$$y_{Ts}^{stick} \geq 0, \quad x_{Ts}^{slip} \geq 0. \tag{4.20}$$

The small sized example presented in Section 5 is solved by using this enumerative procedure. However, the rapid growth of the number of linear eigenproblems with the number of slip particles makes it impossible to use such method in systems with many contact particles. Finally, it is worth noting that the linear eigenproblem corresponding to the sufficient condition (4.6) is just *one* of the 2^{n_s} linear eigenproblems considered in this enumerative procedure (4.19), (4.20).

4 NUMERICAL SOLUTION OF THE COMPLEMENTARITY EIGENPROBLEMS

In the previous section, two complementarity eigenproblems were defined, namely the MCEIP-λ^2 (4.9)-(4.11) and the MCEIP-μ (4.16)-(4.18). In this section, we establish their equivalency to non-linear mixed complementarity problems. This is done by introducing a normalizing constraint and an additional non-negative variable (γ) that is complementary to the eigenvalue (λ^2 or μ); this one, in turn, is also considered as an additional non-negative variable.

Proposition 4.1 *The MCEIP-μ defined in (4.16)-(4.18) has a solution if and only if there is a solution to the following mixed complementarity problem (MCP-μ): find $(x, \mu) \in \mathbb{R}^{N^*+1}$ and $(y, \gamma) \in \mathbb{R}^{N^*+1}$ such that*

$$(K_0 + \mu K_1)x = y, \tag{4.21}$$
$$e^T x_{Ts} = c + \gamma, \tag{4.22}$$
$$y_f = 0, \tag{4.23}$$
$$0 \leq (x_{Ts}, \mu) \perp (y_{Ts}, \gamma) \geq 0, \tag{4.24}$$

where e is a vector of dimension n_s with all components equal to 1, and c is an arbitrary positive real number.

Proof Let x, y and μ solve the MCEIP-μ. Then $x_{Ts} \neq 0$, because $x_{Ts} = 0$ and $x \neq 0$ would imply $x_f \neq 0$, which is impossible, since the equations in (4.16) corresponding to the free degrees of freedom f would reduce to

$$K_{f,f} x_f = 0,$$

$K_{f,f}$ being a SPD matrix. Then (x, μ) and (y, γ), with $\gamma = 0$, solve the MCP-μ (4.21)-(4.24), with $c = e^T x_{Ts} > 0$. Note that this positive value of c is arbitrary, because the norm of the vectors x and y that solve the eigenproblem (4.16)-(4.18) is arbitrary.

Conversely, let (x, μ) and (y, γ) solve the MCP-μ (4.21)-(4.24). Then $e^T x_{Ts} = c + \gamma$ for some $c > 0$, so that $e^T x_{Ts} > 0$, $x_{Ts} \neq 0$, and, consequently, $x \neq 0$. Hence x, y and μ solve the MCEIP-μ (4.16)-(4.18).

□

We can now deduce an equivalent formulation for the MCEIP-λ^2.

Proposition 4.2 *The MCEIP-λ^2 (4.9)-(4.11) has a solution if and only if there is a solution to the following mixed complementarity problem*

(MCP-λ^2): find $(x, \lambda^2) \in \mathbb{R}^{N^*+1}$ and $(y, \gamma) \in \mathbb{R}^{N^*+1}$ such that

$$(\lambda^2 M^* + K^*)x = y, \tag{4.25}$$

$$y_f = 0, \tag{4.26}$$

$$e^T x_{Ts} = c + \gamma, \tag{4.27}$$

$$0 \leq (x_{Ts}, \lambda^2) \perp (y_{Ts}, \gamma) \geq 0, \tag{4.28}$$

where, again, e is a vector of dimension n_s with all components equal to 1, and c is an arbitrary positive constant.

Proof Let x, y and λ^2 be a solution to the MCEIP-λ^2. Then $x_{Ts} \neq 0$, because $x_{Ts} = 0$ and $x \neq 0$ would imply $x_f \neq 0$, which is impossible, since the equations in (4.9) corresponding to the free degrees of freedom f would reduce to

$$(\lambda^2 M_{f,f} + K_{f,f})x_f = 0,$$

with $M_{f,f}$ and $K_{f,f}$ both SPD and $\lambda^2 \geq 0$. Then (x, λ^2) and (y, γ), with $\gamma = 0$, solve the MCP-λ^2 (4.25)-(4.28), with $c = e^T x_{Ts} > 0$. Note that the observation and the arguments presented in the final part of the proof of Proposition 4.1 can also be applied to conclude the present proof. □

Remark 4.3 *Since K_0 is SPD, any solution to the MCEIP-μ has $\mu > 0$ (recall also section 3). Consequently the complementary variable γ vanishes, so that the choice of a positive constant c as data for the MCP-μ specifies the value of $e^T x_{Ts}$, i.e., the l_1 norm of the vector x_{Ts}. In the case of the MCP-λ^2 the same conclusion can be obtained when μ is not a solution to the MCEIP-μ, because in this situation $\lambda^2 \neq 0$ and, again, the complementary variable γ equals zero.*

It may be of interest to search for solutions to the MCEIP-μ with μ below or above a certain prescribed value (μ_0). For that purpose one has to change a non-negative variable in the MCP-μ (4.21)-(4.24): in order to search for solutions with $\mu \leq \mu_0$ (or $\mu \geq \mu_0$), one has to consider the new variable $\zeta = \mu_0 - \mu$ (or $\zeta = \mu - \mu_0$), and then to replace μ in (4.21) by $\mu_0 - \zeta$ (or $\mu_0 + \zeta$), and to replace μ in (4.24) by ζ. Similar procedures can also be followed for the MCEIP-λ^2.

Propositions 4.1 and 4.2 show that the solution of the eigenproblems under study can be obtained by processing two mixed non-linear complementarity problems. There are a number of algorithms for solving these problems. The reader can find in [6] a list of some relevant approaches for non-linear complementarity problems (NCP). More recently, interior-point algorithms [7, 8, 9] and a Newton's method for solving systems of non-differentiable equations [10], based on the so called Fischer function, have also been recommended for this type of problems. The latter algorithms are usually quite efficient, but unfortunately they require the monotonicity of the function, or some similar property, to be useful. The complementarity problems discussed in this section do not share this property. The algorithm PATH described in [4] is an algorithm that can process non-monotone mixed complementarity problems and then has been our choice to process the problems discussed in this section. We recall that PATH is a robust GAMS implementation [11] of a path following technique that was first discussed in Ralph [12] and later improved by Dirkse and Ferris [11]. This algorithm exploits the equivalence of a mixed complementarity problem with a system of non-differentiable equations $F(x) = 0$, where F is the so-called normal map due to Robinson [13]. The zero of the function is computed by a path generation technique that in each iteration pursues a root of a linear approximation of the normal map at the current iterate. This is done by using a pivotal scheme similar to the well-known Lemke algorithm [14, 15]. A non-monotone line-search technique [16] is also included to guarantee sufficient decrease of the Euclidean norm of the normal map. The algorithm possesses strong global convergence properties [4]. As discussed in the next section, the algorithm has been able to process all the non-linear complementarity problems tried so far that were known to have some solution. A theoretical investigation of this behavior is certainly a subject for future research.

5 EXAMPLES AND NUMERICAL RESULTS

The first example involves two particles of mass m each, supported by linear elastic springs of stiffness k, and in contact with an horizontal obstacle (see Fig. 4.1). The same example was discussed earlier by Alart and Curnier [17] in the context of non-uniqueness of solution to incremental quasi-static problems with friction.

The system has four degrees of freedom and the generalized displacements $u = (u_{T1}, u_{N1}, u_{T2}, u_{N2})$ are used. The external applied forces f_1 and f_2 are such that both particles are in a state of impending slip towards the right or towards the left. The mass and stiffness matrices of the system are, respectively,

$$M = mI, \qquad K = k \begin{bmatrix} c^2+1 & -cs & -1 & 0 \\ -cs & s^2+1 & 0 & 0 \\ -1 & 0 & c^2+1 & cs \\ 0 & 0 & cs & s^2+1 \end{bmatrix};$$

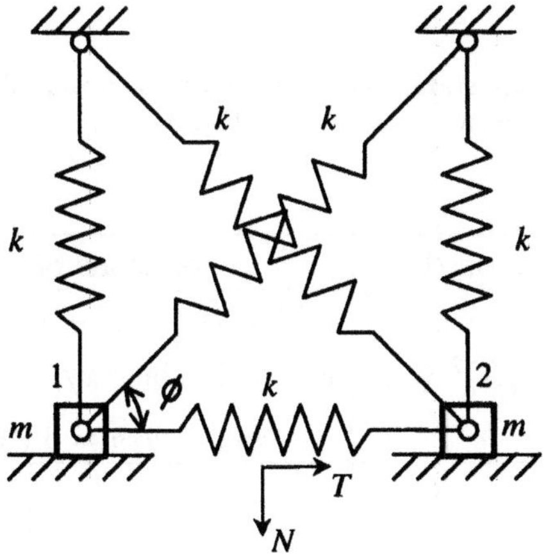

Figure 4.1 A structure with two contact particles leading to multiple solutions of MCEIP-λ^2

where I denotes the 4×4 identity matrix, $s = \sin\phi$, $c = \cos\phi$, and ϕ is the angle between the inclined springs and the horizontal.

For the previously described system, the problem MCEIP-λ^2 is: find $\lambda^2 \geq 0$ and $(x_{T1}, x_{T2}, y_{T1}, y_{T2})$, with $(x_{T1}, x_{T2}) \neq (0,0)$, such that

$$\begin{bmatrix} \lambda^2 m + k(c^2 + 1 - \mu c s S_1) & -S_1 S_2 k \\ -S_1 S_2 k & \lambda^2 m + k(c^2 + 1 + \mu c s S_2) \end{bmatrix} \begin{bmatrix} x_{T1} \\ x_{T2} \end{bmatrix} = \begin{bmatrix} y_{T1} \\ y_{T2} \end{bmatrix}$$

$$0 \leq \begin{bmatrix} x_{T1} \\ x_{T2} \end{bmatrix} \perp \begin{bmatrix} y_{T1} \\ y_{T2} \end{bmatrix} \geq 0,$$

where $S_p = \text{sign}(r^0_{Tp})$, and x_{Tp} and y_{Tp} are defined as in (4.7) and (4.8).

Four different combinations of signs for the tangential reactions exist, each one corresponding to a different static equilibrium state. Table 4.1 contains the solution sets of the above MCEIP-λ^2, for each of those combinations of signs of the static tangential reactions. When both particles are in a state of impending slip to the same side, there is at most one solution; when the left particle is in impending slip towards the left and the right one is in impending slip towards the right, no divergence instability of the type (4.5) is found; when the left particle is in impending slip towards the right and the right particle is in impending slip towards the left, then, depending on the data, there are at most two or three solutions.

As mentioned earlier, all the solutions for this small sized mechanical system could be calculated by the enumerative procedure presented in the end of section 3. But the algorithm PATH was also tested in these small problems. For the sets of numerical data used, the algorithm

Table 4.1 Static equilibrium state, conditions on the data and solutions of MCEIP-λ^2 for the structure with two contact particles of Fig. 4.1

<table>
<tr><th>S_1</th><th>S_2</th><th>Conditions on data</th><th colspan="2">Solutions of MCEIP-λ^2</th></tr>
<tr><td rowspan="6">$+1$</td><td>$+1$</td><td>$c^2+2-(\mu s)^2 \le 0$</td><td colspan="2">$\lambda^2 = \frac{k}{m}\left[-(c^2+1)+\sqrt{1+(\mu cs)^2}\right]$
$x_{T1} > 0$ arbitrary, $x_{T2} = \frac{x_{T1}}{\sqrt{1+(\mu cs)^2}+\mu cs}$
$y_{T1} = y_{T2} = 0$ (SLIP - SLIP)</td></tr>
<tr><td rowspan="5">-1</td><td rowspan="2">$c^2+1-\mu cs \le 0$</td><td rowspan="2">$\lambda^2 =$
$-\frac{k}{m}(c^2+1-\mu cs)$</td><td>$x_{T1} > 0$ arbitrary;
$x_{T2} = 0$;
$y_{T1} = 0$;
$y_{T2} = kx_{T1}$
(SLIP - STICK)</td></tr>
<tr><td>$x_{T1} = 0$;
$x_{T2} > 0$ arbitrary;
$y_{T1} = kx_{T2}$;
$y_{T2} = 0$
(STICK - SLIP)</td></tr>
<tr><td rowspan="3">$c^2+2-\mu cs \le 0$</td><td rowspan="2">$\lambda^2 =$
$-\frac{k}{m}(c^2+1-\mu cs)$</td><td>$x_{T1} > 0$ arbitrary;
$x_{T2} = 0$;
$y_{T1} = 0$;
$y_{T2} = kx_{T1}$
(SLIP - STICK)</td></tr>
<tr><td>$x_{T1} = 0$;
$x_{T2} > 0$ arbitrary;
$y_{T1} = kx_{T2}$;
$y_{T2} = 0$
(STICK - SLIP)</td></tr>
<tr><td>$\lambda^2 =$
$-\frac{k}{m}(c^2+2-\mu cs)$</td><td>$x_{T1} = x_{T2}$
$= x > 0$ arbitrary;
$y_{T1} = y_{T2} = 0$
(SLIP - SLIP)</td></tr>
<tr><td rowspan="2">-1</td><td>$+1$</td><td>—</td><td colspan="2">No Solution</td></tr>
<tr><td>-1</td><td>$c^2+2-(\mu s)^2 \le 0$</td><td colspan="2">$\lambda^2 = \frac{k}{m}\left[-(c^2+1)+\sqrt{1+(\mu cs)^2}\right]$
$x_{T1} = \frac{x_{T2}}{\sqrt{1+(\mu cs)^2}+\mu cs}$, $x_{T2} > 0$ arbitrary
$y_{T1} = y_{T2} = 0$ (SLIP - SLIP)</td></tr>
</table>

PATH was always able to find the corresponding complete solution set presented in Table 4.1. For the particular cases that have two solutions with different values of λ^2, the technique mentioned in Section 4 for searching solutions with λ^2 below or above a certain fixed value of λ^2 was successfully applied.

The second example involves a rectangular polyurethane block sliding on an araldite obstacle that was studied experimentally by Villechaise and Zeghloul ([18, 19]). In the numerical simulations we assume that the elastic block slides on a flat rigid obstacle. The block is discretized with a uniform mesh of 800 linear P1 finite elements that has 21 contact candidate nodes (see Fig. 4.2). The elastic properties are: modulus of elasticity $= 5\,MPa$, Poisson's ratio $= 0.48$. The geometric parameters are length $L = 80\,mm$, height $H = 40\,mm$ and thickness $= 9.6\,mm$. The density of the material is $1.2\,kg/dm^3$.

The block is submitted to a quasi-static loading consisting first of prescribed displacements on the side CD, which is symmetrically pressed against the obstacle until the resultant of the normal reactions on side AB is $-55\,N$. Then the loading proceeds by prescribing an horizontal motion of the side CD towards the left. In this tangential loading phase, the successive equilibrium states have a growing region of nodes in impending slip spreading from right to left.

This same example was studied earlier in [1], also by the finite element method, but using only the necessary (4.14) and the sufficient (4.12), (4.13) conditions for divergence instability recalled in Section 3 of the present paper. For the value of the coefficient of friction ($\mu = 1.1$) identified from the experimental results of [18, 19], the numerical results showed that the necessary condition (4.14) is satisfied very early along the tangential loading process; however, for all the successive equilibrium configurations of the block along that tangential loading, the sufficient condition (4.12), (4.13), that involves slip of all nodes in impending slip, could never be satisfied. The objective of the continuation of that study in the present paper is thus to check if, after the necessary condition is satisfied, there exist or not instability modes of a type different from the all-slip modes of the sufficient condition (4.12), (4.13).

With this purpose, we search first for solutions to the MCEIP-μ at the equilibrium states obtained with $\mu = 1.1$, along the tangential loading of the block, i.e. we search for the values of the coefficient of friction that would originate a transition from stability to instability in that equilibrium configuration. Nontrivial eigenvectors of MCEIP-μ were obtained when 11 or more contact nodes were in a state of impending slip. It is found that the values of μ that solve the MCEIP-μ decrease with the increase of the number of nodes in impending slip in the successive

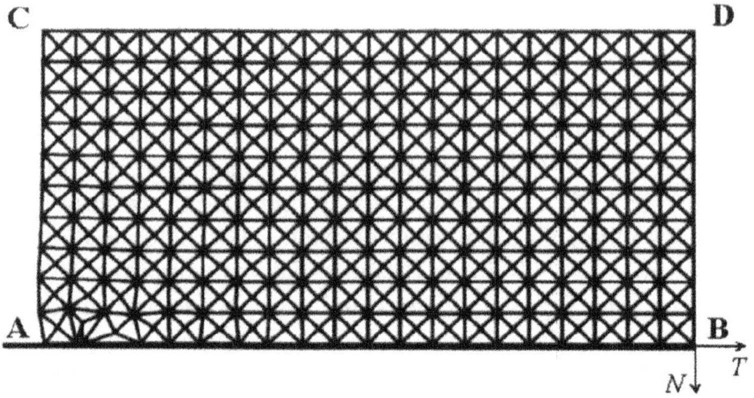

Figure 4.2 An instability mode in the transition between stability and instability for $\mu = 60.81$ (solution of MCEIP-μ)

equilibrium configurations. Moreover, the eigenvectors of the MCEIP-μ associated with higher values of μ correspond to modes having, in average, a larger number of impending slip nodes that get stuck.

For an equilibrium state having the two left nodes stuck, the 15 intermediate nodes in impending slip and the 4 nodes on the right free, the algorithm PATH converged to a solution of the MCEIP-μ. That solution has a very large value of μ (60.81) and a mode represented in Fig. 4.2, where an impending slip node (the fourth from the left) becomes stuck. For the same equilibrium configuration, the classical eigenproblem corresponding to the sufficient condition (4.12), (4.13) was solved, showing that no positive μ exists that corresponds to an admissible non-trivial solution with all impending slip nodes (s) in impending slip or in slip (the sufficient condition (4.12), (4.13) could not be satisfied).

For the final equilibrium state of the loading process, for which the 17 nodes on the left are in impending slip and the 4 nodes on the right are free, a nontrivial eigenvector could be found for a much lower coefficient of friction ($\mu = 1.71$). The corresponding divergence eigenmode is represented in Fig. 4.3. Since all the nodes in impending slip do slide, this mode is given by the sufficient condition (4.12), (4.13) (an all-slip mode). For the same equilibrium configuration and choosing a $\mu > 1.71$ the MCEIP-λ^2 has a similar non-trivial eigenvector and a positive eigenvalue $\lambda^2 > 0$.

For other meshes and other aspect ratios H/L, the same trends were observed in the behavior of the system.

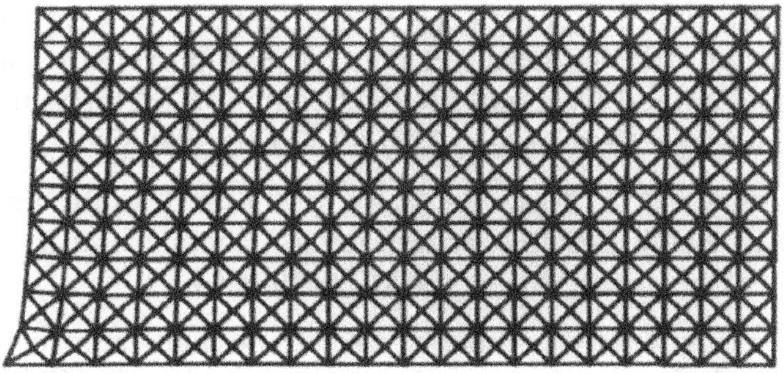

Figure 4.3 An instability mode in the transition between stability and instability for $\mu = 1.71$ (solution of MCEIP-μ)

6 CONCLUSIONS

In this paper a method to solve a mixed complementarity eigenproblem (MCEIP) has been proposed. The motivation to study this mathematical problem was the divergence instability of static equilibrium states of mechanical systems with unilateral frictional contact. The complementarity eigenproblem has been transformed into a non-monotone mixed complementarity problem (MCP), and the algorithm PATH has been applied to solve small sized examples and large finite element problems.

- In all the small sized examples, all the existing solutions could be obtained with the PATH algorithm.
- In the large finite element simulations with the block of Zeghloul and Villechaise [18, 19] it has been observed that:
  - whenever solutions were known to exist [the *all-slip* solutions of the sufficient condition (4.12), (4.13)] the PATH algorithm always converged to one such *all-slip* solution;
  - in some cases where *all-slip* solutions did not exist [the sufficient condition (4.12), (4.13) could not be satisfied] other solutions with slip and stick were obtained, but only for very large values of μ;
  - for reasonably small values of μ, no solutions different from the *all-slip* solutions provided by the sufficient condition (4.12), (4.13) were found.

References

[1] J.A.C. Martins, S. Barbarin, M. Raous, A. Pinto da Costa, Dynamic stability of finite dimensional linearly elastic systems with unilateral contact and Coulomb friction, Computer Methods in Applied Mechanics and Engineering, 177/3-4 (1999), 289-328.

[2] J.A.C. Martins, A. Pinto da Costa, Stability of finite dimensional nonlinear elastic systems with unilateral contact and friction, Int. J. Solids Structures, 37(18) (2000), 2519-2564.

[3] A. Klarbring, Contact, friction, discrete mechanical structures and mathematical programming, Lecture notes for the CISM course Contact Problems: Theory, Methods, Applications, 1997.

[4] S.P. Dirkse, M.C. Ferris, The PATH solver: A non-monotone stabilization scheme for mixed complementarity problems, Optimization Methods and Software, 5 (1995), 123-156.

[5] X. Chateau, Q.S. Nguyen, Buckling of elastic structures in unilateral contact with or without friction, Eur. J. Mech. A/Solids 10(1) (1991), 71-89.

[6] P. Harker and J.S. Pang, Finite-dimensional variational inequalities and nonlinear complementarity problems: a survey of theory, algorithms and applications, Mathematical Programming, 48 (1990) 161-220.

[7] P. Tseng, An infeasible path-following method for monotone complementarity problems, SIAM Journal on Optimization, 7 (1997) 386-402.

[8] S. Bellavia and M. Macconi, An inexact interior point method for monotone nonlinear complementarity problems, Manuscript, Universitá di Firenze, 1999.

[9] M. Kojima, T. Noma and A. Yoshise, Global convergence in infeasible interior-point algorithms, Mathematical Programming, 65 (1994) 43-72.

[10] T. De Luca, F. Facchinei and C. Kanzow, A semismooth equation approach to the solution of nonlinear complementarity problems, Mathematical Programming, 75 (1996) 407-439.

[11] S. Dirkse, M. Ferris, P. Preckel and T. Rutherford, The GAMS callable program library for variational and complementarity solvers, Technical Report 94-07, Computer Sciences Department, Madison, Wisconsin.

[12] D. Ralph, Global convergence of damped Newton's method for non-smooth equations via the path search, Mathematics of Operations Research, 19 (1994), 352-389.

[13] S. Robinson, Normal maps induced by linear transformations, Mathematics of Operations Research, 17 (1992) 691-714.

[14] R.W. Cottle, J.-S. Pang, R.E. Stone, The linear complementarity problem, Academic Press-Computer Science and Scientific Computing, 1992.

[15] K.G. Murty, Linear complementarity, linear and nonlinear programming (Internet edition), 1997.

[16] L. Grippo, F. Lampariello and S. Lucidi, A class of nonmonotone stabilization methods in unconstrained optimization, Numerische Mathematik, 59 (1991) 779-803.

[17] P. Alart, A. Curnier, Contact discret avec frottement: unicité de la solution - Convergence de l'algorithm, École Polytechnique Féderale de Lausanne, 1987.

[18] T. Zeghloul, B. Villechaise, Phénomènes de glissements partiels découlant de l'usage de la loi de Coulomb dans un contat non lubrifié, Materiaux et Techniques - Spécial Tribologie, Décembre (1991) 10-14.

[19] T. Zeghloul, B. Villechaise, Stress waves in a sliding contact, Part 1: experimental study, Proc. 22nd Leeds Lyon Symposium on Tribology, Lyon, 5-8 September (1995).

Chapter 5

VARIATIONAL INEQUALITY MODELS OF RESTRUCTURED ELECTRICITY SYSTEMS

Olivier Daxhelet
ELECTRABEL, R&D Energy Markets, Louvain-la-Neuve, Belgium

Yves Smeers
CORE, Université catholique de Louvain, Louvain-la-Neuve, Belgium

1 INTRODUCTION

Electricity systems are being restructured throughout the world (see [18, 34] for general overviews of the phenomenon). The literature on the subject is now plentiful; it is directed to both academic and professional audiences. It embraces texts on systems descriptions, antitrust and regulatory cases, as well as economic or computational models. Some publications like the Electricity Journal appear to be almost entirely devoted to the various issues generated by the reorganization of the electricity industry. This explosion of interest is due to a major feature of electricity restructuring: the overall process appears to be unusually difficult (see [31] for a discussion of the origins of these difficulties). As a result different paradigms have been proposed and no two systems, even when inspired by the same principles, are identical. Experience with existing systems is now growing but many questions remain. This diversity of paradigms and situations, with no clear cut indication of what is right and wrong, is at the origin of this explosion of interest.

This paper belongs to the computational stream of that literature (see [15] for a recent overview). Its general objective is to survey several restructuring models and to cast them in a single mathematical framework, in this case variational inequalities. It is suggested that this endeavor offers various interesting insights that we briefly discuss now. As indicated above different restructuring models have been proposed

M.C. Ferris et al. (eds.), Complementarity: Applications, Algorithms and Extensions, 85–120.

and extensively discussed in the literature and in practice. The construction of a unifying mathematical framework for formulating them is a first goal that seems worth pursuing. It is certainly well recognized that mathematical formulations will, at least in a first stage, overlook a lot of the market microstructure embedded in restructuring models (see [30, 19] for a discussion of the importance of this market microstructure). Even so, it is submitted that a common mathematical framework still provides a very convenient, albeit abstract, description of the different macro structures proposed for designing electricity systems. Such a common mathematical framework may thus be a useful substitute to verbal discussions and reasoning used for comparing restructuring models; it may also help develop new macro structures. A second point of interest is that this formalism allows one to call upon the theory of variational inequalities to analyze restructuring models. Existence and uniqueness theorems may provide a first body of results to invoke. One can also resort to duality theorems to move from one formulation to another and hence establish relations between restructuring models. A particular point should be made for the capability offered by variational inequalities to explore equilibrium properties of proposed restructuring models. In short, applying the existing theory of variational inequalities produces qualitative results that may be of interest to economists and/or practitioners working in the field of restructuring. Computation is the third goal that we believe justifies the objective of this paper. Eliminating all inefficiencies is almost impossible in market design in general and in electricity restructuring in particular. Trading one inefficiency for another is thus unavoidable in the practice of electricity market design. Being able to computationally assess the magnitude of some of the inefficiencies embedded in certain market designs is therefore useful. Variational inequalities provide this opportunity as they lead to implementable algorithms and powerful codes.

The paper tries to achieve these three goals as follows. Section 2 provides some general background to the problem. It introduces a description of the physical system, basically the electrical network and the load flow equations. A major thrust of electricity restructuring has consisted in unbundling the traditional regulated utility into a vast diversity of agents and services. Section 2 therefore also provides a description of the relevant economic system in terms of goods and services as well as economic agents. Section 3 analyses a bilateral organization of the market. It first considers models for the case where there are no transmission constraints and where there are transmission constraints but no market for transmission services. It then proceeds to introduce transmission services and various congestion charges that have been proposed to price these services. The various possible organizations of the market for transmission services are also discussed in this section. Alternative organizations of the energy market are described in Section 4. A nodal model is first derived from the bilateral model. This allows one to introduce the notion of a pool and some associated models. The relation between the pool and the bilateral model is then analyzed through the variational inequality framework. Section 5 discusses regulatory regimes in transmission markets. The notion of network infrastructure cost is introduced and its coverage through different access pricing charges briefly discussed. Two models, one dealing wih a separate postage stamp charge to recover infrastructure cost, the other, based on zonal prices, are presented. The paper concludes with a numerical illustration. In order to streamline the presentation, all proofs are given in an appendix.

2 BACKGROUND

2.1 THE PHYSICAL SYSTEM

Consider an electrical network (N, L) where $n \in N$ designates the nodes (or bus) and $\ell \in L$ the lines. Lines have physical characteristics (impedances, thermal limits, ...). Flows in the network are commonly described by the DC approximation of Kirchoff's laws. This can be stated as

$$-\overline{f}_\ell \leq \sum_{n \in N} \Gamma_{\ell n} i_n = f_\ell \leq \overline{f}_\ell \quad \ell \in L$$

where $\overline{f}_\ell$ is the thermal limit of line ℓ, $\Gamma_{\ell n}$ is the "distribution factor" into line ℓ of an injection at node n and a withdrawal at some arbitrary but fixed swing bus, and i_n is the net injection at node n (unconstrained). Electrical losses are neglected in this formulation. This implies that $\sum_{n \in N} i_n = 0$. The set of flows satisfying the above constraints is referred to as NC (for network capacity). An intuitive understanding of Kirchoff's laws can be gained by referring to the standard three nodes example [27, 16] depicted on Figure 5.1.

The example assumes that all lines of the network have identical characteristics. Take node 3 as the swing bus. According to Kirchoff's law, a unitary injection in 1 associated to a unitary withdrawal from 3 will generate flows respectively equal to 2/3 on line 1–3 and 1/3 on lines 1–2 and 2–3. Similarly a unitary injection at node 2 and withdrawal at node 3 induce flows of $\frac{2}{3}$ in line 2–3, $-\frac{1}{3}$ in line 1–2 and $\frac{1}{3}$ in line 1–3. For this

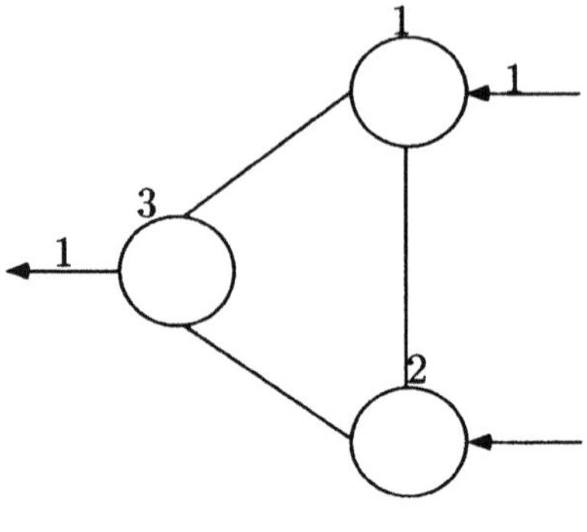

Figure 5.1 A simple three node example

network the matrix Γ has the form

$$\begin{pmatrix} \frac{2}{3} & +\frac{1}{3} \\ \frac{1}{3} & -\frac{1}{3} \\ \frac{1}{3} & \frac{2}{3} \end{pmatrix}$$

For simplicity of notation in the rest of this presentation we only refer to the upper bound constraints and define NC as $NC \equiv \{i \mid \Gamma i \leq \overline{f}\}$.

2.2 GOODS, SERVICES AND ECONOMIC AGENTS

2.2.1 Goods and services. We consider an electrical system where goods and services are limited to energy and transmission services. We neglect ancillary services such as reactive power or frequency control in this paper.

2.2.2 Economic agents. The following agents interact in this system

(i) Generators are located at a subset N_G of nodes G. They generate g_n, $n \in N_G$ according to a marginal cost function $mc_n(g)$ (where $g = (g_n, n \in N_G)$). Note that this marginal cost function is here stated as a function of the whole vector g and not necessarily of the sole g_n. This more general formulation allows one to model certain externalities.[1] It is assumed that generators do not have market power (they sell at their marginal cost $mc_n(g)$). An alternative interpretation is to suppose that $mc_n(g)$ is a supply curve that already encompasses the market power of the generators. The relevant assumption in this paper is that this market power is exogenously specified and does not come as a result of the model.

(ii) Consumers (demand) are located at a subset N_D of nodes N. They consume d_n, $n \in N_D$, according to a marginal willingness to pay function $mw_n(d)$ (where $d = (d_n, n \in N_D)$). Note again that this marginal willingness to pay is specified as a function of the whole demand vector d and not necessarily of the sole d_n.[2] It is again assumed that consumers do not have market power (they buy at marginal willingness to pay $mw_n(d)$ or alternatively their market power is embedded in the representation of $mw_n(d)$). As for generators the relevant assumption for this paper is that this market power is exogenously specified.
(iii) Traders (power marketers) play a fundamental role in this model. A trader p trades between a set of generation and demand nodes, respectively N_G and N_D. This involves the following transactions

- On the energy market; power marketer p trades e_{nmp} from node $n \in N_G$ to node $m \in N_D$. This corresponds to a bilateral representation of energy trade.
- Trader p also pays for transmission services. This will be represented by the generic expression $t^b_{nmp} e_{nmp}$ (or $t^b_{nm} e_{nmp}$) where t^b_{nmp} (or t^b_{nm}) will take on different interpretations depending on the organization of transmission services.

(iv) The network operator or ISO coordinates real time transactions. The description of its (possible) economic role is postponed as long as possible (see [23] for a discussion of these roles). This role is introduced later through some propositions.

It is assumed throughout this paper that mw_m and mc_n are differentiable. This assumption is not strictly necessary but drastically simplifies the presentation.

2.2.3 Traders and market power. Traders are of two types

- Competitive traders, $p \in P^c$ do not exert market power in the sense that they act as price takers and do not consider the effect of their actions on prices. Their *marginal economic loss* on a transaction between n and m is defined as:

$$-mw_m(d) + mc_n(g) + t^b_{nm}$$

- Oligopolistic traders, $p \in P^o$ exert market power and behave à la Cournot both with respect to generators and consumers. Their *marginal loss* on a transaction between n and m accounting for their effect on price is defined as:

$$-\left(mw_m(d) + e_{nmp}\frac{\partial mw_m}{\partial d_m}\right) + \left(mc_n(g) + e_{nmp}\frac{\partial mc_n}{\partial g_n}\right) + t^b_{nm}$$

2.2.4 Notation. Define $P = P^c \cup P^o$ to be the set of (competitive and oligopolistic) traders. The set of transactions conducted by trader p is noted $e_p^T = (e_{nmp}, n \in N_G, m \in N_D)$. The set of all transactions is noted $e^T = (e_p^T, p \in P)$. The net injection at node n is written

$$i_n = \sum_p i_{np} \text{ where } i_{np} = \sum_{j \in N_D} e_{njp} - \sum_{i \in N_G} e_{inp}.$$

The vector of net nodal injections and withdrawals of trader p is noted $i_p^T = (i_{np}, n \in N)$ where $i_{np} = 0$ for $n \notin N_G \cup N_D$. We also note $i_p = S_p e_p$ where S_p is the matrix that converts the vector of bilateral transactions of trader p into the vector i_p of net nodal injections and withdrawals by this trader. Total net nodal injections are noted $i = \sum_p S_p e_p$.

It is sometimes useful to refer to total generation or demand at node n as a function of energy trade. These will respectively be noted $g_n = \sum_{p \in P} \sum_{m \in N_D} e_{nmp} = g_n(e)$ and $d_m = \sum_{p \in P} \sum_{n \in N_G} e_{nmp} = d_m(e)$. We also write $g_n(i)$ and $d_n(i)$ with a similar interpretation.

We complete this notation by restating the marginal economic loss incurred by competitive trader p on transaction nm as

$$L_{nmp}(e; t^b) = -mw_m[d(e)] + mc_n[g(e)] + t^b_{nm}.$$

The mapping of marginal economic losses incurred by competitive trader p is then defined as

$$L_p(e; t^b)^T = (L_{nmp}(e; t^b), n \in N_G, m \in N_D).$$

Similarly the marginal loss of oligopolistic trader p on transaction nm is reformulated as

$$\begin{aligned} L_{nmp}(e; t^b) \quad &= \quad - \left(mw_m[d(e)] + e_{nmp} \frac{\partial mw_m}{\partial d_m} \Big|_{d(e)} \right) \\ &\quad + \left(mc_n[g(e)] + e_{nmp} \frac{\partial mc_n}{\partial g_n} \Big|_{g(e)} \right) + t^b_{nm} \end{aligned}$$

and the mapping of the marginal losses incurred by trader p is defined as

$$L_p(e; t^b)^T = (L_{nmp}(e; t^b), n \in N_G, m \in N_D).$$

3 BILATERAL MODELS OF RESTRUCTURED ELECTRICITY SYSTEMS

3.1 THERE ARE NO TRANSMISSION CONSTRAINTS

3.1.1 The equilibrium.

Definition 5.1 *An equilibrium without transmission constraints is a pair of vectors (g^*, d^*, e^*), $\pi^* = (\pi_n^*, n \in N)$ such that*

- *g_n^* maximizes the profit of generator n at price π_n^*, $n \in N_G$ (this implies $\pi_n^* = mc_n(g^*)$ when $g_n^* > 0$ and $\pi_n^* \leq mc_n(g^*)$ when $g_n^* = 0$)*

- *d_m^* maximizes the surplus of consumer m at price π_m^*, $m \in N_D$ (this implies $\pi_m^* = mw_m(d^*)$ when $d_m^* > 0$ and $\pi_m^* \geq mw_m(d^*)$ when $d_m^* = 0$)*

- *e_p^* maximizes the profit of competitive trader p given the price π^*, $p \in P^c$*

- *e_p^* maximizes the profit of oligopolistic trader p, given the transactions $e_{-p}^* = (e_{p'}^*, p' \in P, p' \neq p)$ of the other traders, $p \in P^o$.*

A competitive equilibrium without transmission constraints is an equilibrium without transmission constraints where oligopolistic traders are inactive ($e_p^* = 0, p \in P^o$) or do not exert market power ($\frac{\partial}{\partial d_m} mw_m(d^*) = \frac{\partial}{\partial g_n} mc_n(g^*) = 0$).

The components of π^* are commonly referred to as the *nodal prices of electricity.*

3.1.2 Computation.

In order to compute this equilibrium we introduce the variational inequality problem

VI_1: Seek $e^* \geq 0$ such that

$$\sum_{p \in P} L_p(e^*; t^b)^T (e_p - e_p^*) \geq 0 \text{ for all } e_p \geq 0.$$

3.1.3 Properties.

Assumption 1. We impose the following assumption already introduced in [20]:

- for all m, the revenue function $[mw_m(e_{nm}+K)]e_{nm}$ obtained by selling energy to consumer m is concave in e_{nm} $\forall\ K > 0$;

- for all n, the procurement cost function $[mc_n(e_{nm}+K)]e_{nm}$ obtained by buying energy from the generator n is convex in e_{nm} $\forall\ K > 0$.

This assumption guarantees the monotonicity of the mappings L_p (see [13] for a definition of monotonicity). Recall that strict monotonicity is equivalent to coercivity when mw_m and mc_n are affine.
Assumption 2. We further bound the set of transactions as follows:

- the set of transaction vectors e is compact.

This assumption can easily be interpreted as resulting from bounded consumption and generation sets. Boundedness of consumption sets is a direct consequence of the budget constraint of the final consumers and of the boundedness of the short term production set of industrial consumers. Boundedness of generation sets results from bounded generation capacities in the short term.

The following propositions can then be established.

Proposition 5.2 *Assume $L_p(e;t^b)$ is monotone for all $p \in P$, then VI_1 has a compact convex, non empty solution set. If $L_p(e;t^b)$ is strictly monotone, the solution is unique.*

Proposition 5.3 *A solution to VI_1 is an equilibrium of the bilateral model without transmission constraints.*

Proposition 5.4 *If competitive traders can engage in the same bilateral transactions as oligopolistic traders, then the equilibrium is competitive.*

3.1.4 Discussion. An obvious question is whether it makes sense to compute equilibria without transmission constraints. This may look doubtful in theory but is relevant in practice. Specifically the constraints imposed by the network have long been ignored explicitly or implicitly by many both in Europe and the US. A commonly heard claim is that the network is overbuild and transmission constraints are irrelevant [2]. The contract path approach which is so common in transmission practice implicitly neglects transmission constraints (see [14] for an extensive discussion of the subject). Also barter deals were commonly used in the old regulatory days to handle occasional congestion at border between control areas. It is sometimes suggested that they can still be used to remedy transmission constraints in the restructured industry ([9] reports that the European Commission held such a position). In brief transmission constraints are often thought to be irrelevant, or are ignored in commercial transactions and only dealt with at the system operation level.

The fallacy of both the diagnostic and the remedy is progressively realized, but the full consequences of network constraints are not for that matter fully taken into consideration. It is indeed sometimes stated that it suffices to prevent access to the network when transmission capacity is limited. The only requirement to fulfill would be that this restriction be non discriminatory. As argued below achieving non discriminatory restriction in access without properly pricing network constraints may be an impossible task in practice. In order to discuss this point we first introduce a bilateral model where transmission constraints are explicitly recognized, even though no market for transmission services is introduced.

3.2 THERE ARE TRANSMISSION CONSTRAINTS BUT NO MARKET FOR TRANSMISSION SERVICES

The following notation will be useful to state the problem. We redefine the network capacity NC as NC^b (network capacity in the bilateral representation) to make the reference to bilateral transactions explicit

$$NC^b \equiv \{e \mid e \geq 0, \sum_p \Gamma S_p e_p \leq \overline{f}\}.$$

The residual network capacity $NC_p^b(\tilde{e}_{-p})$ (network capacity available to p when $\tilde{e}_{-p}$ is given) can then be defined as:

$$NC_p^b(\tilde{e}_{-p}) = \{e_p \mid e_p \geq 0, \Gamma S_p e_p + \sum_{p' \neq p} \Gamma S_{p'} \tilde{e}_{p'} \leq \overline{f}\}.$$

3.2.1 The equilibrium. It is possible to analyze the equilibria achievable when there are transmission constraints but no market for transmission services by resorting to the notion of generalized equilibrium [25].

Definition 5.5 *A generalized equilibrium is a pair of vectors* (g^*, d^*, e^*), π^* *such that*

- g_n^* *maximizes the profit of generator* n *at price* π_n^*, $n \in N_G$
- d_m^* *maximizes the surplus of consumer* m *at price* π_m^*, $m \in N_D$
- e_p^* *maximizes the profit of competitive trader* p *given the price* π^* *and the network capacity* $NC_p^b(e_{-p}^*)$ *available to* p *when the transactions* e_{-p}^* *are given*

- e_p^* *maximizes the profit of oligopolistic trader p, given the transactions* e_{-p}^* *of the other power marketers and the residual network capacity* $NC_p^b(e_{-p}^*)$ *available to p as a result of these transactions.*

A social equilibrium [8] is a generalized equilibrium where oligopolistic traders are inactive or cannot exert market power.

3.2.2 Computation. Define the following quasi variational inequality problem (see [3])
QVI: Seek $e^* \in NC^b$ such that

$$\sum_{p \in P} L_p(e^*; t^b)^T (e_p - e_p^*) \geq 0 \text{ for all } e_p \in NC_p^b(e_{-p}^*).$$

3.2.3 Properties.

Proposition 5.6 *QVI has a solution. This solution will generally not be unique.*

Proposition 5.7 *A solution to QVI is a generalized equilibrium.*

Proposition 5.8 *Suppose that competitive traders can engage in the same transactions as oligopolistic traders. Then both competitive and oligopolistic traders can be active on the same markets.*

3.2.4 Discussion. The relevance of the generalized equilibrium, and hence of the associated QVI problem has been introduced above by referring to general attitudes with respect to network constraints. Various developments both in the theoretical literature and in the restructuring practice provide additional ground to this justification. The importance of tradable transmission rights and of a market for transmission rights in restructured electricity systems has been extensively debated in [22, 29]. It is impossible to do justice to this discussion in a few lines here. It suffices to say that the absence of a market for transmission rights creates a fundamental indeterminacy in the outcome of the energy market and that this uncertainty can exacerbate the market power of some of the agents. Generalized equilibria might thus be a very relevant concept for analyzing systems designed without markets for transmission rights. In practice markets for transmission rights are introduced in many systems in the US (PJM, California, NEPOOL, ...) while EU documents only begin to mention them. An other illustration of the relevance of generalized equilibria can be found in [14, 24, 33] in relation to the definition and uniqueness of the equilibrium in restructured electricity systems.

Because of the importance given to non discriminatory access to the network both in the EU and the US, we believe that the following proposition may contribute to further justify analyzing equilibria without markets for transmission services in case these are effectively implemented in practice. The proposition is stated in this section and discussed in more detail later on.

Proposition 5.9 *Markets without tradable transmission rights lead to implicit discriminatory access prices.*

3.3 THE EQUILIBRIUM WHEN THERE ARE NETWORK CONSTRAINTS AND A MARKET FOR TRANSMISSION SERVICES

3.3.1 Computation. Consider the following variational inequality problem

VI_2: Seek $e^* \in NC^b$ such that
$\sum_{p \in P} L_p(e^*;t^b)^T(e_p - e_p^*) \geq 0$ for all $e \in NC^b$

Because the constraint set is polyhedral, VI_2 is equivalent to (see for instance [13] for the proof of the equivalence):

$\widetilde{VI_2}$: Seek $e^* \geq 0$, $\lambda^* \geq 0$ such that
$\sum_{p \in P} \left[L_p(e^*;t^b) + \lambda^* \Gamma \sum_p S_p\right]^T (e_p - e_p^*) + (\bar{f} - \Gamma \sum_p S_p e_p^*)^T$
$(\lambda - \lambda^*) \geq 0$ for all $e \geq 0, \lambda \geq 0$

where, for a solution λ^* of $\widetilde{VI_2}$, e^* also solves

$\widetilde{\widetilde{VI_2}}$: Seek $e^* \geq 0$ such that
$\sum_{p \in P}(L_p(e^*;t^b) + \lambda^* \Gamma S_p)^T(e_p - e_p^*) \geq 0$ for all $e \geq 0$.

3.3.2 Congestion charges. Let (e^*, λ^*) be a solution to $\widetilde{VI_2}$. $\lambda^* \Gamma S_p$ defines congestion charges which can be given different interpretations:

(i) $\lambda_\ell^*, \ell \in L$ is the congestion charge on line ℓ, $\ell \in L$ [4, 5, 6, 30]. In this interpretation traders pay for the utilization of the lines. This utilization is computed from their energy transactions using the distribution factors of the Γ matrix. λ_ℓ^* is the unit price for the utilization of line ℓ.

(ii) $\lambda^* \Gamma = \tau^*$ is the vector of nodal charges to withdraw (or inject) at node n [27, 16]. In this interpretation, traders pay for the right to inject and withdraw electricity. $-\tau_n^*$ and $+\tau_n^*$ are respectively the charges to inject and withdraw at node n. The total injection and withdrawal services required by trader p are given by $S_p e_p$.

(iii) $\lambda^*\Gamma S_p = \varepsilon_p^*$ where $\varepsilon_{nm}^*(= \varepsilon_{nmp}^*, \forall p)$ is the bilateral charge between n and m [26, 16]. In this interpretation traders pay for a pair of injection and withdrawal rights. Note that $\varepsilon_{nm}^* = \tau_m^* - \tau_n^*$. The vector of transmission services required by trader p is given by e_p.

Setting $t_p^b = \varepsilon_p^* = +\lambda^*\Gamma S_p$, it is easy to see that $\widetilde{VI}_2$ becomes the Variational Inequality problem associated to a bilateral market without transmission constraints but where transmission charges are given by ε_p^*, namely

VI_1: Seek $e^* \geq 0$ such that $\sum_{p\in P} L_p(e^*;\varepsilon_p^*)^T(e_p - e_p^*) \geq 0 \; \forall \; e \geq 0$.

3.3.3 Discussion.

The relationship between the solutions to QVI and VI_2 is easily established from the following result which is a direct adaptation of Theorem 4 in [12].

Lemma 5.10 *e^* is a solution to QVI iff there exists λ_p^*, $p \in P$ such that $e^*, (\lambda_p^*, p \in P)$ is a solution to the following VI problem.*

VI: *Seek $e^* \geq 0$, $\lambda_p^* \geq 0$, $p \in P$ such that*

$$\textstyle\sum_{p\in P}\left[L_p(e^*;t^b)^T + \lambda_p^*\Gamma S_p\right](e_p - e_p^*) + \sum_{p\in P}(\bar{f} - \Gamma\sum_{p'} S_{p'}e_{p'}^*)(\lambda_p - \lambda_p^*) \geq 0 \text{ for all } e \geq 0,\ \lambda_p \geq 0,\ p \in P.$$

This lemma shows that a solution to QVI can be interpreted as an equilibrium of the bilateral market without transmission constraints where power marketers are charged (possibly) different transmission levies $\varepsilon_p^* = \lambda_p^*\Gamma S_p$. In brief a generalized equilibrium is an equilibrium without transmission constraints where access to the network is priced in a discriminatory way except for a very special case. This special case which corresponds to non discriminatory access pricing is characterized in Lemma 5.11.

Lemma 5.11 *A solution e^* to QVI is a solution to VI_2 iff $\lambda_p^* = \lambda^*$ for all p.*

This lemma is an adaptation of Theorem 6 in [12]. It fully characterizes the relationship between solutions to QVI and VI_2.

3.3.4 The equilibrium.

The following property generalizes results found in [16, 4] to the oligopolistic case.

Proposition 5.12 *Let e^* be a solution to VI_2 and i^* and f^* the corresponding net injections and line flows. Let ε^*, τ^* and λ^* be defined as before. Then the following holds*

(i) e^* maximizes $\sum_p \varepsilon_p^* e_p$ s.t. $e_p \geq 0, \Gamma\sum_p S_p e_p \leq \bar{f}$

(ii) i^* *maximizes* $\tau^* \sum_p i^*_p$ *s.t.* $\Gamma \sum_p i_p \leq \overline{f}$
(iii) f^* *maximizes* $\lambda^*_\ell f_\ell$ *s.t.* $f_\ell \leq \overline{f}_\ell$.

Properties (i) and (ii) can be interpreted as the behavior of an ISO which sells transmission rights so as to maximize the value of the network. Notwithstanding its monopoly position, this ISO is supposed to take the price of these services as given. This interpretation is advocated in [16, 14]. Property (iii) can be interpreted as the behavior of a line owner that tries to maximize the value accruing from selling line services. This line owner also takes prices as given. This interpretation is advocated in [4, 5].
Using Proposition 5.12 the following alternative definitions of equilibria can be stated.

Definition 5.13 *An equilibrium of the energy and transmission service markets is a pair of vectors* (g^*, d^*, e^*), $(\pi^*, \lambda^*_p, p \in P)$ *such that*
(i) there is equilibrium on the energy market in the sense introduced below,
(ii) there is equilibrium on the transmission service market in one of the interpretations introduced below.
The energy market. $(g^*, d^*, e^*)(\pi^*, \lambda^*_p, p \in P)$ *is an equilibrium on the energy market iff*

- g^*_n *maximizes the profit of generator* n *at price* π^*_n, $n \in N_G$
- d^*_m *maximizes the surplus of consumer* m *at price* π^*_m, $m \in N_D$
- e^*_p *maximizes the profit of the competitive trader* p *given the price* π^* *and the bilateral transmission charges* $t^*_p = +\lambda^*_p \Gamma S_p$, $p \in P^c$
- e^*_p *maximizes the profit of oligopolistic trader* p*, given the transactions* e^*_{-p} *of the other traders and the transmission charges* $t^*_p = +\lambda^*_p \Gamma S_p$, $p \in P^0$

The transmission market. (e^*_p, λ^*_p), $p \in P$ *is an equilibrium on the transmission service market iff any one of the following alternative definitions holds*

- $\lambda^*_p(\overline{f} - \Gamma \sum_p S_p e^*_p) = 0$ *and* $\lambda^*_p = \lambda^* \geq 0\ \forall p$ *(traders exchange bilateral transmission rights on a bilateral market [1]*
- $\lambda^*_p = \lambda^* \geq 0\ \forall p$ *and* e^* *maximizes* $\sum_p (\lambda^* \Gamma S_p) e_p \mid e_p \geq 0, \Gamma \sum_p S_p e_p \leq \overline{f}$ *(the ISO maximizes the value accruing from sales of bilateral transmission rights compatible with the network capacity [16]*

- $\lambda_p^* = \lambda^* \geq 0 \; \forall p$ *and* $i_p^* = S_p e_p^*$ *maximizes* $\sum_p \lambda^* \Gamma i_p \mid \Gamma \sum_p i_p \leq \overline{f}$ *(the ISO maximizes the value accruing from the sales of nodal transmission rights compatible with the network capacity [16]*

- $\lambda_p^* = \lambda^* \geq 0 \; \forall \; p$ *and* $f_\ell^* = \Gamma \sum_p S_p e_p^*$ *maximizes* $\lambda_\ell^* f_\ell \mid f_\ell \leq \overline{f}_\ell$ *for all* ℓ *(line owners maximize the value of their assets by selling line utilization services [4].*

The relation between these definitions is established through the following propositions.

Proposition 5.14 *All definitions of equilibria are equivalent.*

Proposition 5.15 *An equilibrium of the energy and transmission service market is a generalized equilibrium where congestion charges are priced in a non discriminatory way.*

Finally Proposition 5.16 relates the preceding material to the usual characterization of nodal prices in competitive electricity market [27, 16]. Specifically transmission prices are in these conditions directly derived from nodal electricity prices.

Proposition 5.16 *Suppose that there are no oligopolistic traders or that they are inactive, then* $\tau^* = -\pi^*$.

4 ALTERNATIVE ORGANIZATIONS OF THE MARKET

4.1 PRELIMINARIES: NODAL AND BILATERAL MODELS

A nodal representation of the bilateral model can easily be obtained using the notion of balanced schedules. Let $u = (1, 1, \ldots, 1)$ be a vector of dimension equal to the number of nodes. Because electrical losses are neglected, the vector of net injections of trader p, $i_p = S_p e_p$ satisfies $u i_p = 0$. By definition i_p is a balanced schedule. This notion has played a key role in the Californian reform.

In order to proceed towards the nodal reformulation of the bilateral model we first rewrite the mappings $L_p(e; t)$ of marginal economic losses in terms of net injections. As a preliminary replace the vector t^b of bilateral transmission charges by a vector $t^{no} = (t_n^{no}, n \in N)$ of nodal transmission charges. Marginal economic losses can then be defined as follows.

$$\begin{aligned} L_{np}(i; t^{no}) &= -mw_n(d(i)) + t_n^{no}, \quad n \in N_D, \text{ and} \\ L_{np}(i; t^{no}) &= mc_n(g(i)) + t_n^{no}, \quad n \in N_G, \end{aligned}$$

are respectively the marginal economic losses due to a withdrawal and injection of competitive trader $p \in P^c$ at node n.

Similarly,

$$L_{np}(i;t^{no}) = -\left(mw_n(d(i)) + i_{np}\frac{\partial mw_n}{\partial d_n}\right) + t_n^{no}, \quad n \in N_D, \text{ and}$$
$$L_{np}(i;t^{no}) = \left(mc_n(g(i)) + i_{np}\frac{\partial mc_n}{\partial g_n}\right) + t_n^{no}, \quad n \in N_G$$

are respectively the marginal economic losses due to a withdrawal and injection of the oligopolistic trader $p \in P^0$ at node n.

Finally, $L_p(i;t^{no})^T = (L_{np}(i;t^{no}), n \in N)$ is the mapping of nodal marginal economic losses of trader p.

Using this notation, the bilateral models introduced in Section 3 can be cast in terms of balanced schedules. We accordingly state the following problems

No transmission constraint

VI_1^{no}: Seek $i^*, ui_p^* = 0, p \in P$, such that
$\sum_{p\in P} L_p(i^*;t^{no})(i_p - i_p^*) \geq 0$ for all i such that $ui_p = 0$, $p \in P$.

Transmission constraints, no market for transmission services
Define the residual network capacity $NC_p^{no}(\tilde{i}_{-p}) \equiv \{i_p \mid \Gamma i_p + \sum_{p'\neq p} \Gamma \tilde{i}_{p'} \leq \overline{f}\}$

QVI^{no}: Seek $i^* \in NC^{no}$, $ui_p^* = 0$, $p \in P$ such that
$\sum_{p\in P} L_p(i^*;t^{no})(i_p - i_p^*) \geq 0$ for all i such that $ui_p = 0$ and $i_p \in NC_p^{no}(i_{-p}^*)$.

Transmission constraints and market for transmission services
VI_2^{no}: Seek $i^* \in NC^{no}$, $ui_p^* = 0$, $p \in P$, such that
$\sum_{p\in P} L_p(i^*;t^{no})(i_p - i_p^*) \geq 0$ for all i such that $ui_p = 0$ and $i \in NC^{no}$.

Properties stated in Section 3 carry through with straightforward adaptations to these nodal formulations. Taking stock of these formulations, we introduce a first pool model.

4.2 A FIRST POOL MODEL

In order to define a Pool, we concentrate on the particular situation where there is a single competitive trader hereafter noted c ($P^c = \{c\}$). Consider the problem

VI_2^{no}: Seek $i^* \in NC^{no}$, $ui_p^* = 0$, $p \in P$ such that
$\sum_{p\in P} L_p(i^*;t^{no})(i_p - i_p^*) \geq 0$ for all i such that $ui_p = 0, p \in P$ and $i \in NC^{no}$.

Because its constraint set is polyhedral, VI_2^{no} is equivalent to:

$\widetilde{VI}_2^{no}$: Seek i^*, $ui_p^* = 0$ for $p \in P$, $\lambda^* \geq 0$ such that
$\sum_{p\in P}[(L_p(i^*;t^{no}) + \lambda^*\Gamma]^T(i_p - i_p^*) + (\bar{f} - \Gamma i^*)^T(\lambda - \lambda^*) \geq 0$
for all $i, ui_p = 0$ for $p \in P$, $\lambda \geq 0$.

A solution i^* of $\widetilde{VI}_2^{no}$ is also a solution of the two following problems, where λ^* belongs to a solution pair of $\widetilde{VI}_2^{no}$.

$\widetilde{\widetilde{VI}}_{2,c}^{no}$: Seek i_c^*, $ui_c^* = 0$ such that
$[L_c(i^*;t^{no}) + \lambda^*\Gamma]^T(i_c - i_c^*) \geq 0$ for all i_c, $ui_c = 0$

$\widetilde{\widetilde{VI}}_{2,P^o}^{no}$: Seek i_p^*, $ui_p^* = 0$ for $p \in P^o$, such that
$\sum_{p\in P^o}[L_p(i^*;t^{no}) + \lambda^*\Gamma]^T(i_p - i_p^*) \geq 0$ for all i_p, $ui_p = 0$, $p \in P^o$.

Problem $\widetilde{\widetilde{VI}}_{2,c}^{no}$ describes the behavior of the competitive trader. As before, note $\tau^* = \lambda^*\Gamma$ and define π^* as a vector of nodal prices that satisfy the following complementary slackness conditions (see Section 3.1.1).

$$\begin{array}{llll} \pi_n^* - mc_n(g(i^*)) \leq 0 & \text{and} & [\pi_n^* - mc_n(g(i^*))]g_n(i^*) = 0, & n \in N_G \\ -\pi_n^* + mw_n(d(i^*)) \leq 0 & \text{and} & [-\pi_n^* + mw_n(d(i^*))]d(i^*) = 0, & n \in N_D \end{array}$$

It is easy to see from the statement of problem $\widetilde{\widetilde{VI}}_{2,c}^{no}$ that $\pi^* = \tau^* + \beta_c^*$ (where β_c^* is the dual variable of the constraint $ui_c = 0$ at the solution i_c^*) satisfies these complementary slackness conditions and hence defines a set of nodal prices in the sense of [27, 16] and many others. Also, transmission between two nodes is priced at a level equal to the difference between τ_n^* and hence also between nodal prices π_n^*.

These two properties (the existence of nodal prices and the above relation between transmission and nodal prices) characterize a pool. They are satisfied by the solution to problem VI_2^{no}. To see that they are not satisfied in general, consider problem $\widetilde{\widetilde{VI}}_{2,P^o}^{no}$ when there is no competitive trader. The equilibrium conditions associated to an oligopolistic trader p active at nodes n and m ($i_{np}^* > 0$ and $i_{mp}^* < 0$) can be stated as

$$\begin{array}{rcll} mc_n(g^*) + i_{np}^* \frac{\partial mc_n}{\partial g_n} & = & \tau_n^* + \beta_p^* & n \in N_C \\ mw_m(d^*) + i_{mp}^* \frac{\partial mw_m}{\partial d_m} & = & \tau_m^* + \beta_p^* & m \in N_D \end{array}$$

where β_p^* is the dual variable of the balanced schedule constraint $ui_p = 0$.

Transmission between nodes n and m is priced at

$$\begin{array}{rcl} \tau_m^* - \tau_n^* & = & +mw_m(d^*) - mc_n(g^*) - i_{np}^* \frac{\partial mc_n}{\partial g_n} + i_{mp}^* \frac{\partial mw_m}{\partial d_m} \\ & < & mw_m(d^*) - mc_n(g^*). \end{array}$$

In conclusion a pure oligopolistic market does not satisfy the pool conditions.

The preceding discussion suggests that it suffices to have a competitive trader accessing all generators and consumers to make the system equivalent to a pool. This result is not satisfactory though. Indeed Proposition 5.4 tells us that such a competitive trader makes the system not only a pool by also a competitive market. In fact, the competitive trader drives the oligopolistic traders out of the market. Because pools are generally not exempt from market power in practice (see all the theory of supply curve equilibrium) , it would be interesting to be able to model pools that are not competitive. A first result in this direction is achieved by assuming that the competitive trader(s) do(es) not access the same generators and consumers as the oligopolistic traders. Nodal prices would then be derived from supply and willingness to pay functions of those generators and consumers trading through the competitive marketer(s) while oligopolistic traders would have access to other agents and hence would see different supply and willingness to pay functions. The reasoning leading to Proposition 5.4 would no longer hold and we would have a pool which is not competitive. Such a situation would be achieved in practice if one were to assume that small generators and consumers (or their representative distribution companies) trade through the pool while large generators and consumers trade via bilateral contracts. This system satisfies the property of a pool where nodal prices are determined by the prices paid to and by these small generators and consumers. In this system nodal prices also determine the transmission charges paid by the bilateral transactions that take place outside the pool. Such mixed systems satisfy the definition of a pool and are indeed found in practice. Their mathematical definition remains unsatisfactory though as it requires that the modeler identifies the generation and consumption segments of the market that trade through the pool and those which do not, a daunting task indeed. A more useful construction of a pool is in order. It has recently been provided by Hobbs ([15]). Hobbs model was formulated as a complementarity problem. The following presentation is cast in our variational inequality context.

Before leaving this section, note that a two stage version of the above model has been introduced in [17]. Our one stage version is mathematically much simpler. Whether it is or not more realistic than the two stage model in an empirical question that is not discussed here.

4.3 A COURNOT POOL MODEL

4.3.1 Hobbs' arbitrageur. Consider again the particular case where there is a single competitive trader h hereafter referred to as Hobbs' trader or the arbitrageur ($P^c = \{h\}$). Suppose P^c and P^o cover the same markets. Suppose that the market functions according to a two stage structure similar to [17]. In the first stage, oligopolistic traders trade à la Cournot knowing that there is a second stage. In this latter, Hobbs' trader arbitrages between nodes to take advantage of differences between nodal prices and transmission charges. As seen in Section 4.2, transmission charges in an oligopolistic market are indeed smaller than differences between nodal prices. This creates an opportunity for profit that the arbitrageur takes advantage of. The end result of Hobbs' trader activity is that the difference of nodal prices between two arbitrary nodes at the equilibrium of the second stage game is always equal to the transmission charge between these nodes. This satisfies the characteristics of a pool as all bilateral transactions pay transmission prices equal to differences of nodal prices. Needless to say oligopolistic traders loose some of their market power in the process as they can no longer (price) discriminate between generators and customers. In contrast with the result of the preceding section, the oligopolist and the arbitrageur can both operate on the same segments of the market without making the market competitive. The main differences (and a mathematically essential one) between this model and the one of Section 4.2 is that these traders do not operate in the same stage of the market. We now explore this process in more details.

4.3.2 A two stage equilibrium approach. The preceding discussion can be formalized as an equilibrium problem subject to equilibrium constraints. The following characterizes the second stage equilibrium. For the sake of simplification we drop all references to t^{no} in the rest of this section. We also write $i = (i_{P^o}, i_h)$ where $i_{P^o} = (i_p, p \in P^o)$ to distinguish between the injections of the oligopolistic traders and those of the arbitrageur.

Second stage problem of Hobbs' trader.

- Let $(i_p^*,\ p \in P^o)$ be given from the first stage. Trader h solves the problem:

  Seek i_h^*, $ui_h^* = 0$ such that $[L_h(i_{P^o}^*, i_h^*) + \tau^*]\,(i_h - i_h^*) \geq 0$ for all i_h, $ui_h = 0$.

- Suppose that L_h is affine and coercive in i_h. This variational inequality problem has a unique solution $i_h(i_{P^o})$ which is itself an affine function of i_{P^o}. It is the solution of the affine system of equations

  $L_h(i_{P^o}, i_h) + \tau^* + \beta_h = 0, ui_h = 0$ where β_h(balance) is the dual variable of $ui_h = 0$.

In order to move from the second to the first stage, one needs to model the assumption that trader $p \in P^o$ takes the reaction $i_h(i_P^o)$ of the arbitrageur into account in the first stage. This is done as follows.

First stage problem of an oligopolistic trader.

Consider a trader $p \in P^o$. Note P^o/p the set of oligopolistic traders different from p and define L^o and L^a (respectively o for oligopolistic, a for arbitrage) the respective contributions to the marginal economic loss of trader p from his/her action (i_{np}) and from the arbitrageur's actions (i_{nh}):

$$\begin{array}{lllllll} L^o_{np} & = & -\left(mw_n(d(i)) + i_{np}\frac{\partial mw_n}{\partial i_{np}}\right), & L^a_{np} & = & -i_{np}\frac{\partial mv_n}{\partial i_{nh}} & \text{if } n \in N_D \\ L^o_{np} & = & mc_n(g(i)) + i_{np}\frac{\partial mc_n}{\partial i_{np}}, & L^a_{np} & = & i_{np}\frac{\partial mc_n}{\partial i_{nh}} & \text{if } n \in N_G \end{array}$$

Trader $p \in P^o$ then solves the problem:
Seek $i_p^*, i_h^*, ui_p^* = 0, ui_h^* = 0$ and $L(i_p^*, i_{P^o/p}^*, i_h^*) + \tau^* + \beta_h^* = 0$ such that $\left[L_p^o(i_p^*, i_{P^o/p}^*, i_h^*) + \tau^*\right]^T (i_p - i_p^*) + \left[L_p^a(i_p^*, i_{P^o/p}^*, i_h^*)\right]^T (i_h - i_h^*) \geq 0$ for all $i_p, i_h, ui_p = 0, ui_h = 0$ and $L_h(i_p, i_{P^o/p}^*, i_h) + \tau^* + \beta = 0$.

An equivalent first stage problem.

Combining the behavior of all traders, we state the following variational inequality problem.

VI^{pool}: Seek $i_{P^o}^*, i_h^*, \lambda^* \geq 0$ satisfying $ui_p^* = 0$, $p \in P^o$, $ui_h^* = 0$, $(i_{P^o}^*, i_h^*) \in NC$, $L_h(i_{P^o}^*, i_h^*) + \lambda^*\Gamma + \beta_h^* = 0$ such that

$$\sum_{p \in P^o} \left\{ \left[L_p^o(i_{P^o}^*, i_h^*) + \lambda^*\Gamma\right]^T (i_p - i_p^*) + \left[L_p^a(i_{P^o}^*, i_h^*)\right]^T (i_h - i_h^*)\right]\right\}$$
$$+[\overline{f} - \lambda^*\Gamma(\textstyle\sum_{p \in P^o \cup \{h\}} i_p^*)]^T(\lambda - \lambda^*) \geq 0$$

for all $\lambda \geq 0$ and i_{P^o}, i_h, satisfying $ui_p = 0$, $p \in P^o$, $ui_h = 0$, $L_h(i_{P^o}, i_h) + \lambda\Gamma + \beta_0 = 0$.

Proposition 5.17 *Suppose the mappings L_p^o, L_p^a, $p \in P^o$ are strictly monotone and L_c is affine and coercive, then VI^{pool} has a single solution. This solution is a Pool with Cournot players in the sense of Hobbs.*

4.4 THE POOL AND BILATERAL MODELS

The relation between the bilateral and pool organizations of the market received considerable attention in the literature. The discussion probably peaked at the time of the restructuring in California. The general conclusion of this discussion is that the two systems are equivalent when there is no market power. To the best of our knowledge this conclusion was never proved formally and no analysis of the relation between the two systems is available for the case where there is market power. This subsection takes stock of the preceding variational inequality formulations to formally establish the equivalence between the Pool and the bilateral model when there is no market power. It also establishes that this equivalence does not hold when there is market power. This is stated in the following proposition.

Proposition 5.18 *The bilateral and pool models are equivalent when there is no market power. They are not equivalent when there is market power.*

A standard result of imperfect competition theory is that Cournot equilibria converge to perfect competition equilibria when the number of Cournot players increases. This result suggests that the equilibria in the bilateral and pool models under the Cournot assumption might converge to a competitive equilibrium when the number of Cournot players increases. Such a proposition was first hinted in [28]. It is stated here in general form.

Proposition 5.19 *Consider the sequence of bilateral models obtained by replacing each trader by n identical traders (n going to ∞) and the corresponding sequence of equilibria generated by this sequence of problems. Then any accumulation point of this sequence of equilibria is a competitive equilibrium. A similar property holds for pool models.*

This proposition can be interpreted by saying that even though the bilateral and pool models are not equivalent when there is market power, they become almost equivalent to a competitive equilibrium when the number of oligopolistic traders is large enough.

5 REGULATED TRANSMISSION MARKETS

5.1 INFRASTRUCTURE COSTS: BACKGROUND

Congestion charges are due to limitations of transmission capacities. It is commonly admitted that congestion charges do not provide revenue sufficient to cover infrastructure costs. These costs, or their fraction that is not covered by congestion charges, must then be recovered through other levies.

A wide variety of charges have been introduced (access charge, network services ...) to recover infrastructure costs. A natural question is whether one can tackle both congestion and these other charges in a single computational framework. The answer is that computational models can still be devised even though solution existence and uniqueness become much more difficult if not impossible to establish.

For the sake of simplification in this presentation, we assume a single network service charge collected through a postage stamp, that is a uniform, per Kwh, charge.

5.2 SEPARATE CONGESTION AND NETWORK SERVICE CHARGES

5.2.1 Computation. The following model considers the situation where two types of charges are recovered from the users of the network (here the power marketers). The first charge is the standard congestion charge discussed in the preceding sections. The second levy is a postage stamp which is supposed to cover the revenue requirement of the network. Altogether, these charges add up to more than the total costs incurred by the transmission activity. This decomposition of charges is common. It can be found, for instance, in [10] where network infrastructure costs and measures for dealing with congestion are indeed seen as two separate matters. This model is constructed on the basis of the nodal version of the equilibrium problem.

Consider the nodal model where we initially assume $t^{no} = 0$.

VI_2^{no}: Seek $i^* \in NC^{no}$, $ui_p^* = 0$, $p \in P$, such that
$\sum_{p \in P} L_p(i^*; 0)(i_p - i_p^*) \geq 0$ for all i such that $ui_p = 0$
and $i \in NC^{no}$

Let RRN be an amount (revenue requirement of the network) to be collected through a postage stamp t^s (a unique charge independent of the node). It is assumed that this postage stamp is only charged at the consumption sites. The covering of the revenue requirement is expressed by the constraint $-t^s \sum_{n \in N_D} \sum_p i_{np} = RRN$ (recall $i_{np} \leq 0$ for $n \in N_D$). Introducing this constraint in VI_2^{no} leads to the following mixed variational inequality problem.

MVI: Seek $i^* \in NC^{no}$, t^{*s}, $ui_p^* = 0$, $p \in P$ such that
$\sum_{p \in P} [L_p(i^*; t^{s*})]^T (i_p - i_p^*) \geq 0$ for all i such that $ui_p = 0$, $p \in P$,
and $i \in NC^{no}$ and $-t^{s*} \sum_{n \in N_D} \sum_p i_{np}^* = RRN$.

The difference between problems VI_2^{no} and MVI is twofold. First, the constraint set of MVI is obtained by intersecting the constraint set of VI_2^{no} with an additional restriction that imposes a certain revenue requirement. Second, the postage stamp t^s is no longer exogenous but is a variable appearing both in the mapping and in the constraints.

5.2.2 The equilibrium. Transmission charges at some node thus consist of two parts: t^{s*} which is only collected (in this version of the model) at the demand node and τ_n^* which is the congestion charge. The definitions of the equilibrium presented in Section 3 are readily transposed to this new context. We only mention the adaptation to the case where the ISO maximizes the revenue accruing from nodal congestion charges.

Definition 5.20 *An equilibrium of the energy and transmission service markets with coverage of infrastructure cost is a pair of vectors* (g^*, d^*, i^*), $(\pi^*, \lambda^*, t^{s*})$ *such that*
(i) there is equilibrium on the energy market in the sense introduced below
(ii) there is equilibrium on the transmission service market in the sense introduced below

The energy market. $(g^*, d^*, i^*), (\pi^*, \lambda^*, t^*)$ is an *equilibrium on the energy market* iff

- g_n^* maximizes the profit of generator n at price π_n^*, $n \in N_G$
- d_m^* maximizes the profit of consumer m at price π_m^*, $m \in N_D$
- i_p^* maximizes the profit of the competitive trader p, given the price π^* and the nodal transmission charges $\tau^* = \lambda^*\Gamma$ and the infrastructure charge t^{s*}
- i_p^* maximizes the profit of oligopolistic trader p, given the transactions i_{-p}^* of the other traders, the nodal transmission charges $\tau^* = \lambda^*\Gamma$ and the infrastructure charge t^{s*}.

Transmission market. (i_p^*, τ^*, t^{s*}) is an equilibrium on the transmission service market iff the following holds

- i^* maximizes $\{\tau^* \sum_p i_p \mid \Gamma \sum_p i_p \leq \bar{f}\}$ (the ISO maximizes the value accruing from the sales of nodal transmission rights compatible with the network capacity)
- $t^{s*} \sum_{n \in N_D} \sum_p i_{np}^* = -RRN$ (the infrastructure costs are covered).

While the existence and uniqueness of the solution of the models of Sections 3 and 4 can easily be derived from standard results, this is no longer the case here. Indeed, the mapping appearing in problem VI^{no} after substituting t by its expression derived from the revenue requirement constraint does not have any monotonicity property. Solution may thus fail to exist or multiple solutions may also exist. This is stated in the following proposition.

Proposition 5.21 *MVI may have no, one or several solutions. Any solution of MVI is an equilibrium in the above sense.*

5.3 NODAL AND ZONAL SYSTEMS

5.3.1 Background. Distinguishing electricity prices on a nodal basis is not a readily accepted idea. The common wisdom has been for a long time that nodal systems are too complex to be implemented in practice (even though such systems were already operating in some countries). This attitude prevailed in the early days in the US and the issue is still heavily debated today. This belief is certainly the rule in Europe as no nodal system is envisaged at this stage on the continent. Zonal prices have been proposed as an alternative to nodal prices. By definition zonal pricing associates prices, not to individual nodes but to zones of nodes. The limit case of the zonal system, when zones are made smaller, is the single node zone or nodal system.

The following shows that the variational inequality framework allows one to model zonal systems. As an example we consider the case where congestion is charged according to a zonal pattern and no other charge (such as infrastructure cost) is recovered. At least two approaches can be envisaged for zonal prices. The first one consists of distributing zonal congestion charges through each zone. In the second approach the ISO engages is countertrading and distributes the cost accruing from this countertrading activity. We limit ourselves to the first case.

5.3.2 Computation. Consider a partitioning $N = \cup_{z \in Z} N_z$ (z for zones) of the nodes into zones. Zonal charges are obtained by perturbing nodal prices so as to make all nodal prices in a zone equal. Zonal charges should also cover nodal congestion charges in each zone. We achieve both requirements by appropriately choosing the additional nodal charge t_n^{no}. Consider the nodal problem VI_2^{no}.

VI_2^{no}: Seek $i^* \in NC$, $ui_p^* = 0$, $p \in P$ such that
$\sum_{p \in P} L_p[(i^*; t^{no})](i_p - i_p^*) \geq 0$ for all i such that $ui_p = 0$, $p \in P$, and $i \in NC^{no}$

At equilibrium i^* solves the problem

$\widetilde{\widetilde{VI}}_2^{no}$: Seek $i^*, ui_p^* = 0, p \in P$ such that
$\sum_{p \in P}[L_p(i^*; t^{no}) + \lambda^* \Gamma](i_p - i_p^*) \geq 0$ for all i such that $ui_p = 0, p \in P$.

A zonal pricing scheme is achieved for a t^{no*} such that

(i) $t_n^{no*}+(\lambda^*\Gamma)_n = at_z^{no*}$ (where at_z^{no} stands for average nodal transmission charge) for all nodes n, $n \in N_z$, and

(ii) $at_z^{no*}\sum_{n\in N_z}\sum_p i_{np}^* = \sum_{n\in N_z}(\lambda^*\Gamma)_n(\sum_p i_{np}^*)$ (all congestion charges are covered).

Under these conditions all nodal charges in a zone are identical and turn out to be averages of nodal congestion charges.

Define $t^{no} = (t_n^{no}, n \in N)$ and $at^{no} = (at_z^{no}, z \in Z)$. These conditions can be cast in the following single variational inequality problem

VI_2^{zo}: Seek $i^* \in NC$, $ui_p^* = 0$, $p \in P$, t^{no*}, at^{no*} such that
$\sum_{p\in P}[L_p(i^*;t^{no*})]^T(i_p - i_p^*) + \sum_{z\in Z}\sum_{n\in N_z}[t_n^{no*} - (\lambda^*\Gamma)_n - at_z^{no*}]^T(t_n^{no} - t_n^{no*})$
$+\sum_{z\in Z}[at_z^{no*}\sum_{n\in N_z} i_n^* - \sum_{n\in N_z}(\lambda^*\Gamma)_n i_n^*]^T(at_z^{no} - at_z^{no*}) \geq 0$
for all $i \in NC$, $ui_p = 0$, $p \in P$, t^{no}, at^{no}.

As for the problem MVI discussed in Section 5.2, the mapping appearing in problem VI^{zo} does not have any monotonicity property. Solutions may thus fail to exist or multiple solutions may also exist. As for Section 5.2, the algorithmic situation is also much bleaker even though procedures for handling the problems of Sections 5.2 and 5.3 may be found in the literature (and seem to be working quite well). Two approaches coming from quite different origins [7, 11] can be used and lead to similar approaches. Specifically both will require to solve the following sequence of problems

$VI_2(t^{no,k})$: Seek $i^{k*} \in NC$, $ui_p^{k*} = 0$, $p \in P$ such that
$\sum_{p\in P} L_p(i^{k*};t^{no,k})(i_p^k - i_p^{k*}) \geq 0$ for all i^k such that $ui_p^k = 0, p \in P, i^k \in NC$

Update $t^{no,k}$ into $t^{no,k+1}$ by solving the system

$$\begin{aligned} at_z^{no,k+1}\sum_{n\in N_z} i_n^{k*} &= \sum_{n\in N_z}(\lambda^{k*}\Gamma)_n i_n^{k*} \\ t_n^{no,k+1} &= (\lambda^{k*}\Gamma)_n + at_z^{no,k+1}. \end{aligned}$$

Convergence conditions for this approach can be found in [7].

5.3.3 The equilibrium. The definition of the equilibrium associated with the solution of this variational inequality problem is readily derived from the bilateral equilibria discussed in Section 3. An illustration of this type of adaptation is given in section 5.2.2. A similar procedure applies here and will not be repeated. The definition of the overall equilibrium and of the equilibrium on the energy markets are identical to those of Section 5.2. The equilibrium on the transmission market is expressed by saying that transmission prices are identical in each zone. A proposition very similar to Proposition 5.21 can also be stated.

6 A NUMERICAL ILLUSTRATION

The following example illustrates the relevance of being able to model alternative market designs and to simulate their operation.

We consider a problem with 13 nodes and 21 lines. Four generators are located on these nodes. They are supposed to be owned by 4 power marketers. This formulation is slightly different from the one adapted previously in the paper in two respects. First, because the power marketers own the generators, they only exert market power towards consumers. Second, two markets segments (residential/tertiary and industrial) are located at each node. Each is represented by an affine demand curve. In order to illustrate the impact of market design, we consider two alternative assumptions of competition and three alternative regimes of transmission pricing. Specifically we model the case where power marketers do not exert market power and the one where they behave à la Cournot. As to transmission we consider two main situations. In the first one nodal congestion charges constitute the only payment for transmission services (case 1), in the second situation both congestion charges and the cost of the infrastructure are recovered in a single transmission charge. In this latter situation we further distinguish between the case where the total transmission charges only cover infrastructure cost (case 2) and the one where they cover the sum of congestion and infrastructure costs (case 3).

The outcome of a market structure can be judged on the basis of different criteria. The following are reported below.

AVER _P	is the quantity averaged price of electricity computed over all demand nodes and sectors. Price levels constitute one of the most commonly reported results in restructured electricity systems. It is here expressed in US$ per MWh.
TOT_QN	is the total energy traded. It is equal to the sum of the generated or consumed energy. It is measured in TWh.
UTILITY	measures the total willingness to pay of the consuming sectors. It is directly integrated from the affine demand curves of both the final and industrial consumers. It is measured in millions of US$.
TOTAL INCOME	is the revenue of the power marketers. It is measured in millions of US$.
VARCST	is the total variable cost of generation plants. It is measured in millions of US$.
FIXCST	is the fixed production cost of generation plants. It is measured in millions of US$.
PROFIT	is the difference between the total revenue and the sum of the variable and fixed costs.
WELFARE	is the global welfare of the producers and consumers. It is the sum of producers and consumer surplus and is measured in millions of US$.
MARGIN	is the addup to congestion charges necessary to cover infrastructure costs. This increase only covers the difference between the revenue requirement and congestion costs in case 2. It covers the whole revenue requirement in case 3.

The numerical results agree with the common wisdom. First, prices drastically increase when introducing market power. Second, for each assumption of market power (that is to say with and without market power) prices also increase when departing from nodal prices to include a higher recovery of fixed infrastructure costs. Recall that infrastructure costs are recovered through a single levy that includes congestion charge in case 2, while infrastructure costs and congestion are recovered separately through two different charges in case 3.

Even though market power and inefficient transmission pricing both contribute to decreasing total welfare, as reflected in row WELFARE, market power plays the dominant role. As can be seen, imposing the worst transmission-pricing scheme (case 3) when there is no market power is still better in terms of welfare than an efficient (nodal and congestion based only) transmission price (case 1) when there is market power. The overall decrease of welfare throughout these cases is relatively smooth though. This is in contrast with the allocation of this welfare among generators/power marketers and consumers (see columns UTILITY and PROFIT). Consumers lose a lot while generators/power marketers correspondingly gain significantly when market power is assumed.

Models of this type only measure a potential for market power. They do not indicate the extent to which the latter is exerted. It may thus be relevant to relate these results to empirical studies of existing systems. Wolfram in [32] assessed the extent to which generators were exerting market power in England and Wales. She found that, even though prices included considerable margin with respect to marginal costs, these were significantly below the Cournot levels or even below the levels that would result from supply curve equilibrium. In fact none of the existing theories of market power accounted for the observed price levels. Her assessment of market power is conducted through a coefficient very much akin to the conjectural variation appearing in the standard theory of oligopolistic pricing. It may be useful to note that it is quite easy to introduce a conjectural variation in the above models. This means that even though the models only measure the potential for market power, they can also include information obtained from empirical studies about the extent to which market power is indeed exerted. Considerable attention has been given recently to market power in restructured electricity systems. The results obtained in the illustrative example fully justify this interest; as indicated in the table market power is certainly detrimental to global welfare but much more so to consumers welfare. This is quite in line with the conclusion in [21] that shareholders benefited while consumers lost from the restructuring in England and Wales.

	Competitive			Cournot		
	Case 1	Case 2	Case 3	Case 1	Case 2	Case 3
AVER_P	28	36	36	158	164	169
TOT_QN	124	122	122	94	92	91
UTILITY	34048	33153	33065	19952	19752	19082
INCOME	3418	3170	3150	14839	15105	15329
VARCST	1848	1789	1783	1419	1379	1344
FIXCST	955	955	955	955	955	955
PROFIT	615	426	412	12465	12770	13030
WELFARE	34663	33579	33477	32417	32522	32112
MARGIN	0	9	10	0	7	13

We suggested in the introduction that the models presented in this paper could only capture part of the features of real restructured systems. This is particularly true for the market microstructure that has been found quite important by various authors [30, 19]. It is not clear at this stage that the microstructure of the market can be represented in computational models. We would suggest that it is equally unclear that it can not. It is certainly a reasonable objective for further work to take advantage of the possibility to develop multistage equilibrium models to try to take market microstructure on board.

Notes

1. Suppose for the three nodes example that plants 1 and 2 are located along a river. Plant 1 is upstream and plant 2 downstream. Suppose further that both plants are thermal units using the water stream as cold source. The efficiency of each plant and hence its marginal cost depends on the temperature of the stream. While the temperature of the water used by unit 1 is unaffected by the operation of plant 2, the converse is not true. Indeed a higher operations level of plant 1 increases the temperature of the water used by plant 2. This decreases the efficiency of plant 2 and hence increases its marginal cost. This externality is modeled by letting the marginal cost of plant 2 be a function of both g_1 and g_2.

2. This allows to represent cross price elasticities between demand at different nodes. This possibility may be relevant to model activity relocation phenomena.

References

[1] Boucher, J. and Y. Smeers (1999). Alternative models of restructured electricity systems. Part 1: No market power. Mimeo, CORE Discussion Paper 9950, Université catholique de Louvain, Louvain-la-Neuve, Belgium. To appear in Operations Research.

[2] Boucher, J., Ghilain, B. and Y. Smeers (1998). Security-constrained dispatch gives financially and economically significant nodal prices. *The Electricity Journal*, 53–59.

[3] Chan, D. and J.S. Pang (1989). The Generalized quasi-variational inequality problem. *Mathematics of Operations Research*, 212–222.

[4] Chao, H.-p. and S. Peck (1996). A market mechanism for electric power transmission. *Journal of Regulatory Economics*, 10, 25–60.

[5] Chao, H.-p. and S. Peck (1997). An institutional design for an electricity contract market with central dispatch. *The Energy Journal*, 18(1), 85–110.

[6] Chao, H.-p. and S. Peck (1998). Reliability management in competitive electricity markets. *Journal of Regulatory Economics*, 3, 189–200.

[7] Dafermos, S. (1983). An iterative scheme for variational inequalities. *Mathematical Programming*, 26, 40–47.

[8] Debreu, G. (1952). A social equilibrium existence theorem. *Proceedings of the National Academy of Sciences U.S.A.*, 38, 886–893.

[9] EC Energy Monthly (1999). Commission, industry rift over transmission pricing. *Financial Times*, 122/9.

[10] Eurelectric (1998). Draft report on national transmission tariffs. Unipede/Eurelectric, Brussels, November 18, 1998.

[11] Greenberg, H.J. and F.H. Murphy (1985). Computing regulated market equilibria with mathematical programming. *Operations Research*, 35(5), 935–955.

[12] Harker P.T. (1991). Generalized Nash Games and Quasi-Variational Inequalities. *European Journal of Operations Research*, 54, 81–94.

[13] Harker P. T. and J.S. Pang (1990). Finite-dimensional variational inequality and non-linear complementarity problems: a survey of theory, algorithm and applications. *Mathematical Programming*, 48B, 161–220.

[14] Harvey, S.M., Hogan, W.W. and S. Pope (1997). Transmission capacity reservations and transmission congestion contracts. Center for Business and Government, Harvard University, Cambridge (Mass.).

[15] Hobbs, B.F. (1999). Linear complementarity models of Nash Cournot competition in bilateral and POOLCO power markets. Submitted to *IEEE Transactions on Power Systems*, June 1999.

[16] Hogan, W. (1992). Contract networks for electric power transmission. *Journal of Regulatory Economics*, 4, 211-242.

[17] Hogan, W. (1997). A market power model with strategic interaction in electricity network. *The Energy Journal*, 18(4), 107–141.

[18] Hunt, S. and G. Shuttleworth (1996). *Competition and choice in electricity.* New York, John and Wiley and Sons.

[19] Joskow, P. and J. Tirole (2000). Transmission rights and market power in electric power networks. MIT-CEEPR 2000-002 Working Paper. Center for Energy and Environmental Policy Research, Massachusetts Institute of Technology.

[20] Murphy, F.H., Sherali, H.D. and A.L. Soyster (1983). Stackelberg-Nash-Cournot equilibria: Characterisation and computation. *Operations Research*, 31(3), 253–276.

[21] Newbery, D.M. and M.G. Pollit (1997). The restructuring and privatisation of the CEGB – was it worth it. *Journal of Industrial Economics*, 45(3), 269–303.

[22] Oren, S.S. (1997). Economic inefficiency of passive transmission rights in congested electricity systems with competitive generation. *The Energy Journal*, 18(1), 63–84.

[23] Oren, S.S. (1998). Authority and responsibility of the ISO: objectives, options and tradeoffs. In H.-p. Chao and H.Huntington (eds.),

Designing Competitive Electricity Markets. Kluwer's International Series.

[24] Oren, S., Spiller, P. Varariya, P. and F. Wu (1995). Nodal prices and transmission rights: a critical appraisal. *The Electricity Journal*, 8, 24–35.

[25] Rosen, J.B. (1965). Existence and uniqueness of equilibrium points for concave n-person game. *Econometrica*, 33, 520–534.

[26] Schweppe, F.C., Caramanis, M.C. and R.E. Bohm (1986). The costs of wheeling and optimal wheeling rates. *IEEE Transactions on Power System*, Volume PWRS-1, No. 1.

[27] Schweppe, F.C., Caramanis, M.C. Tabors, R.E. and R.E. Bohn (1988). *Spot pricing of electricity.* Boston, Kluwer Academic Publishers.

[28] Smeers, Y. and J.-Y. Wei (1997). Do we need a power exchange if there are enough power marketers. CORE Discussion Paper 9760, Université catholique de Louvain, Louvain-la-Neuve, Belgium.

[29] Stoft, S. (1999). Financial transmission rights meet Cournot: How TCCS curb market power. *The Energy Journal*, 20(1), 1–23.

[30] Wilson, R.B. (1998). Design principles. In H.-p. Chao and H.Huntington (eds.), *Designing Competitive Electricity Markets.* Kluwer's International Series.

[31] Wilson, R.B. (1999). Market architecture. Mimeo, Stanford University, Stanford.

[32] Wolfram, C.D. (1999). Measuring duopoly in the British electricity spot market. *American Economic Review*, 89(4), 805–826.

[33] Wu, F. Varaiya, P., Spiller, P. and S. Oren (1996). Folk theorems on transmission access: Proofs and counterexample. *Journal of Regulatory Economics*, 10(1), 5–24.

[34] Yajima Masajyuki (1997). *Deregulatory reforms of the electricity supply industry.* Quorum books, Westport Connecticut.

Appendix: Proofs

Proposition 5.2 The proposition is an adaptation of standard results (e.g. see Harker and Pang (1990)).
Proposition 5.3 The proof is a special case of the proof of Proposition 5.14.
Proposition 5.4 Suppose that $p^c \in P^c$ and $p^o \in P^o$ both trade between nodes n and m, that is $e^*_{nmp^c} > 0$ and $e^*_{nmp^o} \geq 0$. Being a solution to

VI_1 implies

$$
\begin{aligned}
0 &= -mw_m[d(e^*)] + mc_n[g(e^*)] + t^b_{nm} \text{ (trader } p^c) \\
0 &= -mw_m[d(e^*)] + mc_n[g(e^*)] + t^b_{nm} + e^*_{nmp^o}\left(-\tfrac{\partial mw_m}{\partial d_m} + \tfrac{\partial mc_n}{\partial g_n}\right) \\
&\quad \text{(trader } p^o).
\end{aligned}
$$

Because of the assumption, $-\frac{\partial mw}{\partial d_m}$ and $\frac{\partial mc_n}{\partial g_n}$ are both ≥ 0. These, combined with the two above equalities imply either $e^*_{nmp^o} = 0$ or $-\frac{\partial mw_n}{\partial d_m} = \frac{\partial mc_n}{\partial g_n} = 0$.

Proposition 5.6 The proof follows from Theorem 2 of Harker (1991) where the compactness required in this theorem is guaranteed by Assumption 2.

Proposition 5.7 The proof follows from well established reasoning (See Theorem 4 in Harker (1991) applied in the context of the proof of Proposition 5.14).

Proposition 5.8 The following example proves the proposition.

Take a single line network (line 1–2 between nodes 1 and 2). There is one generator of marginal cost $20 + q$ at node 1 and one consumer of demand curve $120 - q$ at node 2. The capacity of the line is 40. There are two power marketers trading between these two nodes. Trader 1 is competitive and buys and sells at marginal cost. Trader 2 is oligopolistic. Consider the situation where power marketers 1 and 2 respectively trade 30 and 10 between nodes 1 and 2. Nodal prices are respectively 60 and 80 at nodes 1 and 2. One easily checks that 30 maximizes the profit

$$(80 - 60)q_1 \text{ subject to } 0 \leq q_1 \leq 40 - 10$$

of competitive trader 1 given the residual capacity 30 of line 1–2. Similarly 10 maximizes the profit

$$[120 - (30 + q_2)]q_2 - [20 + (30 + q_2)]q_2 \text{ subject to } 0 \leq q_2 \leq 40 - 30$$

of oligopolistic (here monopolistic) trader 2 given the residual capacity 10 of line 1–2.

Proposition 5.9 The proof of this proposition is directly derived from Lemma 5.10, as explained in Section 3.3.3.

Proposition 5.12 We prove (i) (the proof of (ii) and (iii) are established in a similar way). Consider the pair of primal and dual problems

$$
\begin{aligned}
&\max_{e\geq 0}\left\{\lambda^*\Gamma\textstyle\sum_p S_p e_p \mid \Gamma\sum_p S_p e_p \leq \overline{f}\right\} \\
&\min_{\lambda\geq 0}\left\{\lambda^T\overline{f} \mid \lambda(\Gamma\textstyle\sum_p S_p) \geq \lambda^*(\Gamma\sum_p S_p)\right\}.
\end{aligned}
$$

e^* is feasible for the primal and λ^* is feasible for the dual. Because $\overline{f}^T\lambda^* = \lambda^*\Gamma\sum_p S_p e^*_p$ their objective functions are equal. Therefore e^*

is an optimal solution of the primal and hence maximizes the revenue accruing from the network.

Proposition 5.14 Denote $i^T_{N_D} = (i_n, n \in N_D)$ and $i^T_{N_G} = (i_n, n \in NG)$. Define the mapping $H(i)$ and $\Lambda_p(e)$, $p \in P$ as

$$\begin{aligned} H_n(i) &= mw_n(-i_{N_D}) \quad \text{if } n \in N_D \\ &= mc_n(i_{N_G}) \quad \text{if } n \in N_G. \end{aligned}$$

(Recall $i_n = 0$ when $n \notin N_G \cup N_D$).

$$\begin{aligned} \Lambda_{nmp}(e) &= e_{nmp}\left(-\frac{\partial mw_m}{\partial d_m} + \frac{\partial mc_n}{\partial g_n}\right) + t^b_{nm} \quad \text{if } p \in P^o \\ &= t^b_{nm} \quad \text{if } p \in P^c \end{aligned}$$

Note that

$$\sum_p [L_p(e^*; t^b)]^T (e_p - e^*_p) = H(i^*)(i - i^*) + \sum_p \Lambda_p(e^*)(e_p - e^*_p)$$
$$\text{where } i = \sum_p S_p e_p \text{ and } i^* = \sum_p S_p e^*_p$$

VI_2 can then be restated as

VI_2': Seek $e^* \geq 0$, i^*, $e^* \in NC^b$, $i^* = \sum_p S_p e^*_p$ such that $H(i^*)(i - i^*) + \sum_p \Lambda_p(e^*)(e_p - e^*_p) \geq 0$ for all $e \geq 0$, $i, e \in NC^b, i = \sum_p S_p e_p$.

Let π denote the dual variable of the constraints $-i + \sum_p S_p e_p = 0$ and λ be defined as before. VI_2' is equivalent to

VI_2'': Seek $e^* \geq 0$, i^*, π^*, $\lambda^* \geq 0$ such that

$$[H(i^*) - \pi^*](i - i^*) + \sum_p [\Lambda_p(e^*) + \pi^* S_p + \lambda^* \Gamma S_p]^T (e_p - e^*_p)$$
$$+ (-i^* + \sum_p S_p e^*_p)^T (\pi - \pi^*) + (\bar{f} - \Gamma \sum_p S_p e^*_p)^T (\lambda - \lambda^*) \geq 0$$

for all $e \geq 0, i, \pi, \lambda \geq 0$.

This reformulation of the problem is used to prove the proposition.

A. We first show that a solution to VI_2'' is an equilibrium. A solution to VI_2'' satisfies the following properties

(i) $[H(i^*) - \pi^*](i - i^*) \geq 0 \; \forall \; i$ or $H_n(i^*) = \pi^*_n$, $n \in N_g \cup N_D$.
This shows that generators maximize their profit and consumers their surplus at the nodal prices π^*_n for net injections/withdrawals i^*.

(ii) $[\Lambda_p(e^*) + \pi^*_p S_p + \lambda^* \Gamma S_p]^T (e_p - e^*_p) \geq 0 \; \forall \; e_p \geq 0$. This shows that power marketers, whether competitive or oligopolistic maximize their profit after paying transmission charges $\lambda^* \Gamma S_p$ (use (i) to prove the proposition for oligopolistic traders).

(iii) $(-i^* + \sum_p S_p e^*_p)^T (\pi - \pi^*) \geq 0 \; \forall \pi$ implies that $i^* = \sum_p S_p e^*_p$ and hence that quantities produced, traded and consumed balance.

(iv) $(\overline{f} - \Gamma \sum_p S_p e_p^*)^T (\lambda - \lambda^*) \geq 0$ implies that

$$\overline{f} \geq \Gamma \sum_p S_p e_p^* \text{ and } \overline{f}^T \lambda^* = \lambda^{*T} \Gamma \sum_p S_p e_p^*.$$

The inequality implies that e^* is feasible for the network. Consider the pair of primal and dual problems

$$\begin{array}{l} \max_{e \geq 0} \left\{ \lambda^{*T} \Gamma \sum_p S_p e_p \mid \Gamma \sum_p S_p e_p \leq \overline{f} \right\} \\ \min_{\lambda \geq 0} \left\{ \lambda^T \overline{f} \mid \lambda^T (\Gamma \sum_p S_p) \leq \lambda^{*T} (\Gamma \sum_p S_p) \right\}. \end{array}$$

e^* is feasible for the primal and λ^* is feasible for the dual. Because $\overline{f}^T \lambda^* = \lambda^* \Gamma \sum_p S_p e_p^*$ their objective functions are equal. Therefore e^* maximizes the revenue accruing from the network. This proves equilibrium on the transmission system. It is easy to show that any of the three other statements of equilibrium on the transmission system also hold true.

B. Consider now the reverse statement. Using the same notation as before we have

(i) Generators and consumers respectively maximize their profit and surplus. This implies

$$\begin{array}{rcl} mc_n(i_{N_G}^*) & = & \pi_n^* \text{ if } i_n^* > 0 \text{ and } mc_n(i^*) \geq \pi_n^* \text{ if } i_n^* = 0 \\ mw_n(-i_{N_D}^*) & = & \pi_n^* \text{ if } i_n^* < 0 \text{ and } mw_n(i^*) \geq \pi_n^* \text{ if } i_n^* = 0 \end{array}$$

(ii) Competitive traders maximize their profit at the current prices π^* after paying for transmission. This implies

$$(\pi_n^* - \pi_m^* + \lambda^{*T} \Gamma S_p + t_{nm})(e_{nmp} - e_{nmp}^*) \geq 0 \text{ for all } e_{nmp},$$
$$n \in N_G, m \in N_D.$$

$e_{nmp}^* > 0$ implies $i_n^* > 0$ and $i_m^* < 0$. Using (i) one can replace π_n^* by $mc_n(i_{N_G}^*)$, $n \in N_G$ and π_m^* by $mw_m(-i_{N_D}^*), m \in N_D$. This leads to

$$(-mw(d(e^*)) + mc_n(g(e^*)) + \lambda^* \Gamma S_p + t_{mn})(e_{nmp} - e_{nmp}^*) \geq 0.$$

Suppose now $e_{nmp}^* = 0$. This implies $mc_n(i^*) \geq \pi_n^*$ and $mw_n(i^*) \leq \pi_n^*$. The same replacement of the π^* by mc and mw can thus be still performed while keeping the inequality. Therefore one has for a competitive trader

$$L_p(e^*; t^b)(e_p - e_p^*) \geq 0.$$

(iii) Oligopolistic trader. By definition an oligopolistic trader at optimality satisfies

$$L_p(e^*; t^b)(e_p - e_p^*) \geq 0.$$

(iv) It is easy to see that any of the condition defining the equilibrium in transmission implies

$$\lambda^* \geq 0, (\bar{f} - \Gamma \sum_p e_p^*) \geq 0 \text{ and } (\bar{f} - \Gamma \sum_p S_p e_p^*)\lambda^* = 0$$

and hence

$$(\bar{f} - \Gamma \sum_p e_p^*)^T (\lambda - \lambda^*) \geq 0.$$

Summing these different variational inequalities one obtain that e^*, λ^* solve the variational inequality problem $\widetilde{VI}_2$ and hence is also a solution to VI_2.
Proposition 5.15 This proposition is a restatement of Lemma 5.11.
Proposition 5.16 Take an electricity system where competitive traders also cover the markets of the oligopolistic traders. Then oligopolistic traders are either absent or do not exert market power. Consider now the behavior of the competitive traders in the proof of Proposition 5.14. One has

$$\begin{array}{l} \sum_{p \in P} [\pi^* S_p + \lambda^* \Gamma S_p]^T (e_p - e_p^*) \geq 0, \text{ or} \\ (\pi^* + \lambda^* \Gamma)^T (i - i^*) \geq 0, \ \forall\, i. \end{array}$$

This implies that $\pi_n^* = -\tau_n^*$ for all $n \in N_D \cup N_G$. One can complete the definition of π^* by setting $\pi_n^* = -\tau_n^*$ for $n \notin N_G \cup N_D$.
Proposition 5.17 First note that the solution of VI^{pool} satisfies

$$L_h(i_{p^o}^*, i_h^*) + \lambda^* \Gamma + \beta_h^* = 0.$$

The nodal prices can be defined as

$$\pi^* = -\lambda^* \Gamma - \beta_h^*.$$

The transmission charge between two nodes is thus equal to the difference between two nodal prices. The solution of VI^{pool} is thus a pool. To see that it also satisfies the definition of a Cournot equilibrium in the sense of Hobbs, set $i_p = i_p^*$ for $p \neq \tilde{p}$. VI^{pool} then reduces to the optimal behavior of oligopolistic trader $\tilde{p}$.

Proposition 5.18 Consider a system where there are only competitive traders. The equivalence between the pool and the bilateral model comes from the fact that VI_2^{no} boils down to a rewriting of VI_2 and VI_2^{no} is a pool when there is no market power.

To see that the Pool and the bilateral model are not equivalent when there is market power consider a single node network with a single generator and two customers. Assume a single power marketer that behaves as a monopoly. The bilateral model boils down to a discriminating monopoly charging different monopoly prices to the two consumers. The pool boils down to a non discriminating monopoly charging a single price to both segments of the market. As is well know from standard economic theory, these are not equivalent.

Proposition 5.19 Consider a system with only oligopolistic traders and define the following sequence of problems VI_2^k. To each oligopolistic trader p of the original problem VI_2 associate k identical oligopolistic traders. Keep the rest of the system (generators, consumers, network) as such. Define $e^{\ell}_{nmp}, \ell = 1, \dots, k$ and e^{ℓ}_p, $\ell = 1, \dots, k$ the trades and vector of trades of power marketer ℓ of type p. Let $e^T = \{[(e^{\ell}_p)^T, \ell = 1, \dots, k], p \in P\}$. Let $L^{\ell}_p(e)$ be the loss function of trader (p, ℓ).

Define the problem

VI_2^k: Seek $e^* \geq 0$, $e^* \in NC^b$ such that

$$\sum_p \sum_{\ell=1}^k L^{\ell}_p(e^*)(e^{\ell}_p - e^{\ell *}_p) \geq 0 \text{ for all } e \geq 0, e \in NC^b.$$

By symmetry the $e^{\ell *}_p$ are identical for all $\ell = 1, \dots, k$. Let e^{o*}_p be this common value and let $e^{k*}_p = ke^{o*}_p$. The $L^{\ell}_p(e^{k*})$ are also identical for all $\ell = 1, \dots, k$. Let $L^o_p(e^{k*})$ be this common expression. VI_2^k can be rewritten as

VI_2^k: Seek $e^{k*} \geq 0$, $e^{k*} \in NC^b$ such that

$$\sum_p L^o_p(e^{k*})(e^k_p - e^{k*}_p) \geq 0 \text{ for all } e^k \geq 0, e^k \in NC^b.$$

Denote $L^c_p(e^{k*})$ the mapping of a competitive trader if it existed in this system. One can verify that

$$L^o_p e^{k*}) = L^c_p(e^{k*}) + \frac{1}{k}\text{diag}\left\{e^{k*}_{nmp}\left[-(\frac{\partial mw_m}{\partial d_n})_{d_n(e^{k*}} + (\frac{\partial mc_n}{\partial g_n})_{g_n(e^{k*}}\right]\right\}$$

Suppose now that $e^{k*} \to \tilde{e}^*$ when $k \to \infty$. At the limit, $\tilde{e}^* \geq 0$ and $\tilde{e}^* \in NC^b$. Moreover $\tilde{e}^*$ satisfies

$$\sum_p L^c_p(\tilde{e}^*)(e_p - \tilde{e}^*_p) \geq 0 \text{ for all } e \geq 0, e \in NC^b.$$

Therefore $\tilde{e}^*$ satisfies the variational inequality problem associated to a market where there are only competitive traders. Selecting k large enough makes thus the oligopolistic market arbitrary close to the competitive market.

The proof the the proposition for the pool model follows similar lines. Consider a sequence of problems $VI^{\text{pool},k}$ where oligopolistic traders are subject to the same treatment as in the definition of VI_2^k. No change is imposed to Hobbs trader which remains unique. The reaction of Hobbs trader to a given set of injections from the oligopolistic traders is thus invariant whatever the number of oligopolistic traders. The same reasoning as before shows that the first stage problem becomes at the limit identical to the problem of a competitive trader. The two stage problem becomes one stage where the first stage and second stage problems are identical. Because the first stage already satisfies the Pool condition, the reaction of the arbitrageur vanishes and the second stage problem becomes redundant. The limit is thus a single stage problem VI_2 with only competitive traders.

Proposition 5.21 Consider the simple system consisting of a single producer of marginal cost $mc = 0$, a single consumer of demand curve $mw = 1 - q$, and a single line network of infrastructure cost RRN. Suppose that this line has an infinite capacity and hence that there is no congestion. A solution of MVI must satisfy $t^*q = RRN$. Because there are no network capacity limit and the marginal generation cost is zero, there are no congestion costs. Therefore the solution of MVI also satisfies $1 - q = mw = t$. This implies that $q(1 - q) = RRN$, an equation that may have zero, one or two solutions. Consider now a solution of MVI. By construction, it satisfies the revenue requirement constraint and hence the second requirement of the equilibrium on the transmission market. This solution is also a solution of the variational inequality problem VI_2^{no} where t^{no} is equal to t^*. The other equilibrium conditions therefore follow immediately from the properties of Section 3 restated in the nodal formulation.

Chapter 6

OPTIMIZATION APPROACHES TO SEMI-SUPERVISED LEARNING

Ayhan Demiriz

Department of Decision Sciences and Engineering Systems, Rensselaer Polytechnic Institute. Troy, NY 12180

demira@rpi.edu

Kristin P. Bennett

Department of Mathematical Sciences, Rensselaer Polytechnic Institute. Troy, NY 12180

bennek@rpi.edu

Abstract We examine mathematical models for semi-supervised support vector machines (S^3VM). Given a training set of labeled data and a working set of unlabeled data, S^3VM constructs a support vector machine using both the training and working sets. We use S^3VM to solve the transductive inference problem posed by Vapnik. In transduction, the task is to estimate the value of a classification function at the given points in the working set. This contrasts with inductive inference which estimates the classification function at all possible values. We propose a general S^3VM model that minimizes both the misclassification error and the function capacity based on all the available data. Depending on how poorly-estimated unlabeled data are penalized, different mathematical models result. We examine several practical algorithms for solving these model. The first approach utilizes the S^3VM model for 1-norm linear support vector machines converted to a mixed-integer program (MIP). A global solution of the MIP is found using a commercial integer programming solver. The second approach uses a nonconvex quadratic program. Variations of block-coordinate-descent algorithms are used to find local solutions of this problem. Using this MIP within a local learning algorithm produced the best results. Our experimental study on these statistical learning methods indicates that incorporating working data can improve generalization.

M.C. Ferris et al. (eds.), Complementarity: Applications, Algorithms and Extensions, 121–141.

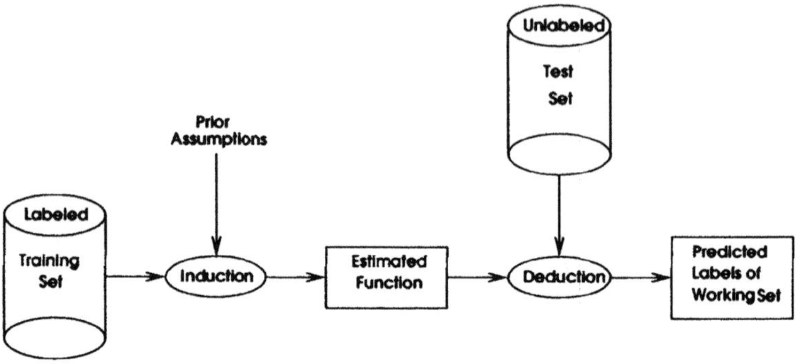

Figure 6.1 Inductive Learning

1 INTRODUCTION

The focus of this paper is mathematical programming approaches to semi-supervised learning for classification tasks. The main idea of semi-supervised learning is to construct a classifier using both a training set of labeled data and a working set of unlabeled data. If none of the labels are known then the problem becomes clustering. If some of the labels are known, then the problem is classification. There are many practical domains in which unlabeled data are abundant but labeled data are expensive to generate and therefore relatively scarce (e.g. medical diagnosis, web search, drug design, and database marketing). When the training data consist of relatively few labeled data points in a high-dimensional space, something must be done to prevent the classification or regression function from overfitting the training data. The key idea is that by exploiting the unlabeled data we hope to be able to provide additional information about the problem that can be used to improve accuracy on data with unknown labels (generalization).

By including the unlabeled data in the testing set, semi-supervised learning can be used to perform transductive learning instead of the more typical inductive learning. In induction, the task is to construct a good discriminant function valid everywhere. This function is fixed and applied to any future test data (Figure 6.1). In transduction, the labeled training data and unlabeled testing data are given, then the discriminant function is constructed based on all the available data. The learning task is to predict the labels of only those specific test data points, not all possible future points. This simpler task can result in theoretically better bounds on the generalization error [30], thus reducing the amount of required labeled data for good generalization (Figure 6.2).

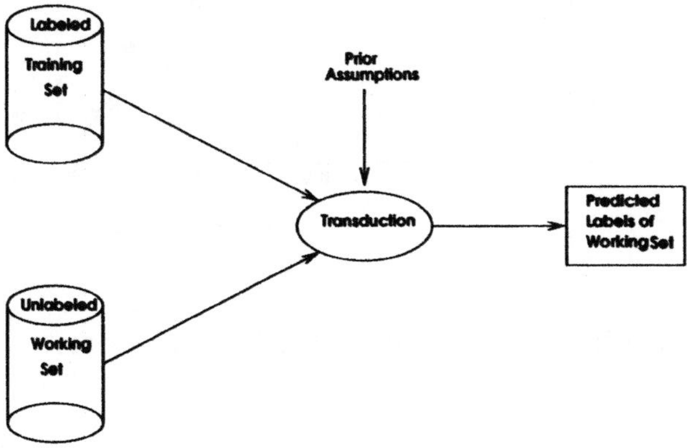

Figure 6.2 Transductive Learning

Our semi-supervised support vector machine approach can be illustrated by a simple example. Consider the two-class problem shown in Figure 6.3(a). Since the labeled training sets are linearly separable, there exists an infinite number of possible separating planes that correctly classify the two sets. Intuitively, the best linear classifier is the middle plane shown that separates the two sets with greatest margin. The margin is the sum of distances from the closest points (the support vectors) in each set to the plane or equivalently the distance between the supporting planes for each set. The supporting planes are shown using dotted lines. Statistical Learning Theory proves that for a given misclassification error, maximizing the margin of separation minimizes a bound on the expected misclassification error on future unseen data [30]. Maximizing the margin reduces the capacity of the function to fit data. Intuitively, a "fat" plane with wide margin has less capacity to fit data than a "skinny" one. In SVM, the optimal plane can be found using quadratic or linear programming depending on the metric used to measure the margin distance [30, 29, 3]. Consider now the additional unlabeled test data shown in Figure 6.3(b). The SVM performs poorly on this particular test set in terms of classification accuracy of the testing data. Note also that the resulting margin for the combined labeled training data and unlabeled testing data is very small. If we construct the SVM margin that correctly classifies the training data and achieves the widest margin based on all the data, the results found by our semi-supervised SVM are significantly improved and the preferable plane is shown in Figure 6.3(c). Results in statistical learning theory show that, for a fixed misclassification error, maximizing the margin based on all

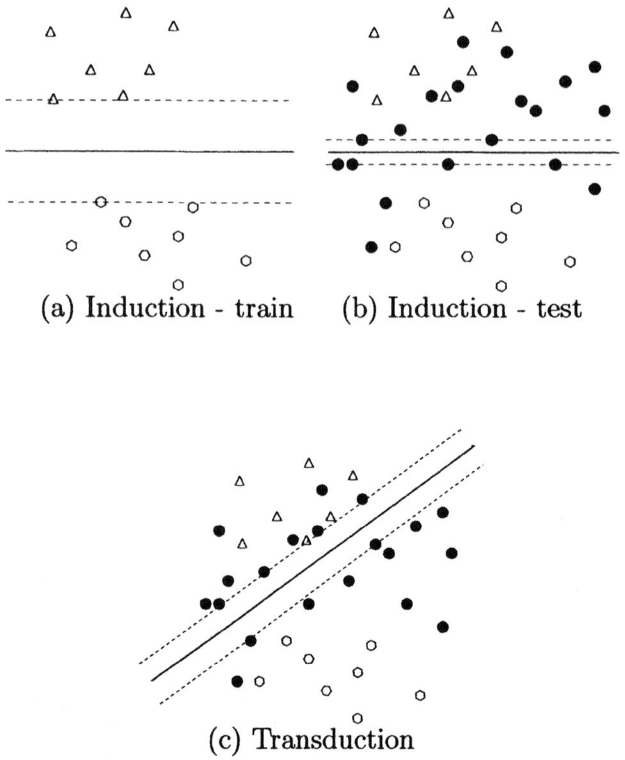

Figure 6.3 Traditional SVM (a, b) versus Semi-Supervised SVM (c)

the data (train and test) can lead to better bounds on the expected generalization error[30].

For semi-supervised SVM we consider all possible labels of the test data and assign the labels that produce the best SVM with maximum margin based on all the available data, both labeled and unlabeled. For the purpose of this paper we limit our discussion to linear SVM, but these methods can be extended to nonlinear support vector machines using the standard SVM approach of including kernel functions [30, 22]. In Section 2 we review support vector machines. In Section 3 we provide a general framework for viewing the semi-supervised support vector machine problem. Depending on how we penalize unlabeled data appearing in the margin the problem can be formulated as a linear or convex quadratic program with additional equilibrium constraints, mixed-integer constraints, or nonconvex objective terms. In Section 4 we examine practical approaches using the linear mixed integer program (MIP) formulation first introduced in [5]. By incorporating the MIP within a local learning approach, performance is greatly enhanced. In Section 5 we examine practical algorithms for a nonconvex quadratic formulation. Finally, we conclude our paper with a brief summary and discussion of optimization issues in semi-supervised learning.

Other researchers have reported favorable results on semi-supervised methods on web-based text classification problems for example using an EM (Expectation-Maximization) [26, 23], co-training in Bayesian networks [10], and a transductive version of SVM-Light [19]. Cataltepe and Magdon-Ismail [13] propose augmented error, which has components from both labeled and unlabeled data. They provide an analytical solution in the case of linear, noisy targets and linear hypothesis functions. They also show some results for the non-linear case. Theoretical results exist [12] on the relative value of labeled and unlabeled data.

2 REVIEW OF SVM

The underlying problem of interest is to estimate a classification function $f : R^N \to \{\pm 1\}$ using input-output training data from two classes

$$(x_1, y_1), \ldots, (x_\ell, y_\ell) \in R^n \times \{\pm 1\}. \tag{6.1}$$

The function f should correctly or almost correctly classify unseen examples $(\boldsymbol{x}, y)$, i.e. $f(\boldsymbol{x}) = y$ if $(\boldsymbol{x}, y)$ is generated from the same underlying probability distribution as the training data. In this work we limit discussion to linear classification functions. If the points are linearly separable, then there exist an n-vector $\boldsymbol{w}$ and scalar b such that

$$\begin{array}{ll} \boldsymbol{w} \cdot x_i - b \geq 1 & if\ y_i = 1,\ and \\ \boldsymbol{w} \cdot x_i - b \leq -1 & if\ y_i = -1,\ i = 1, \ldots, \ell \end{array} \tag{6.2}$$

or equivalently

$$y_i[\boldsymbol{w} \cdot x_i - b] \geq 1,\ i = 1, \ldots, \ell. \tag{6.3}$$

The "optimal" separating plane, $\boldsymbol{w} \cdot x = b$, is the one that is furthest from the closest points in the two classes. Geometrically this is equivalent to maximizing the separation margin or distance between the two parallel planes $\boldsymbol{w} \cdot x = b + 1$ and $\boldsymbol{w} \cdot x = b - 1$ (see Figure 6.4).

The "margin of separation" in Euclidean distance is $2/\left\|\boldsymbol{w}\right\|_2$ where $\left\|\boldsymbol{w}\right\|_2 = \sqrt{\sum_{i=1}^n \boldsymbol{w}_i^2}$ is the 2-norm. To maximize the margin, we minimize $\left\|\boldsymbol{w}\right\|_2/2$ subject to the constraints (6.3). According to structural risk minimization, for a fixed empirical misclassification rate, larger margins should lead to better generalization and prevent overfitting in high-dimensional attribute spaces[29]. The classifier is called a support vector

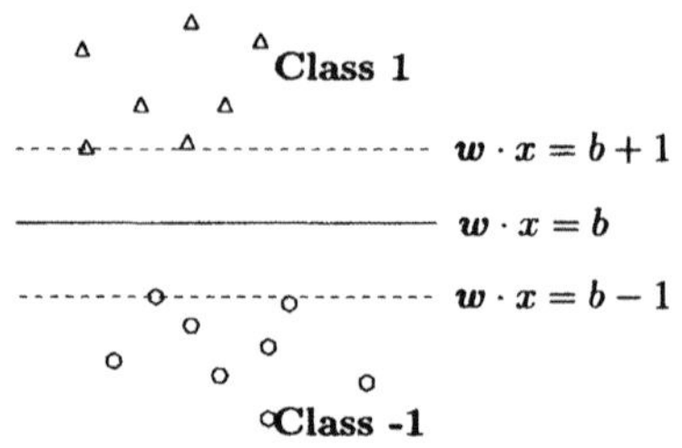

Figure 6.4 Optimal Plane Maximizes Margin

machine because the solution depends only on the points (called support vectors) located on the two supporting planes $\boldsymbol{w} \cdot x = b - 1$ and $\boldsymbol{w} \cdot x = b + 1$.

In general the classes will not be linearly separable, so the generalized optimal plane problem (6.4) [14, 29] is used. A slack term η_i is added for each point such that if the point is misclassified, $\eta_i \geq 1$. The quadratic programming formulation is (SVM-QP):

$$\begin{array}{ll} \min\limits_{\boldsymbol{w},b,\eta} & C\sum\limits_{i=1}^{\ell} \eta_i + \frac{1}{2}\|\boldsymbol{w}\|^2 \\ s.t. & y_i[\boldsymbol{w} \cdot x_i - b] + \eta_i \geq 1 \\ & \eta_i \geq 0, \quad i = 1, \ldots, \ell \end{array} \tag{6.4}$$

where $C > 0$ is a fixed penalty parameter. The capacity control provided by the margin maximization can greatly improve generalization [31, 28]. Typically, the following dual form of (6.4) is solved in practice:

$$\begin{array}{ll} \min\limits_{\alpha} & \frac{1}{2}\sum\limits_{i=1}^{\ell}\sum\limits_{j=1}^{\ell} y_i y_j \alpha_i \alpha_j x_i \cdot x_j - \sum\limits_{i=1}^{\ell} \alpha_i \\ s.t. & \sum\limits_{i=1}^{\ell} \alpha_i y_i = 0 \\ & 0 \leq \alpha_i \leq C \quad i = 1, \ldots, \ell \end{array} \tag{6.5}$$

The Robust Linear Programming approach to SVM is identical to SVM-QP except the margin term is changed from the 2-norm $\|\boldsymbol{w}\|_2$ to the 1-norm, $\|\boldsymbol{w}\|_1 = \sum_{j=1}^{n} |w_j|$. The problem becomes the following robust linear program (SVM-RLP) [6, 11, 4]:

$$\begin{aligned} \min_{\boldsymbol{w},b,s,\eta} \quad & C\sum_{i=1}^{\ell}\eta_i + \sum_{j=1}^{n} s_j \\ s.t. \quad & y_i[\boldsymbol{w}\cdot x_i - b] + \eta_i \geq 1 \\ & \eta_i \geq 0, \quad i = 1,\dots,\ell \\ & -s_j <= \boldsymbol{w}_j <= s_j, \quad j = 1,\dots,n. \end{aligned} \tag{6.6}$$

The RLP formulation is a useful variation of SVM with some nice characteristics. The 1-norm weight reduction still provides capacity control. The results in [21] can be used to show that minimizing $\|\boldsymbol{w}\|_1$ corresponds to maximizing the separation margin using the infinity norm. Statistical learning theory could potentially be extended to incorporate alternative norms. One major benefit of SVM-RLP over SVM-QP is dimensionality reduction. Both SVM-RLP and SVM-QP minimize the magnitude of the weights $\boldsymbol{w}$. But RLP forces more of the weights to be 0 due to the properties of the 1-norm. This results in dimensionality reduction since variables with 0 weights can be removed fro the model. Another benefit of SVM-RLP over SVM-QP is that it can be solved using linear programming instead of quadratic programming.

SVM are easily generalized to nonlinear discriminants through the introduction of kernel functions [30, 22]. The basic idea is that the data are mapped nonlinearly to a higher dimensional space and a linear SVM is constructed in the transformed space corresponding to a nonlinear classifier in the original space. We limit our formulation to the linear classification problem and leave computational studies of these approaches extended with kernels to future work.

3 SEMI-SUPERVISED SVM

The basic idea of semi-supervised support vector machines is that we want the best support-vector machine on the labeled data that has no or very few unlabeled points in the margin. Thus we want to penalize the support vector machine if unlabeled points fall in the margin. Specifically, we define the semi-supervised support vector machine problem (S^3VM) as:

$$\begin{aligned} \min_{\boldsymbol{w},b,\eta,\xi,z} \quad & C\left[\sum_{i=1}^{\ell}\eta_i + \sum_{j=\ell+1}^{\ell+k} g(\boldsymbol{w}\cdot x_j - b)\right] + \|\ \boldsymbol{w}\ \| \\ s.t. \quad & y_i(\boldsymbol{w}\cdot x_i - b) + \eta_i \geq 1 \quad \eta_i \geq 0 \quad i = 1,\dots,\ell \end{aligned} \tag{6.7}$$

where $C > 0$ is a fixed misclassification penalty parameter and $g(\alpha)$ is the margin penalty function on unlabeled data.

The question then is how to define g. For a hard margin approach in which no unlabeled points are allowed in the margin, the margin penalty function is defined as

$$g_{\infty}(\alpha) := \begin{array}{ll} \infty & for \ -1 < \alpha < 1 \\ 0 & otherwise \end{array} \tag{6.8}$$

If an unlabeled point falls outside the margin, it is considered well-classified and no penalty is incurred.

We can transform the hard margin g_{∞} problem into a linear or quadratic program with an additional equilibrium constraint. We start with either SVM formulation, (6.4) or (6.6), and then add two constraints for each point in the working set. One constraint calculates the misclassification error as if the point were in class 1 and the other constraint calculates the misclassification error as if the point were in class -1. We add a constraint that forces one of the two misclassification errors per point to be zero. This produces the following mathematical programming problem with equilibrium constraints:

$$\begin{array}{lll} \min\limits_{w,b,\eta,\xi,z} & C\left[\sum\limits_{i=1}^{\ell} \eta_i\right] + \| w \| & \\ s.t. & y_i(w \cdot x_i - b) + \eta_i \geq 1 & \eta_i \geq 0 \;\; i = 1, \dots, \ell \\ & w \cdot x_j - b + \xi_j \geq 1 & \xi_j \geq 0 \;\; j = \ell+1, \dots, \ell+k \\ & -(w \cdot x_j - b) + z_j \geq 1 & z_j \geq 0 \\ & \xi_j \cdot z_j = 0 & j = \ell+1, \dots, \ell+k \end{array} \tag{6.9}$$

The requirement that no unlabeled points may fall in the margin may be too strong. A natural relaxation of the problem would be to move the equilibrium constraint into the objective and use it as the margin penalty function g. This results in the following nonconvex quadratic optimization problem:

$$\begin{array}{lll} \min\limits_{w,b,\eta,\xi,z} & C\left[\sum\limits_{i=1}^{\ell} \eta_i + \sum\limits_{j=\ell+1}^{\ell+k} \xi_j \cdot z_j\right] + \| w \| & \\ s.t. & y_i(w \cdot x_i + b) + \eta_i \geq 1 & \eta_i \geq 0 \;\; i = 1, \dots, \ell \\ & w \cdot x_j - b + \xi_j \geq 1 & \xi_j \geq 0 \;\; j = \ell+1, \dots, \ell+k \\ & -(w \cdot x_j - b) + z_j \geq 1 & z_j \geq 0 \end{array} \tag{6.10}$$

Close examination of this choice of error function shows that it has attractive properties. If the unlabeled point x_j falls outside or on the margin then ξ_j or z_j is 0, and there is no error associated with that point. If the point falls in the margin, then for $\gamma = w \cdot x_j - b$, $\xi_j = 1 - \gamma$ and $z_j = 1 + \gamma$ by construction of the support vector machine. The

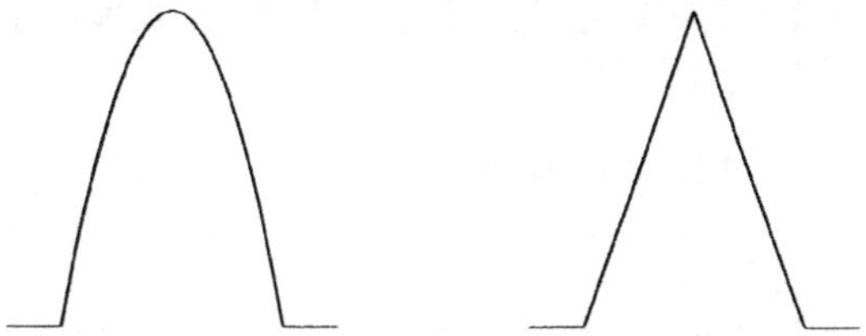

Figure 6.5 Margin Penalty Functions

following piecewise quadratic margin penalty function is produced (see Figure 6.5(a)):

$$g_2(\gamma) := \begin{array}{ll} 1-\gamma^2 & for\ -1<\gamma<1 \\ 0 & otherwise. \end{array} \tag{6.11}$$

Another natural choice would be a margin penalty function that calculates the minimum of the two possible misclassification errors. The final class of a point corresponds to the one that results in the smallest error. This is the transductive idea as proposed by Vapnik [30]. It has the advantage that if the correct labels are found, the resulting SVM will be identical to the one produced if the points were known. The minimum error formulation is [5]:

$$\begin{array}{ll} \min\limits_{\boldsymbol{w},b,\eta,\xi,z} & C\left[\sum\limits_{i=1}^{\ell}\eta_i + \sum\limits_{j=\ell+1}^{\ell+k}\min(\xi_j,z_j)\right] + \|\boldsymbol{w}\| \\ s.t. & y_i(\boldsymbol{w}\cdot\boldsymbol{x}_i+b)+\eta_i \geq 1 \quad \eta_i \geq 0 \quad i=1,\ldots,\ell \\ & \boldsymbol{w}\cdot\boldsymbol{x}_j - b + \xi_j \geq 1 \quad \xi_j \geq 0 \quad j=\ell+1,\ldots,\ell+k \\ & -(\boldsymbol{w}\cdot\boldsymbol{x}_j - b) + z_j \geq 1 \quad z_j \geq 0 \end{array} \tag{6.12}$$

The resulting margin penalty function is shown in Figure 6.5(b)

$$g(\gamma) = g_1(\gamma) := \begin{array}{ll} 1-|\gamma| & for\ -1<\gamma<1 \\ 0 & otherwise \end{array} \tag{6.13}$$

For our experimental study of practical methods for solving these problems we focused on the minimum error formulation (6.12). But this is not to say that other formulations are not possible or preferable. In the next two sections we explore two different approaches to practically solving this problem.

4 MIXED-INTEGER PROGRAMMING FORMULATION

Integer programming can be used to exactly solve S^{3}VM (6.12). The basic idea is to add a 0 or 1 decision variable, d_j, for each point $\boldsymbol{x}_j$ in the working set. This variable indicates the class of the point. If $d_j = 1$ then the point is in class 1 and if $d_j = 0$ then the point is in class -1. This results in the following mixed integer program (S^{3}VM-MIP):

$$\begin{array}{lrl} \min\limits_{\boldsymbol{w},b,\eta,\xi,z,d} & C\left[\sum\limits_{i=1}^{\ell}\eta_i + \sum\limits_{j=\ell+1}^{\ell+k}(\xi_j + z_j)\right] + \parallel \boldsymbol{w} \parallel & \\ \text{s.t.} & y_i(\boldsymbol{w}\cdot x_i - b) + \eta_i \geq 1 & \eta_i \geq 0 \quad i = 1,\dots,\ell \\ & \boldsymbol{w}\cdot x_j - b + \xi_j + M(1-d_j) \geq 1 & \xi_j \geq 0 \quad j = \ell+1,\dots,\ell+k \\ & -(\boldsymbol{w}\cdot x_j - b) + z_j + Md_j \geq 1 & z_j \geq 0 \quad d_j = \{0,1\} \end{array} \tag{6.14}$$

The constant $M > 0$ is chosen sufficiently large such that if $d_j = 0$ then $\xi_j = 0$ is feasible for any optimal $\boldsymbol{w}$ and b. Likewise if $d_j = 1$ then $z_j = 0$. In this paper we use the 1-norm of $\boldsymbol{w}$ in the objective. A globally optimal solution to this problem can be found using CPLEX or other commercial mixed integer programming codes [15] provided computer resources are sufficient for the problem size. Using the mathematical programming modeling language AMPL [16], we were able to express the problem in approximately thirty lines of code plus a data file and solve it using CPLEX. One practical limitation of this approach is the capacity of the MIP solver used. Using CPLEX 4.0 on a Sun Ultra 1 with 700MB RAM we found it was practical to include about 50 unlabeled data points due to the CPU time limitation.

4.1 LOCAL SEMI-SUPERVISED SUPPORT VECTOR MACHINES

To get around the practical restriction on the number of integer variables and thus unlabeled data handled by our MIP solver, we utilized the S^{3}VM-MIP as part of a local learning algorithm. In local learning, a point is classified based on points in its "neighborhood". For example, in the K-Nearest-Neighbor algorithm (K-NN), the K nearest neighbors to a point (by Euclidean distance or some other metric) are found and then the point is assigned the majority class of the K nearest neighbors. Local learning methods are often called memory-based methods, because training examples are kept in "memory" and used to classify new points. Since the local models have fewer training examples, it takes much less computational time to optimize each local S^{3}VM than to train one global one at the expense of many local models. Previous empirical studies have shown that the generalization ability of local methods often exceeds that of global ones since the local models include only the points which are related to the query point (interested unlabeled data) in a given learning task. Many variations exist for both selecting the neighborhoods and determining the output class based on the neighbor. For example, Discriminant Adaptive Nearest Neighbor [17] uses local discriminant analysis to estimate the class within K-NN classification. Lawrence et al. [20] use local neural network models for function approximation. See [1] for a survey of approaches.

4.2 LOCAL S^{3}VM AND EXPERIMENTAL RESULTS

Local S^{3}VM is nothing but an application of S^{3}VM in a local neighborhood of each unlabeled point as determined by the K-NN algorithm using Euclidean distance. This neighborhood includes both labeled and unlabeled examples. In order to have enough labeled examples in each neighborhood, we arbitrarily pick K as 10% of all available data points. Further study is needed on how to best select the neighborhood of a point. We can summarize the method (Local-S^{3}VM) to classify a given unlabeled point in the following steps:

1. Find K-NN for a given unlabeled point.
2. If all the labeled points in the neighborhood are in one class, then label the unlabeled point as in that class and end. Otherwise continue.
3. Solve the S^{3}VM-MIP (6.14) in the neighborhood.
4. Label the point according to the result of S^{3}VM .

There are many advantages to using Local S^{3}VM over using a single global S^{3}VM. In transduction for any data, we need to construct a new model. So the fact that local S^{3}VM must compute a new model for each point is also true for any transductive algorithm. Although there are as many models as unlabeled points to solve in Local S^{3}VM , the overall computational time of the algorithm including time to find the local neighborhood is generally less than the global S^{3}VM algorithm. This is because we have fewer unlabeled points in each local model which means we have fewer binary variables in each model. Having fewer binary variables results in less running time for each local model. Another advantage is that the overall classification function by Local S^{3}VM is

Table 6.1 Dataset Summary Statistics

Data Set	Dim	Points	Test-size
Bright	14	2462	50*
Cancer	9	699	70
Diagnostic	30	569	57
Dim	14	4192	50*
Heart	13	297	30
Housing	13	506	51
Ionosphere	34	351	35
Musk	166	476	48
Sonar	60	208	21
Pima	8	769	50*

nonlinear (piecewise linear to be exact) when a linear S^3VM is used locally.

Determining nearest neighbors of a point can become problematic for large datasets. One must consider an appropriate metric and method to find K-NN. Since we use datasets which have relatively small dimensions, we use Euclidean distance combined with a partial sort algorithm [25] to find the local neighborhood. As mentioned in the outlines of the algorithm, for each unlabeled point, a related data file is created and the S^3VM model is solved using AMPL. Then the output of AMPL is analyzed to find the label of the point.

Our computational study of S^3VM consisted of 10 trials using the ten real-world data sets described in Table 6.1 (eight from [24] and the bright and dim galaxy sets from [27]) [1]. The basic properties of the datasets are summarized in Table 6.1. Each dataset is sampled randomly 10 times and each working set is composed of 10% of the data except the Bright, Dim, and Pima datasets in which the size of the working set is set to 50 points and rest of the data are used as the training set. We use the following formula to pick the penalty parameter: $C = \frac{(1-\lambda)}{\lambda(\ell+k)}$ with $\lambda = 0.001$, ℓ is the size of the training set, and k is the size of the working set. The average working set errors are reported in Table 6.2. The best result from different models is underlined for each dataset.

Columns two and three of Table 6.2 provide a comparison of the inductive linear 1-norm support vector machine (SVM-RLP 6.6) with the transductive linear 1-norm SVM optimized used mixed integer programming (S^3VM-MIP 6.14). On all ten datasets, the transductive S^3VM-MIP results are either slightly better or not significantly different than

Table 6.2 Average Error Results for Inductive and Transductive SVM Methods

Data Set	SVM-RLP	S^3VM	Local SVM	Local S^3VM	3-NN
Bright	0.02	0.018	0.008	<u>0.006</u>	0.028
Cancer	0.036	<u>0.034</u>	0.06	0.059	<u>0.034</u>
Diagnostic	0.035	<u>0.033</u>	0.039	0.039	0.039
Dim	0.064	0.054	<u>0.042</u>	0.044	0.074
Heart	0.173	<u>0.16</u>	0.257	0.253	0.17
Housing	0.155	0.151	<u>0.118</u>	0.124	0.177
Ionosphere	0.109	<u>0.106</u>	0.117	0.109	0.129
Musk	0.173	0.173	0.092	<u>0.085</u>	0.208
Sonar	0.281	0.219	0.181	<u>0.143</u>	0.171
Pima	0.22	0.222	0.22	<u>0.218</u>	0.264

the inductive results found using SVM-RLP. Note that all parameters of the formulations are identical; the only difference between the two formulations is the use of unlabeled data for the transductive case. For this formulation, unlabeled data seems to help and never hurt generalization.

Columns 4 and 5 of Table 6.2 compare an inductive version of Local SVM and the transductive version of Local S^3VM . In our study, the neighborhoods of points used by both Local SVM and Local S^3VM are identical. Thus for each testing set point the optimization problem solved by local S^3VM is identical to the one solved by local SVM once the terms involving the unlabeled data are removed. This was done to ensure that the introduction of unlabeled data was the only change in the experiment. But in fact, it means the unlabeled data are being used to determine the effective size of the neighborhood for Local SVM which in itself is a form of transduction. Column 6 of Table 6.2 gives results for the 3-nearest neighbor algorithm. This was done to examine improvements that occur by simply switching to a local algorithm. Local S^3VM outperformed or did as well as Local SVM on eight of the ten datasets, once again supporting the transductive hypothesis. The improvements cannot be simply attributed to a local learning strategy since 3-NN did worse than both Local SVM and S^3VM on nine of ten datasets.

Overall, Local S^3VM was consistently the best or almost the best in our experiments. Either S^3VM or the Local S^3VM obtained the best results on most of the datasets except Dim and Housing datasets. The results indicate that using the labeled and unlabeled points in a transduction model can improve accuracy. Local S^{3}VM resulted in better accuracy than S^{3}VM on six datasets. One noteworthy point is that in some cases (Sonar, Musk, Housing, Bright) Local S^{3}VM improved accuracy notably. On Cancer, Diagnostic, Heart, and Ionosphere the fact that S^{3}VM performed best indicates that if the neighborhood of Local S^{3}VM is increased, Local S^{3}VM could perform better. The best method of choosing neighborhoods for local methods is still very much an open question. The proposed algorithm in this section takes into consideration only one unlabeled point at a given time. Although there might be many unlabeled points in a given neighborhood, the algorithm returns the results only on the test point of interest. The results for other points are basically discarded. One extension would be keeping these results for a final vote at the end of the algorithm. In this case, we can assign a probability of class membership for a certain point. The results from one point can also be used as starting points to improve the solution time of Local S^{3}VM on nearby points.

5 NONCONVEX QUADRATIC APPROACH

An alternative approach to solving the minimum error S^{3}VM problem (6.12) is to convert it into a nonconvex quadratic program. We adapt the approach used previously to handle disjunctiveness of classification labels within the bilinear separability [7] and global tree optimization problems [9, 2, 7]. Once again a decision variable d_j is introduced for each point such that at optimality if $d_j = 1$ then the predicted class of x_j is 1 and if $d_j = 0$ then the x_j is predicted as class -1. The resulting problem is (S^{3}VM-QP)

$$
\begin{array}{lll}
\displaystyle\min_{\boldsymbol{w},b,\eta,\xi,z,d} & C\left[\displaystyle\sum_{i=1}^{\ell}\eta_i + \sum_{j=\ell+1}^{\ell+k}(d_j\xi_j + (1-d_j)z_j)\right] + \| \boldsymbol{w} \| & \\
s.t. & y_i(\boldsymbol{w}\cdot x_i - b) + \eta_i \geq 1 \quad \eta_i \geq 0 \quad i = 1,\dots,\ell & \\
 & \boldsymbol{w}\cdot x_j - b + \xi_j \geq 1 \quad \xi_j \geq 0 \quad j = \ell+1,\dots,\ell+k & \\
 & -(\boldsymbol{w}\cdot x_j - b) + z_j \geq 1 \quad z_j \geq 0 \quad 0 \leq d_j \leq 1 &
\end{array}
\tag{6.15}
$$

An intuitively simple approach is to adapt a block coordinate descent algorithm (e.g. [8]) which alternates between fixing d and estimating the SVM weights $\boldsymbol{w}, b$ and other dependent variables, and optimizing d with the other SVM variables fixed. In [9], it was shown for a class of problems that includes S^{3}VM-QP (6.15), using 2-norm $\| \boldsymbol{w} \|_2$ such an approach will converge in a finite number of iterations to a solution satisfying the minimum principal necessary optimality conditions. No linesearch is required. The proof in [9] does require each subproblem be solved to optimality, but this condition can be relaxed to require only a strict decrease in the objective function. On the global tree optimization problem [2, 9] , the block coordinate descent algorithm was found to be very prone to local minima so a tabu search method was used. When applied to transduction, we also found this simple algorithm to be very prone to local minima and thus do not report the results here. To improve the results, we developed a heuristic variation of the block coordinate descent algorithm. We introduce this algorithm in the following section.

5.1 A DESCENT ALGORITHM FOR TRANSDUCTIVE SVM

The essential idea behind our heuristic approach is that we start by heavily penalizing solutions with points falling within the margin and then relax this requirement in order to find solutions with wider margin. Just as in the basic block coordinate descent method, we first estimate the labels (d_j, $j = \ell+1, \ldots, \ell+k$) based on our current estimate of the SVM, and then solve S^{3}VM-QP with d fixed. Note that in practice and for easy introduction of nonlinearity via kernels we solve the dual of Problem (6.15) which for fixed d reduces to the usual dual SVM problem (Eq. 6.5) tailored for transduction :

$$
\begin{array}{ll}
\min\limits_{\alpha} & \frac{1}{2}\sum\limits_{i=1}^{\ell+k}\sum\limits_{j=1}^{\ell+k} y_i y_j \alpha_i \alpha_j K(x_i, x_j) - \sum\limits_{i=1}^{\ell+k} \alpha_i \\
s.t. & \sum\limits_{i=1}^{\ell+k} \alpha_i y_i = 0 \\
 & 0 \leq \alpha_i \leq C \quad i = 1, \ldots, \ell+k
\end{array}
\qquad (6.16)
$$

where $y_j = 2*(d_j - \frac{1}{2})$ for $j = \ell, \ldots, \ell+k$. This process is repeated until a local minimum is reached. Then the weight on the misclassification error C is decreased allowing wider margins. In order to escape from local minima, the algorithm switches the labels of unlabeled data close to the separating hyperplane, if necessary. For this purpose, we check the consecutive solutions to track local minima. If 10 consecutive solutions are the same we assign the opposite labels to the points satisfying $|\boldsymbol{w} \cdot x_{i+1} - b| < S$. Occassionally a local minima is found with all points classified in one class ($\boldsymbol{w} = 0$). In this case, the algorithm restarts using the same initial conditions except for a reduced margin penalty parameter C for the unlabeled data. We empirically picked $C = \frac{\lambda}{100*(1-\lambda)}$, because it performed well in most cases. To ensure a good starting solution, the initial label assignments are made based on the closest class center for each unlabeled point. The resulting algorithm can be summarized as follows:

Algorithm 6.1 *S^3VM-IQP*

- *Find class centers from training points*
- *Assign labels d_0 to working set according to the closest class center*
- *Initialization:* $i = 0$, $\lambda = 0.9$, $C = \frac{\lambda}{100*(1-\lambda)}$, *counter* $= 0$, $S = 0.2$.
- *While* $i \leq max\_iteration$
  1. *Fix* d_i *and solve Problem 6.15 (or its dual (6.16)) to find* $(\boldsymbol{w}_{i+1}, b_{i+1}, \eta_{i+1}, \xi_{i+1}, z_{i+1})$.
  2. *Fix* $(\boldsymbol{w}_{i+1}, b_{i+1}, \eta_{i+1}, \xi_{i+1}, z_{i+1})$ *and solve Problem 6.15 for* d_{i+1}.
  3. *Check convergence criteria*
     - *If solution is same as the previous one*
       then counter=counter +1 and $\lambda = \lambda * 0.9$
       else if there exists no point within margin
       then stop
       else let counter $= 0$
     - *if counter* > 10 *then let counter* $= 0$ *and assign the opposite labels to the points satisfying* $|\boldsymbol{w}_{i+1} \cdot x - b_{i+1}| < S$
     - *if solution is all-in-one-class then reassign initial conditions*
       except i *and let* $\lambda * = 0.9$
  4. $i = i + 1$

As a benchmark for transductive SVM, we report results from SVM-Light proposed by Joachims in [19, 18]. Transductive SVM-Light also can be viewed as a block coordinate descent algorithm that alternates between estimating the class labels and optimizing the SVM based on those labels. Transductive SVM-Light has an inner and an outer loop. The outer loop adjusts the penalty parameters on misclassification errors. Different errors are used for the unlabeled data according to their estimated class labels. After initial inductive iteration, the algorithm starts with low penalty terms for unlabeled data. Two penalty terms (C^*_-, C^*_+) are used in transductive SVM-Light, each for classifying an

Table 6.3 Average Error Results for Transductive and Inductive Methods

Data Set	SVM-QP	SVM-Light	S^3VM-IQP
Heart	<u>0.16</u>	0.163	0.1966
Housing	0.1804	<u>0.1608</u>	0.1647
Ionosphere	<u>0.0857</u>	0.1572	0.0943
Sonar	0.1762	0.2524	<u>0.1572</u>

unlabeled point as a class -1 or a class 1 object respectively. Then it uniformly increases the influence of unlabeled data up to a user-defined penalty level. During this phase, the algorithm tunes these penalty terms in a way to satisfy a user-defined bias in data. The inner loop optimizes the SVM for the given penalties. The inner loop switches the labels of two given points, if such an action reduces the overall error. Like S^3VM-IQP, SVM-Light alternates the labels to avoid local minima. The primary difference is that SVM-Light changes the signs of at most two points at a time. Another difference is SVM-Light uses different margin penalty parameters for class 1 and class -1 objects. In addition, unlike S^3VM-QP, it starts with lower values for margin penalty parameters. Details of SVM-Light and successful results on large datasets can be found in [19] We use the default parameter options in our experiments with SVM-Light.

5.2 S^3VM-IQP RESULTS

In this section we compare S^3VM-IQP with SVM-QP (Eq. 6.5) and transductive SVM-Light. We use same datasets as in the previous section. Due to the long computational times for S^3VM-IQP and transductive SVM-Light, we limit our experiments to only the Heart, Housing, Ionosphere, and Sonar datasets. Linear kernel functions are used for all methods used in this section. The results given in Table 6.3 show that using unlabeled data in the case of datasets Heart and Ionosphere affects generalization ability slightly but the difference between the best transductive result and SVM-QP (Eq. 6.5) is not statistically significant. In the other two cases (Housing and Sonar), the best transductive method outperforms SVM-QP significantly. On two datasets S^3VM-IQP performs significantly better than transductive SVM-Light and in one case (Housing) the difference between two methods is not statistically significant.

As indicated above, the results from both S^3VM-IQP and SVM-Light are inconclusive. Both algorithms are much more expensive than their inductive versions. From the results on the Mixed Integer Programming Approaches we know that transduction can improve learning. We speculate that the reason that these improvements were not found using S^{3}VM-IQP and SVM-Light is that the optimization problem is very difficult and that the methods are failing to find the global minima. We know from the prior experiments that there is very little room for improvement on these specific learning tasks. Very few local minima will lead to better generalization. S^{3}VM -MIP and its local version are finding globally optimal solutions that are better. From the results on SVM-Light reported in [19] we know that on larger problems in text categorization, transductive inference using SVM-Light did lead to significant improvements. So on different learning tasks S^{3}VM-IQP may perform better as well. We speculate that on problems where there are many local minima that improve generalization, it is not as essential that the global minimum be found. Further studies are needed to identify when methods that find good but not globally optimal solutions are sufficient. Note that nonlinear kernels also might result in better generalization.

6 CONCLUSION

We examined mathematical models for semi-supervised support vector machines (S^{3}VM). We proposed a general S^{3}VM model that minimizes both the misclassification error and the function capacity based on all the available data. Three different functions for penalizing unlabeled points falling in the margin were discussed. Our computational investigation focused on the minimum error formulation for the transductive inference problem. We converted this problem to a mixed-integer program that can be exactly solved using commercial integer programming packages. By using the MIP formulation with a local learning algorithm, a powerful scalable transductive inference method was created. Our computational experiments found that the local learning method was the most effective overall. Further studies are needed to determine how to best select neighborhoods and to choose the parameters within the local S^{3}VM-MIP. In addition, very efficient computational methods for the local S^{3}VM-MIP are needed. One possibility is to use the estimated labels and models for one point as a starting point for other points. We also examined a nonconvex quadratic optimization approach to S^{3}VM. Our computational studies were less conclusive using this approach. The best optimization approach for solving this problem is still very much an open question.

Acknowledgments

This work was partially supported by NSF IRI-9702306, NSF IIS-9979860 and NSF DMS-9872019. Thanks to the referees and Scot Vandenberg for helpful comments.

Notes

1. The continuous response variable in Housing dataset was categorized at 21.5

References

[1] C. G. Atkeson, A. W. Moore, and S. Schaal. Locally weighted learning. *Artificial Intelligence Review*, 11:11–73, 1997.

[2] K. P. Bennett. Global tree optimization: a non-greedy decision tree algorithm. *Computing Science and Statistics*, 26:156–160, 1994.

[3] K. P. Bennett. Combining support vector and mathematical programming methods for classification. In B. Schölkopf, C. Burges, and A. Smola, editors, *Advances in Kernel Methods – Support Vector Machines*, pages 307–326, Cambridge, MA, 1999. MIT Press.

[4] K. P. Bennett and E. J. Bredensteiner. Geometry in learning. Web manuscript, Rensselaer Polytechnic Institute, http://www.rpi.edu/~bennek/geometry2.ps, 1996. Accepted for publication in Geometry at Work, C. Gorini et al, editors, MAA Press.

[5] K. P. Bennett and A. Demiriz. Semi-supervised support vector machines. In D. Cohn M. Kearns, S. Solla, editor, *Advances in Neural Information Processing Systems*, pages 368–374, Cambridge, MA, 1999. MIT Press.

[6] K. P. Bennett and O. L. Mangasarian. Robust linear programming discrimination of two linearly inseparable sets. *Optimization Methods and Software*, 1:23–34, 1992.

[7] K. P. Bennett and O. L. Mangasarian. Bilinear separation in n-space. *Computational Optimization and Applications*, 4(4):207–227, 1993.

[8] D. P. Bertsekas. *Nonlinear Programming.* Aethena Scientific, Cambridge, MA, 1996.

[9] J. Blue. *A hybrid of tabu search and local descent algorithms with applications in artificial intelligence.* PhD thesis, Rensselaer Polytechnic Institute, Troy, NY, 1998.

[10] A. Blum and T. Mitchell. Combining labeled and unlabeled data with co-training. In *Proceedings of the 1998 Conference on Computational Learning Theory*, Madison WI, 1998. ACM Inc.

[11] E. J. Bredensteiner and K. P. Bennett. Feature minimization within decision trees. *Computational Optimization and Applications*, 10:110–126, 1997.

[12] V. Castelli and T. M. Cover. On the exponential value of labeled samples. *Pattern Recognition Letters*, 16:105–111, 1995.

[13] Z. Cataltepe and M. Magdon-Ismail. Incorporating test inputs into learning. In *Proceedings of the Advances in Neural Information Processing Systems, 10*, Cambridge, MA, 1997. MIT Press.

[14] C. Cortes and V. N. Vapnik. Support vector networks. *Machine Learning*, 20:273–297, 1995.

[15] CPLEX Optimization Incorporated, Incline Village, Nevada. *Using the CPLEX Callable Library*, 1994.

[16] R. Fourer, D. Gay, and B. Kernighan. *AMPL A Modeling Language for Mathematical Programming.* Boyd and Frazer, Danvers, MA, 1993.

[17] T. Hastie and R. Tibshirani. Discriminant adaptive nearest neighbor classification. *IEEE PAMI*, 18:607–616, 1996.

[18] T. Joachims. Text categorization with support vector machines: Learning with many relevant features. In *European Conference on Machine Learning(ECML)*, 1998.

[19] T. Joachims. Transductive inference for text classification using support vector machines. In *International Conference on Machine Learning*, 1999.

[20] S. Lawrence, A. C. Tsoi, and A. D. Back. Function approximation with neural networks and local methods: Bias, variance and smoothness. In Peter Bartlett, Anthony Burkitt, and Robert Williamson, editors, *Australian Conference on Neural Networks, ACNN 96*, pages 16–21. Australian National University, 1996.

[21] O. L. Mangasarian. Arbitrary norm separating plane. *Operations Research Letters*, 24(1-2), 1999.

[22] O. L. Mangasarian. Generalized support vector machines. In A. Smola, P. Bartlett, B. Schölkopf, and D. Schuurmans, editors, *Advances in Large Margin Classifiers*, pages 135–146, Cambridge, MA, 2000. MIT Press. ftp://ftp.cs.wisc.edu/math-prog/tech-reports/98-14.ps.

[23] A. McCallum and K. Nigam. Employing em and pool-based active learning for text classification. In *Proceedings of the 15th International Conference on Machine Learning (ICML-98)*, 1998.

[24] P.M. Murphy and D.W. Aha. *UCI repository of machine learning databases.* Department of Information and Computer Science, University of California, Irvine, California, 1992.

[25] D. R. Musser and A. Saini. *STL Tutorial and Reference Guide: C++ Programming with the Standard Template Library.* Addison-Wesley, 1996.

[26] K. Nigam, A. McCallum, S. Thrun, and T. Mitchell. Learning to classify text from labeled and unlabeled documents. In *Proceedings of the 15th National Conference on Artificial Intelligence (AAAI-98)*, 1998.

[27] S. Odewahn, E. Stockwell, R. Pennington, R Humphreys, and W Zumach. Automated star/galaxy discrimination with neural networks. *Astronomical Journal*, 103(1):318–331, 1992.

[28] V. N. Vapnik. *Estimation of dependencies based on empirical Data.* Springer, New York, 1982. English translation, Russian version 1979.

[29] V. N. Vapnik. *The Nature of Statistical Learning Theory.* Springer Verlag, New York, 1995.

[30] V. N. Vapnik. *Statistical Learning Theory.* Wiley Inter-Science, 1998.

[31] V. N. Vapnik and A. Ja. Chervonenkis. *Theory of Pattern Recognition.* Nauka, Moscow, 1974. In Russian.

Chapter 7

PREPROCESSING COMPLEMENTARITY PROBLEMS

Michael C. Ferris
Computer Sciences Department
University of Wisconsin at Madison
Madison, WI 53706
ferris@cs.wisc.edu

Todd S. Munson
Computer Sciences Department
University of Wisconsin at Madison
Madison, WI 53706
tmunson@cs.wisc.edu

Abstract Preprocessing techniques are extensively used in the linear and integer programming communities as a means to improve model formulation by reducing size and complexity. Adaptations and extensions of these methods for use within the complementarity framework are detailed and shown to be effective on practical models. The preprocessor developed is comprised of two phases. The first recasts a complementarity problem as a variational inequality over a polyhedral set and exploits the uncovered structure to fix variables and remove constraints. The second discovers information about the function and utilizes complementarity theory to eliminate variables. The methodology is successfully employed to preprocess several models.

Keywords: complementarity problems, preprocessing, variational inequalities

1 INTRODUCTION

Complementarity is becoming a widely used format in a variety of application areas, ranging from finance and economics to frictional contact and engineering [11]. Some of the increase is due to the availability of

M.C. Ferris et al. (eds.), Complementarity: Applications, Algorithms and Extensions, 143–164.

general purpose codes, such as PATH [5], both as subroutine libraries [9] and as solvers for models written with languages such as AMPL and GAMS.

General purpose codes for solving complementarity problems have previously lacked one significant feature: a powerful preprocessor. The benefits of preprocessing have long been known to the linear [1, 2] and integer [20] programming communities, yet have not been studied from a complementarity perspective. In many cases, a preprocessor just uncovers structure from a model that may not have been explicitly communicated to the solver. The reasons for this are varied. The modeler may be completely unaware of the structure or unwilling or unable to pass the known structure to a solver. In other cases, the structure arises from simplifications due to particular data values. For example, if a general inequality constraint can be reduced to use only one variable, the solver would prefer to treat it as a bound constraint.

In general, the purpose of a preprocessor is to reduce the size and complexity of a model to achieve improved performance by the main algorithm. For example, by fixing variables, some nonlinear constraints may become linear, and smaller linear systems are typically solved with less memory and in less time. Another benefit of the analysis performed is the detection of model infeasibilities. Most nonlinear complementarity solvers continue attempting to solve a model until a time or iteration limit is reached. In cases where the model has no solution, this is obviously wasteful. Furthermore, although some progress has been made to furnish the modeler with useful information regarding the source of the difficulty [8], most codes just terminate with a resource limit message. A preprocessor can sometimes process the model enough to determine (and report) the source of the difficulty. In this paper, a comprehensive preprocessor is developed for the mixed complementarity problem and computational experience with an implementation is reported.

The AMPL and GAMS modeling environments enable a practitioner to specify and solve their application by providing a mixed complementarity problem (MCP) format. The models generated, along with auxiliary information such as gradient and Jacobian evaluations, are then communicated directly to available solvers [6, 10]. The mixed complementarity problem supported by these modeling languages and accepted by most solvers is the (box-constrained) variational inequality:

$$0 \in F(x) + N_{[L,U]}(x), \tag{7.1}$$

where $F : \mathbb{R}^n \to \mathbb{R}^n$ is continuously differentiable and $[L, U]$ represents a Cartesian product of closed, not necessarily compact, intervals. In this definition, $N_X(\cdot)$ is the normal cone [19] of X defined by

$$N_X(x) := \begin{cases} \{y \mid \langle \bar{x} - x, y\rangle \leq 0,\ \forall \bar{x} \in X\} & \text{if } x \in X \\ \emptyset & \text{otherwise,} \end{cases}$$

under the assumption that X is a closed convex set. Note that (7.1) is just the standard variational inequality

$$x \in [L, U] \text{ and } \langle F(x), \bar{x} - x\rangle \geq 0 \text{ for all } \bar{x} \in [L, U].$$

When $X = [L, U]$, the normal cone can be determined as a Cartesian product, namely

$$N_{[L,U]}(x) = \prod_{i=1}^{n} N_{[L_i,U_i]}(x_i)$$

Each of the components in this product depends on x_i, L_i and U_i but is either $\mathbb{R}$, $\mathbb{R}_+$, $\mathbb{R}_-$, $\{0\}$ or $\emptyset$. As an example, suppose

$$[L, U] = [-2, \infty) \times \{3\} \subseteq \mathbb{R}^2.$$

In this case,

$$N_{[L,U]}(x) = \begin{cases} \mathbb{R}_- \times \mathbb{R} & \text{if } x_1 = -2, x_2 = 3 \\ \{0\} \times \mathbb{R} & \text{if } x_1 > -2, x_2 = 3 \\ \emptyset & \text{otherwise,} \end{cases}$$

and hence (7.1) is the mixed system of inequalities and equations

$$\begin{aligned} F_1(x) &\geq 0 \\ x_1 &\geq -2 \\ (x_1 + 2)F_1(x) &= 0 \\ x_2 &= 3 \end{aligned}$$

Note that F_2 is irrelevant, and so the example reduces to the standard complementarity problem

$$F_1(z - 2, 3) \geq 0,\ z \geq 0\ zF_1(z - 2, 3) = 0$$

in the single variable z.

The preprocessor for complementarity problems works upon two equivalent representations of the same model. To understand the basic methodology developed, consider a standard convex quadratic programming problem:

$$\begin{array}{ll} \min & \frac{1}{2}x^T Q x + c^T x \\ \text{s.t.} & Ax \geq b \\ & x \geq 0, \end{array} \tag{7.2}$$

where $Q \in \mathbb{R}^{n\times n}$ is a symmetric positive semi-definite matrix, $A \in \mathbb{R}^{m\times n}$, $b \in \mathbb{R}^m$, and $c \in \mathbb{R}^n$. (7.2) can be posed as a variational inequality in one of two ways. First, when dual variables, λ, are introduced, the complementary slackness conditions for quadratic programs form the box constrained variational inequality:

$$0 \in \begin{bmatrix} Q & -A^T \\ A & 0 \end{bmatrix} \begin{bmatrix} x \\ \lambda \end{bmatrix} + \begin{bmatrix} c \\ -b \end{bmatrix} + \begin{bmatrix} N_{\mathbb{R}^n_+}(x) \\ N_{\mathbb{R}^m_+}(\lambda) \end{bmatrix}. \tag{7.3}$$

Alternatively, the first order conditions can be succinctly written as a polyhedrally constrained variational inequality:

$$0 \in Qx + c + N_C(x), \tag{7.4}$$

where $C = \{x \mid Ax \geq b, x \geq 0\}$. Since C is a geometric object, a computationally attractive algebraic representation can be chosen for C. Exploiting two representations of the mixed complementarity problem (analogous to (7.3) and (7.4) for quadratic programs) is a key concept in the preprocessor developed.

As mentioned above, the MCP is communicated as a box-constrained variational inequality (7.1). The preprocessor automatically rewrites the problem in the form

$$0 \in H(x) + N_C(x). \tag{7.5}$$

Both formulations are used in distinct phase of the preprocessor. The majority of the preprocessor reductions documented involve exploiting the polyhedral set C in (7.5). From the problem description, the polyhedral structure in C will need to be recovered before it can be used. Once this is achieved, the general inequalities in the set C can be used to modify the bounds L_i and U_i on a variable x_i. In particular if $x_i = L_i = U_i$ then (see above) $N_{[L_i,U_i]}(x_i) = \mathbb{R}$. Hence, fixing a variable x_i means that the corresponding constraint

$$0 \in F_i(x) + N_{[L_i,U_i]}(x_i) \iff 0 \in F_i(x) + \mathbb{R}$$

is trivially satisfiable. This was shown in the example above, where the variable x_2 was fixed, thus removing the constraint involving F_2. Thus, preprocessing in the complementarity case attempts to fix variables and thus remove constraints.

Section 2 begins by detailing the process used to uncover and exploit polyhedral structure in an MCP. The general idea is to reformulate (7.1) in a form similar to (7.4), with a general polyhedral set C replacing $[L, U]$. The representation of the set C can then be modified by either removing constraints or bounding variables. When converted back to a mixed complementarity problem a reduction in the number of variables is realized. Note that the process developed in this section recovers most checks done by traditional linear programming codes[1] when given the complementary slackness necessary and sufficient conditions for linear programs, but is applicable to a larger class of problems.

Further reductions to the MCP can be made by utilizing information about F and its Jacobian, ∇F, as developed in Section 3. In particular, the range of F is used to eliminate variables from the model. Row and column duplicates are also removed. By detecting special block structure, a sequence of smaller problems can be solved to find an answer to the original problem.

Finally, both phases are incorporated into a complete preprocessor for mixed complementarity problems in Section 4. Computational results for some test problems are presented indicating the success of the procedure outlined.

More information about the problem must be provided to the preprocessor than is necessary to solve it. The basic requirement is a listing of the linear and nonlinear elements in the Jacobian of F. This knowledge is sufficient to find and utilize special structures. The AMPL [12] and GAMS [3] environments already provide this information. Users of other interfaces, such as MATLAB and NEOS [7], will need to develop the appropriate routines. Some checks in Section 3 based on the nonlinear functions need to know the range of F over X. Routines to calculate these intervals are not currently provided by any of the interfaces.

2 POLYHEDRAL CONSTRAINTS

The first stage of the preprocessor detects polyhedral structure in a mixed complementarity problem. The structure is exploited by transforming the source problem into a model of lower dimension where C is the intersection of a closed product of intervals and a polyhedral set. The representation of C is then modified by removing constraints and changing bounds with the resultant MCP containing fewer variables. After the preprocessed model has been solved, a solution to the original MCP is recovered with a postsolve step.

2.1 PRESOLVE

Polyhedral structure can be exploited when given a special type of complementarity problem. Suppose the variables can be split into (x, y)

and (7.1) has the form:

$$0 \in \begin{bmatrix} F(x) - A^T y \\ Ax - b \end{bmatrix} + \begin{bmatrix} N_X(x) \\ N_Y(y) \end{bmatrix}, \tag{7.6}$$

where $F : \mathbb{R}^n \to \mathbb{R}^n$ is continuously differentiable, $A \in \mathbb{R}^{m \times n}$, $b \in \mathbb{R}^m$, $X \subseteq \mathbb{R}^n$ is a Cartesian product of closed intervals, and $Y \subseteq \mathbb{R}^m$ is a Cartesian product of $\mathbb{R}$, $\mathbb{R}_+$, or $\mathbb{R}_-$. Note that if $Y_i = \{0\}$ then y_i can be fixed at zero and the corresponding $A_{i,\cdot}x - b_i$ removed. Further, if $Y_i = [L_i, \infty)$ or $Y_i = (-\infty, U_i]$ for some finite L_i or U_i then an appropriate change of variables, possibly adding constant vectors to $F(x)$ and b, replaces Y_i with $\mathbb{R}_+$ and $\mathbb{R}_-$ respectively.

Theorem 7.1 *(Propositions 1 and 2 of [18]) Under the assumptions placed on X, Y, and the structure of the problem given above the following hold:*

1. *If $(\bar{x}, \bar{y})$ solves (7.6) then $\bar{x}$ is a solution to*

$$0 \in F(x) + N_{X \cap \{x | b - Ax \in Y^o\}}(x), \tag{7.7}$$

where Y^o denotes the polar cone of Y which is defined as

$$Y^o := \{y \mid \langle y, \bar{y} \rangle \leq 0, \ \forall \bar{y} \in Y\}.$$

2. *If $\bar{x}$ solves (7.7) then the linear optimization problem*

$$\begin{array}{ll} \min_{y \in Y} & \langle A\bar{x} - b, y \rangle \\ s.t. & 0 \in F(\bar{x}) - A^T y + N_X(\bar{x}). \end{array} \tag{7.8}$$

has a nonempty solution set. Further, for any $\bar{y}$ solving (7.8), $(\bar{x}, \bar{y})$ solves (7.6).

Theorem 7.1 provides the machinery used by the first stage of the preprocessor. The Jacobian matrix, ∇F is stored in both row- and column-oriented data structures. Utilizing the information provided about the types of the elements in the Jacobian, a row and column possessing the necessary skew symmetric structure of (7.6) can be quickly identified. Theorem 7.1 is then applied to this single row and column to create a problem of the form (7.7). The polyhedral set, $X \cap \{x \mid b - Ax \in Y^o\}$, is then checked for possible reductions. That is, we determine whether $X \cap \{x \mid b - Ax \in Y^o\}$ can be replaced with an identical set $\tilde{X} \cap \{x \mid b - Ax \in \tilde{Y}^o\}$ in such a way that the MCP recovered by using Theorem 7.1 (in the reverse mode) has a reduced size. The identification and modification is repeated until no further simplifications are made.

Note that Theorem 7.1 is only applied to a small number of constraints at a time during preprocessing while [18] uses the machinery to ready a problem for solution by a polyhedrally constrained variational inequality solver [21]. Automatically finding a set of polyhedral constraints with maximum size from (7.1) is a harder problem and is not considered here.

Information about any modifications performed are pushed onto a stack. The stack is a convenient data structure with two basic operations: pushing an element onto the top and popping an element from the top. Changes are pushed onto the stack in the order performed and are popped off the stack in the reverse order during the postsolve.

The complete algorithm for the first phase of the preprocessor is as follows:

A.1 Mark all rows and columns with the skew symmetric structure as *eligible* candidates, excluding any rows complementary to a variable with finite lower and upper bounds.

A.2 Using some ordering, pick one of the candidates and transform the problem into a polyhedral-constrained equation using Theorem 7.1.

A.3 Analyze the polyhedral set and modify the representation as detailed below. Push any changes on top of the stack.

A.4 Transform the modified problem back to box-constrained form.

A.5 Repeat steps A.2–A.4 until there are no reductions possible.

The implementation performs all simple reductions (Section 2.1.1 and Section 2.1.2) first. Once all of these are completed, forcing constraints (Section 2.1.3) and redundant rows (Section 2.1.4) are checked. In a nonlinear model, additional rows and columns can become linear when variables are fixed. Therefore, after all tests are completed on the current list of eligible candidates, another pass is made through the Jacobian to mark new eligible rows and columns which are checked using A.2–A.5. When no new eligible rows and columns are created the process stops.

2.1.1 Simple Reductions. The simplest reduction made is when an eligible row contains zero elements. This corresponds to the case where the polyhedral set in the transformed problem is:

$$X \cap \{x \mid b \in Y^o\}. \tag{7.9}$$

If (7.9) is empty, then the original problem has no solution. Otherwise, the polyhedral component is irrelevant, because it does not impose any constraints on x, and (7.9) can be replaced with:

$$X \cap \{x \mid b \in \{0\}^o\}.$$

Note that $\{0\}^o = \mathbb{R}$, ensuring the constraint $b \in \{0\}^o$ is always satisfied. When transformed back to the original space, the corresponding multiplier is fixed at 0 and removed from the problem, resulting in a reduction of one variable and the corresponding constraint.

Another simple reduction occurs when the eligible row contains a single element. The polyhedral set in this case is:

$$X \cap \{x \mid b - ax_i \in Y^o\}. \tag{7.10}$$

Since Y is $\mathbb{R}_+$, $\mathbb{R}_-$, or $\mathbb{R}$, the constraint will be either $b - ax_i \leq 0$, $b - ax_i \geq 0$, or $b - ax_i = 0$ respectively. Each of these imply simple bounds on x_i, which can be explicitly incorporated in X. Therefore, (7.10) is replaced by

$$\tilde{X} \cap \{x \mid b - ax_i \in \{0\}^o\},$$

where $\tilde{X}$ includes the tightened bounds on x_i. This modification results in a reduction of at least one variable.

2.1.2 Doubleton Rows. Doubleton rows having the skew symmetric property can also be preprocessed. Consider a row, i, which is an equation of the form

$$ax_j + bx_k = c$$

where either x_j or x_k is a column singleton. Assume x_k is the column singleton. If row k also has the skew symmetric property, then X can be modified by changing the bounds on x_j to make x_k free. Immediately following this change, row k (which must be a singleton row complementary to the free variable x_k) is preprocessed out of the model using the singleton check described above.

2.1.3 Forcing Constraints. Forcing constraints are constraints for which, given the bounds on the variables, there is exactly one feasible point. Once it is known that only one solution is possible, all variables appearing in the constraint can be fixed, potentially leading to a large reduction in problem size.

Let the polyhedral constraint be written in the form:

$$X \cap \{x \mid b - a^T x \in Y^o\}. \tag{7.11}$$

Without loss of generality, assume that $Y = \mathbb{R}_+$ which means that $Y^o = \mathbb{R}_-$. (7.11) can then be explicitly stated as:

$$X \cap \{x \mid b \leq a^T x\}.$$

Using X, bounds, $\underline{a}$ and $\bar{a}$, can be implied such that $\underline{a} \leq a^T x \leq \bar{a}$ for all $x \in X$. The ranges are determined as follows:

$$\begin{aligned} \underline{a} &= \sum_{\{i|a_i>0\}} a_i L_i + \sum_{\{i|a_i<0\}} a_i U_i \\ \bar{a} &= \sum_{\{i|a_i>0\}} a_i U_i + \sum_{\{i|a_i<0\}} a_i L_i, \end{aligned}$$

where L_i and U_i are the lower and upper bounds on variable i respectively. If $\bar{a} < b$ the problem is infeasible. Otherwise, if $\bar{a} = b$, there is only one feasible point and the values of the variables are fixed at U_i for all i with $a_i > 0$ and L_i for all i with $a_i < 0$. Set (7.11) is then replaced with:

$$\tilde{X} \cap \{x \mid b - a^T x \in \{0\}^o\}, \tag{7.12}$$

where $\tilde{X}$ contains the fixed variables. The net result is that the forcing constraint and a number of variables are removed from the original problem.

2.1.4 Redundant Rows. Redundancy in the Jacobian matrix can cause difficulty for many algorithms. Therefore, it is advantageous to remove as much redundancy as possible. The algorithm given in [22] is used to identify duplicate rows. All eligible constraints are checked simultaneously with the algorithm. After finding two duplicate rows, any inconsistencies are uncovered (meaning that the model is unsolvable) and one of the constraints is removed wherever possible. Without loss of generality, let the constraint set be written as

$$X \cap \left\{ x \,\middle|\, \begin{array}{l} b - ax \in Y_1^o \\ c - ax \in Y_2^o \end{array} \right\}. \tag{7.13}$$

Several cases are presented in Table 7.1 along with the associated action taken. The other cases are symmetric to those given in the table.

2.1.5 Extensions. The requirements in Theorem 7.1 can be slightly weakened. Let $D \in \mathbb{R}^{n \times n}$ be a positive diagonal matrix. Then the following form will suffice instead of (7.6):

$$0 \in \begin{bmatrix} F(x) - DA^T y \\ Ax - b \end{bmatrix} + \begin{bmatrix} N_X(x) \\ N_Y(y) \end{bmatrix}. \tag{7.14}$$

Table 7.1 Redundant Rows Cases

Case	Action
$Y_1 = \mathbb{R}$ and $Y_2 = \mathbb{R}$	If $b = c$ remove one of the constraints. Otherwise the problem is infeasible.
$Y_1 = \mathbb{R}$ and $Y_2 = \mathbb{R}_+$	If $b \geq c$ remove the inequality constraint. Otherwise the problem is infeasible.
$Y_1 = \mathbb{R}_+$ and $Y_2 = \mathbb{R}_+$	If $b \geq c$ remove the constraint associated with Y_2. Otherwise remove the Y_1 constraint.
$Y_1 = \mathbb{R}_-$ and $Y_2 = \mathbb{R}_+$	If $b < c$, the problem is infeasible. Otherwise if $b = c$ make one of the constraints an equation and remove the other. Otherwise, it is a range constraint; nothing is done by the preprocessor.

(7.14) can be reduced to (7.6) by applying a diagonal row scaling of $\begin{bmatrix} D & 0 \\ 0 & 1 \end{bmatrix}^{-1}$ and recalling that the normal cone does not change under multiplication by a positive diagonal matrix.

Furthermore, free variables imply that the corresponding function is an equation because $N_{\mathbb{R}}(\cdot) \equiv \{0\}$. Therefore, the equation can be negated to obtain the required structure.

2.2 POSTSOLVE

Once the algorithm has solved the preprocessed model, all of the presolve steps must be undone in the reverse order to recover a solution to the original model. The stack of presolve records is used for this purpose. The following steps are performed:

B.1 Remove a presolve record from the top of the stack.

B.2 Transform the problem into the polyhedral-constrained setting.

B.3 Undo the changes made to the model.

B.4 Solve the optimization problem (7.8) using $\bar{x}$ to obtain $\bar{y}$. The generated $(\bar{x}, \bar{y})$ solves the model before the presolve step was performed.

B.5 Repeat until the presolve stack is empty.

The optimization problem (7.8) is typically only in one dimension and is trivial to solve. Care must be taken when calculating $N_X(\bar{x})$ because the algorithm may only find a solution to within a prespecified tolerance.

Therefore, variables within some tolerance of their bounds should be treated as if they are exactly on their bounds when constructing the normal cone. The current default used in the implementation is the convergence tolerance as specified by the algorithm.

The only case where two variables are involved in the optimization problem is when two inequalities are replaced by one equation. The optimization problem in this case has an objective function equal to zero because at the solution $A\bar{x} - b = 0$. Therefore, a feasible point need only be generated. In the presolved model, $0 \in F(x) - a\hat{y} + N_X(\bar{x})$ for the solution $(\bar{x}, \hat{y})$ given. Without loss of generality, assume $Y_1 = \mathbb{R}_+$ and $Y_2 = \mathbb{R}_-$. Select $y_1 \in Y_1 = \mathbb{R}_+$ and $y_2 \in Y_2 = \mathbb{R}_-$ such that $y_1 + y_2 = \hat{y}$. These conditions can always be trivially met. Because the inclusion holds at $\hat{y}$, it also holds for the y_1 and y_2 selected, which is then a feasible point as required.

In the unfortunate case that the algorithm fails to solve the preprocessed model, the optimization problem may have no solution either because it is infeasible or unbounded. In this case, a value for the multiplier is chosen in Y such that the norm of the error in the box constrained representation is minimized given $\bar{x}$. This greedy heuristic will lead to the best possible value in terms of the residual at each stage in the unrolling of the preprocessing steps, but not necessarily the least residual solution overall.

3 STRUCTURAL IMPLICATIONS

The second phase of the preprocessor utilizes complementarity theory to eliminate variables from the MCP. The reductions documented are based on the rows and columns of the Jacobian, ∇F. The main ingredient for the row-based rules is uniqueness. If the value for a variable can be uniquely determined prior to solving the remainder of the problem, it is fixed and removed. The column-based rules rely upon existence arguments. Once a solution to the reduced model is known, a solution to the original problem always exists and can be calculated. Mechanisms developed include using interval evaluations, uncovering duplicate rows and columns, and exploiting special structure. Note that when a row with zero elements and corresponding zero column are present in a model, the variable can always be fixed at an appropriate value and removed.

3.1 INTERVALS

An interval evaluator determines the tightest possible $\underline{F}$ and $\bar{F}$ such that for all $x \in X$, $\underline{F} \leq F(x) \leq \bar{F}$. For example, with a linear constraint,

$F_i(x) = a^T x - b$, the bounds

$$\underline{F}_i = \sum_{\{j|a_j>0\}} a_j L_j + \sum_{\{j|a_j<0\}} a_j U_j - b$$
$$\bar{F}_i = \sum_{\{j|a_j>0\}} a_j U_j + \sum_{\{j|a_j<0\}} a_j L_j - b$$

can be used where L_j and U_j are the lower and upper bounds on variable j respectively. The range of a nonlinear function is dependent upon the model and the bounds must be computed by a user supplied routine.

Using the ranges, variables in the model can be fixed. If $\underline{F}_i > 0$, then x_i must be fixed at its finite lower bound or the problem is infeasible. Furthermore, if $\bar{F}_i < 0$, then x_i must be fixed at its finite upper bound or the problem is infeasible.

Some of the constraints in the model will imply tighter bounds on the variables; i.e. a linear constraint. These can be used by the interval evaluator to strengthen the ranges on other constraints, potentially leading to more variables being fixed. The tightened bounds should only be used when calculating intervals. Modifying the bounds on the complementarity problem using this information could change the solution set of the problem.

3.2 DUPLICATES

Duplicate rows and columns can be very problematic for a solver. By applying the same algorithm used in the polyhedrally constrained case (Section 2.1.4), two such linear rows or columns can be identified. First consider the case of two duplicate rows in the problem. Without loss of generality, the model can be written as:

$$0 \in \begin{bmatrix} F(x,y,z) \\ a^T(x,y,z)+b \\ a^T(x,y,z)+c \end{bmatrix} + \begin{bmatrix} N_X(x) \\ N_Y(y) \\ N_Z(z) \end{bmatrix}.$$

Table 7.2 discusses the reductions that can be made.

To remove column duplicates, one of the variables needs to be free and the other must have two finite bounds. The problem in this case is:

$$0 \in F(x) + ay + az + N_{X \times \mathbb{R} \times [L,U]}(x,y,z),$$

where L and U are the finite lower and upper bounds on z. The reduction removes the z variable from the problem and solves the reduced system to obtain $(\bar{x},\bar{y})$. If $F_z(\bar{x}) + a_z\bar{y} > 0$, then $\hat{z} = L$. Otherwise, $\hat{z} = U$. Set $\hat{y} = \bar{y} - \hat{z}$. Then $(\bar{x},\hat{y},\hat{z})$ solves the original problem as can be easily verified.

Table 7.2 Duplicate Rows Cases

Case	Action
$Y = \mathbb{R}$ and $Z = \mathbb{R}$	If $b \neq c$ the problem is infeasible. Otherwise, nothing is done.
$Y = \mathbb{R}$ and $Z = \mathbb{R}_+$	If $b > c$ fix z at its lower bound. Otherwise, if $b < c$, the problem is infeasible.
$Y = \mathbb{R}_+$ and $Z = \mathbb{R}_+$	If $b > c$ fix z at its lower bound. If $b < c$ fix y at its lower bound.
$Y = \mathbb{R}_-$ and $Z = \mathbb{R}_+$	If $b < c$, the problem is infeasible. Otherwise, nothing is done.

3.3 SPECIAL STRUCTURE

For a system of nonlinear equations, if the problem has the form:

$$\begin{bmatrix} F(x) \\ G(x,y) \end{bmatrix} = \begin{bmatrix} 0 \\ 0 \end{bmatrix}$$

with $F : \mathbb{R}^k \to \mathbb{R}^k$, then $F(x) = 0$ can be solved giving $\bar{x}$ and then a $\bar{y}$ solving $G(\bar{x}, y) = 0$ can be found. If $F(x) = 0$ has multiple solutions, this procedure may fail by finding an $\bar{x}$ for which $G(\bar{x}, y) = 0$ has no solution. For example, consider $F(x) = x^2 - 1$ and $G(x, y) = x + y^2$. $F(x) = 0$ has two solutions $\bar{x} = 1$ and $\bar{x} = -1$. Choosing $\bar{x} = 1$ leads to the case where $G(\bar{x}, y)$ has no solution. If $F(x) = 0$ has at most one solution, this case is precluded provided the original problem has a solution. Similarly, the difficulty is alleviated if $G(\bar{x}, y)$ has a solution $\bar{y}$ for all $\bar{x}$, since whatever $\bar{x}$ is found, the system $G(\bar{x}, y) = 0$ is solvable. This section applies similar block reduction schemes to the mixed complementarity problem.

Consider a problem of the form:

$$0 \in \begin{bmatrix} F(x) \\ G(x,y) \end{bmatrix} + \begin{bmatrix} N_X(x) \\ N_Y(y) \end{bmatrix},$$

where, as usual, X and Y are Cartesian products of intervals. There are two sets of reductions that can be made. If $0 \in F(x) + N_X(x)$ has a unique solution, $\bar{x}$, then x can be fixed and the algorithm will only work on the reduced problem $0 \in G(\bar{x}, y) + N_Y(y)$. If $F(x)$ is an affine function, i.e. $F(x) = Ax - b$, then it is known that $0 \in F(x) + N_X(x)$ has a unique solution if A_X, the normal map associated with this variational inequality, is coherently oriented [17]. For example, when $X = \mathbb{R}^k_+$, this condition is equivalent to A being a P-matrix. For simple cases, coherent orientation can be checked; e.g. when $k = 1$ or 2. In particular,

Table 7.3 Coherent Orientation Conditions. X is assumed to be a Cartesian product of intervals with L and U being two finite numbers.

Case	Coherent Orientation Condition
$X = \mathbb{R}$	$A_{1,1} \neq 0$
$X = [L, \infty)$	$A_{1,1} > 0$
$X = (-\infty, U]$	$A_{1,1} > 0$
$X = [L, U]$	$A_{1,1} > 0$
$X = \mathbb{R} \times \mathbb{R}$	$det(A) \neq 0$
$X = \mathbb{R} \times [L, \infty)$	$det(A) \neq 0$ and $sign(det(A)) = sign(A_{1,1})$
$X = \mathbb{R} \times (-\infty, U]$	$det(A) \neq 0$ and $sign(det(A)) = sign(A_{1,1})$
$X = \mathbb{R} \times [L, U]$	$det(A) \neq 0$ and $sign(det(A)) = sign(A_{1,1})$
All other cases	$det(A) > 0$ and $sign(A_{1,1}) = sign(A_{2,2}) = 1$

when $F(x) = ax - b$ is a row singleton with a linear element on the diagonal, then coherent orientation is $a \neq 0$ when $X = \mathbb{R}$ and $a > 0$ in all other cases. Satisfaction of this condition guarantees uniqueness of the solution. When $k = 2$, the condition is again that A is a P-matrix, unless one or more of the intervals defining X is $\mathbb{R}$. In these later cases, the conditions are weaker. Table 7.3 summarizes all of the checks for coherent orientation. The preprocessor identifies double blocks by finding a linear row with two elements, one of which is on the diagonal. A check of the row corresponding to the other variable is performed to see if there is a doubleton block.

The other reduction to consider is where $0 \in G(x, y) + N_Y(y)$ has a solution for all $x \in X$. In this case, $0 \in F(x) + N_X(x)$ is solved to find $\bar{x}$ and then a $\bar{y}$ satisfying $0 \in G(\bar{x}, y) + N_Y(y)$ is found. Assume that $G(x, y)$ is linear in y, i.e. $G(x, y) = H(x) + By$. The coherent orientation conditions outlined above applied to B suffice in this case as well, since they guarantee existence as well as uniqueness. However, to guarantee only existence, weaker conditions are sufficient. For $k = 1$ it is necessary and sufficient to have coherent orientation or Y compact. When $k = 2$, the conditions are outlined in Table 7.4 and are derived from Theorem 2 in [15] and [14]. The conditions given for the cases where there is at least one free variable are necessary and sufficient to guarantee existence for all $x \in X$.

In the nonlinear setting, intervals on the Jacobian elements can be used to verify conditions related to uniqueness and existence of a solution. For example in the single element case, if the value of the Jacobian element is positive and uniformly bounded away from zero, i.e. it is a

Table 7.4 Existence Conditions. Y is assumed to be a Cartesian product of intervals with L, $\tilde{L}$, U, and $\tilde{U}$ being finite numbers.

Case	Condition
$Y = \mathbb{R}$	$B_{1,1} \neq 0$
$Y = [L, \infty)$	$B_{1,1} > 0$
$Y = (-\infty, U]$	$B_{1,1} > 0$
$Y = [L, U]$	nothing
$Y = \mathbb{R} \times \mathbb{R}$	$det(B) \neq 0$
$Y = \mathbb{R} \times [L, \infty)$	$det(B) \neq 0$ and $sign(det(B)) = sign(B_{1,1})$
$Y = \mathbb{R} \times (-\infty, U]$	$det(B) \neq 0$ and $sign(det(B)) = sign(B_{1,1})$
$Y = \mathbb{R} \times [L, U]$	$B_{1,1} \neq 0$
$Y = [L, \infty) \times [\tilde{L}, \infty)$	$B > 0$ or
	$(det(B) > 0$ and $sign(B_{1,1}) = sign(B_{2,2}) = 1)$
$Y = [L, \infty) \times (-\infty, \tilde{U}]$	$B > 0$ or
	$(det(B) > 0$ and $sign(B_{1,1}) = sign(B_{2,2}) = 1)$
$Y = [L, \infty) \times [\tilde{L}, \tilde{U}]$	$B_{1,1} \neq 0$
$Y = (-\infty, U] \times (-\infty, \tilde{U}]$	$B > 0$ or
	$(det(B) > 0$ and $sign(B_{1,1}) = sign(B_{2,2}) = 1)$
$Y = (-\infty, U] \times [\tilde{L}, \tilde{U}]$	$B_{1,1} \neq 0$
$Y = [L, U] \times [\tilde{L}, \tilde{U}]$	nothing

uniform P-function, then the existence and uniqueness is always guaranteed and the same substitutions can be performed. Finding the solution becomes more difficult, as it involves solving a nonlinear problem.

4 RESULTS

The preprocessing algorithm implemented alternates between exploiting the polyhedral structure and the functions. Initially all possible reductions based on the polyhedral constraints are made. Then all reductions based on the functional implications are made. These two steps are repeated until no changes are made to the model.

The potential for preprocessing is mainly limited to finding and exploiting linear parts of the problem because interval evaluators are not available at present in the modeling language environments. The majority of the reductions made come from exploiting polyhedral structure. However, the reductions from the second stage can also be significant to the success of the algorithm.

The preprocessor was tested on three different sets of problems. The first test compares the performance of the MCP preprocessor to the one used by the commercial CPLEX code [16]. Using the linear programs contained in NETLIB [13], the first order conditions from linear programming were constructed and given to the MCP preprocessor. CPLEX was given the original linear program. Reported in Tables 7.5 and 7.6 are the sizes of the preprocessed models. CPLEX is capable of performing aggregations, while the MCP preprocessor currently does not. Therefore, in the tables, the size of the model produced by CPLEX both with aggregations (With) and without aggregations (Without) are stated. As evidenced by the table, the MCP preprocessor is competitive with CPLEX on linear programs when aggregations are not allowed. One interesting point to note is that the problems `fit*p` and `fit*d` are primal-dual pairs - the MCP preprocessor generates an identical system in both cases. Exploiting dual information in the `fit*p` problems significantly reduces the size of the preprocessed models.

A second test was performed using quadratic programming problems reformulated using the complementary slackness conditions (7.3). Some artificial quadratic programs were created for testing purposes from the NETLIB collection. A term of $\frac{1}{2}x^T x$ was added to the objective function and the resulting complementary slackness conditions were given to the preprocessor and the PATH algorithm [5]. Table 7.7 reports the size reductions and compares the solution times on the original and preprocessed models. On the problems successfully preprocessed, the reductions in time are significant. Some other quadratic programs from

Table 7.5 Comparison of CPLEX and MCP Preprocessors on NETLIB problems

Model	Size	CPLEX With	CPLEX Without	MCP Preprocessor
adlittle	153	147	147	146
afiro	59	48	52	56
agg	651	271	275	433
agg2	818	530	538	743
agg3	818	531	541	743
bandm	777	398	483	467
beaconfd	435	55	220	205
blend	157	108	140	149
bnl1	1807	1443	1668	1670
bnl2	5769	3031	4226	4341
boeing1	909	711	713	720
boeing2	320	281	281	292
bore3d	547	105	182	191
brandy	431	265	311	311
capri	608	383	547	547
cycle	4743	2700	3416	3884
czprob	4221	2904	3349	3430
d2q06c	7338	6450	6871	6286
d6cube	6588	5844	5867	6423
degen2	978	855	977	974
degen3	3321	3125	3321	3310
dfl001	18301	13062	17091	16915
e226	505	397	411	414
etamacro	1006	754	850	821
fffff800	1378	933	983	1284
finnis	1066	739	786	808
fit1d	1050	1048	1048	1050
fit1p	2304	2054	2054	1050
fit2d	10525	10388	10388	10525
fit2p	16525	16525	16525	10525
forplan	554	466	476	483
ganges	2990	1202	2177	2466
gfrd-pnc	1708	1116	1656	1656
greenbea	7691	4055	5900	5763
greenbeb	7679	4044	5892	5738
grow15	945	945	945	945
grow22	1386	1386	1386	1386
grow7	441	441	441	441
israel	316	304	304	304

Table 7.6 Comparison of CPLEX and MCP Preprocessors on NETLIB (cont.)

Model	Size	CPLEX With	CPLEX Without	MCP Preprocessor
kb2	84	67	79	82
lotfi	461	399	399	408
nesm	3585	3325	3373	3440
perold	1937	1571	1769	1757
pilot4	1380	1111	1200	1210
sc105	207	117	207	207
sc205	407	231	405	405
sc50a	97	57	97	97
sc50b	96	56	96	96
scagr25	971	591	841	734
scagr7	269	159	229	194
scfxm1	787	612	694	698
scfxm2	1574	1228	1388	1396
scfxm3	2361	1844	2082	2094
scorpion	746	172	590	453
scrs8	1659	913	1429	1438
scsd1	837	837	837	817
scsd6	1497	1497	1497	1481
scsd8	3147	3147	3147	3135
sctap1	780	608	608	660
sctap2	2970	2303	2303	2500
sctap3	3960	3111	3111	3340
share1b	342	297	315	310
share2b	175	168	172	172
shell	2061	1427	1935	1935
ship04l	2478	2174	2182	2208
ship04s	1818	1426	1482	1500
ship08l	4995	3569	3569	3611
ship08s	3099	1760	1858	1890
ship12l	6469	4756	4756	4790
ship12s	3805	2114	2258	2288
stair	741	512	740	740
stocfor1	228	113	190	188
stocfor2	4188	2474	3822	3825
tuff	878	514	738	788
wood1p	2838	1898	1898	1971
Total	181745	137202	155734	150753

Table 7.7 Comparison of PATH solution times in seconds on QP models with and without preprocessing.

Model	Original Size	Original Solution Time	Preprocessed Size	Preprocessed Solution Time
agg	651	6.6	454	1.0
beaconfd	435	1.1	283	0.7
finnis	1066	9.1	918	1.5
lotfi	461	5.9	434	0.9
nesm	3585	57.6	3481	53.0
scorpion	746	1.1	617	0.8
ship08s	3099	6.3	1966	3.3
tuff	878	4.3	849	3.9

other sources were also tested. In one of the models, hwayoung, over 70% of the variables were removed by the preprocessor reducing the size from 46123 variables to 13655, leading to a significant reduction in the total time needed to solve the problem.

Finally, the models in MCPLIB [4] were given to the preprocessor. The results on these models are less encouraging than the other two tests. This stems from a lack of linear problems in the test set and the inability to obtain interval evaluations for the nonlinear functions. Many of the models did not benefit from preprocessing. However, some successes are reported in Table 7.8. Note that the `explcp` model that is supposed to display exponential behavior for Lemke's algorithm is completely solved in the preprocessor. The preprocessor for the `golanmcp` model removes 18 redundant rows. The remaining problem solves without any proximal perturbation, leading to the substantial reduction in solution time.

In our initial testing, some of the preprocessing performed was detrimental. For example, the force* models became harder to solve after preprocessing even though they were significantly reduced in size. We added an option to the PATH solver `check_skew_symmetry` that reports extra information to the modeler regarding orientation of equality constraints. This uncovered four models in MCPLIB that have equality constraints not formulated in a skew symmetric sense, namely the electric, force* and lincont models. Rewriting to change this orientation gave the improved timings noted in parentheses in the table.

Overall the results of the preprocessor are very encouraging. Unfortunately, many of the models that are currently in MCPLIB do not have large amounts of exploitable linear structure, so the benefits of prepro-

Table 7.8 Comparison of PATH solution times in seconds on MCP models with and without preprocessing.

	Original		Preprocessed	
Model	Size	Solution Time	Size	Solution Time
electric	158	1.3	149(143)	0.5 (0.1)
explcp	16	0.0	0	0.0
forcebsm	184	0.1	72	0.2 (0.1)
forcedsa	186	0.1	70	0.1 (0.1)
golanmcp	4321	80.9	4303	25.0
merge	9536	2254.6	8417	1954.2

cessing them are limited. Further work in exploiting range constraints and aggregations is warranted. An interval evaluator is planned for the GAMS modeling language which will enable the nonlinear models to be more effectively preprocessed.

Acknowledgments

This material is based on research supported by National Science Foundation Grant CCR-9972372 and Air Force Office of Scientific Research Grant F49620-98-1-0417.

References

[1] E. Andersen and K. Andersen. Presolving in linear programming. *Mathematical Programming*, 71:221–245, 1995.

[2] A. Brearley, G. Mitra, and H. Williams. Analysis of mathematical programming problems prior to applying the simplex algorithm. *Mathematical Programming*, 8:54–83, 1975.

[3] A. Brooke, D. Kendrick, and A. Meeraus. *GAMS: A User's Guide.* The Scientific Press, South San Francisco, CA, 1988.

[4] S. P. Dirkse and M. C. Ferris. MCPLIB: A collection of nonlinear mixed complementarity problems. *Optimization Methods and Software*, 5:319–345, 1995.

[5] S. P. Dirkse and M. C. Ferris. The PATH solver: A non-monotone stabilization scheme for mixed complementarity problems. *Optimization Methods and Software*, 5:123–156, 1995.

[6] M. C. Ferris, R. Fourer, and D. M. Gay. Expressing complementarity problems and communicating them to solvers. *SIAM Journal on Optimization*, 9:991–1009, 1999.

[7] M. C. Ferris, M. P. Mesnier, and J. Moré. NEOS and Condor: Solving nonlinear optimization problems over the Internet. *ACM Transactions on Mathematical Software, forthcoming*, 1999.

[8] M. C. Ferris and T. S. Munson. Case studies in complementarity: Improving model formulation. In M. Thera and R. Tichatschke, editors, *Ill–Posed Variational Problems and Regularization Techniques*, number 477 in Lecture Notes in Economics and Mathematical Systems, pages 79–98. Springer Verlag, Berlin, 1999.

[9] M. C. Ferris and T. S. Munson. Interfaces to PATH 3.0: Design, implementation and usage. *Computational Optimization and Applications*, 12:207–227, 1999.

[10] M. C. Ferris and T. S. Munson. Complementarity problems in GAMS and the PATH solver. *Journal of Economic Dynamics and Control*, 24:165–188, 2000.

[11] M. C. Ferris and J. S. Pang. Engineering and economic applications of complementarity problems. *SIAM Review*, 39:669–713, 1997.

[12] R. Fourer, D. M. Gay, and B. W. Kernighan. *AMPL: A Modeling Language for Mathematical Programming*. Duxbury Press, 1993.

[13] D. M. Gay. Electronic mail distribution of linear programming test problems. *COAL Newsletter*, 13:10–12, 1985.

[14] M. S. Gowda. Applications of degree theory to linear complementarity problems. *Mathematics of Operations Research*, 18:868–879, 1993.

[15] M. S. Gowda. An analysis of zero set and global error bound properties of a piecewise affine function via its recession function. *SIAM Journal on Matrix Analysis and Applications*, 17:594–609, 1996.

[16] ILOG CPLEX Division, 889 Alder Avenue, Incline Village, Nevada. *CPLEX Optimizer*. http://www.cplex.com/.

[17] S. M. Robinson. Normal maps induced by linear transformations. *Mathematics of Operations Research*, 17:691–714, 1992.

[18] S. M. Robinson. A reduction method for variational inequalities. *Mathematical Programming*, 80:161–169, 1998.

[19] R. T. Rockafellar. *Convex Analysis*. Princeton University Press, Princeton, New Jersey, 1970.

[20] M. W. P. Savelsbergh. Preprocessing and probing techniques for mixed integer programming problems. *ORSA Journal on Computing*, 6:445–454, 1994.

[21] H. Sellami and S. M. Robinson. Implementation of a continuation method for normal maps. *Mathematical Programming*, pages 563–578, 1997.

[22] J. Tomlin and J. Welch. Finding duplicate rows in a linear programming model. *Operations Research Letters*, 5(1):7–11, 1986.

Chapter 8

ON THE CONNECTEDNESS OF SOLUTION SETS OF PARAMETRIZED EQUATIONS AND OF SOLUTION SETS IN LINEAR COMPLEMENTARITY PROBLEMS

M. Seetharama Gowda

Department of Mathematics and Statistics, University of Maryland, Baltimore County, Baltimore, MD 21250, USA

gowda@math.umbc.edu

G.S.R. Murthy

Indian Statistical Institute, Street No. 8, Habsiguda, Hyderabad 500 007, India

isihyd@ap.nic.in

T. Parthasarathy

tps@isid.ac.in

Indian Statistical Institute, 7, SJS Sansanwal Marg, New Delhi 110 016, India

Abstract In this article, we prove, under certain conditions, the connectedness of sets of the form $\{x : f(x,y) = 0, y \in E\}$ where f is a function with x varying over an open set in R^n and the parameter y varying over a topological space. Based on this, we show that the partitioned matrix

$$M = \begin{bmatrix} A & B \\ C & D \end{bmatrix}$$

is (LCP) connected (i.e., for all q, the solution set of LCP(q, M) is connected) when $A \in \boldsymbol{P_0} \cap \boldsymbol{Q}$, $C = 0$, and D is connected. We also show that (a) any nonnegative $\boldsymbol{P_0} \cap \boldsymbol{Q_0}$-matrix is connected and (b) any matrix M partitioned as above with C and D nonnegative, and $A \in \boldsymbol{P_0} \cap \boldsymbol{Q}$ is connected.

Keywords: Linear complementarity problem, solution sets, connectedness, weak univalence.

M.C. Ferris et al. (eds.), Complementarity: Applications, Algorithms and Extensions, 165–177.

1 INTRODUCTION

In the first part of the paper, we consider a continuous function $f : \Omega \times Y \to R^n$ where Ω is an open subset of R^n and Y is a topological space. Under the assumptions that the function $x \mapsto f(x,y)$ is weakly univalent and the set $S_y := \{x : f(x,y) = 0\}$ is nonempty and compact for each parameter $y \in Y$, we show that the sets

$$\cup_{y \in E} S_y = \{x : f(x,y) = 0,\ y \in E\} \quad \text{and} \quad \{(x,y) : f(x,y) = 0,\ y \in E\}$$

are connected whenever E is connected in Y. This result extends a result of Gowda and Sznajder [7] proved for a singleton set E. We note that among other things, connectedness results are important and useful in studying the structural and stability properties of equations arising from complementarity and variational inequality problems [11].

The second part of the paper deals with the connectedness of solution sets in linear complementarity problems. Recall that for a given matrix $A \in \boldsymbol{R}^{n \times n}$ and a vector $q \in \boldsymbol{R}^n$, the *Linear Complementarity Problem* LCP(q, A) is to find a vector $z \in \boldsymbol{R}^n$ such that

$$z \geq 0, \quad Az + q \geq 0, \quad \text{and} \quad z^t(Az + q) = 0.$$

In order to present our results in this setting, we shall briefly introduce the notation and background. The reader may refer to [4] for further details. Let $F(q, A) := \{ z \in \boldsymbol{R}^n_+ : Az + q \geq 0 \}$ denote the so-called feasible set and

$$S(q, A) := \{ z \in F(q, A) : z^t(Az + q) = 0 \}$$

denote the solution set of LCP(q, A).

The following are the well known classes of real square matrices in the literature of LCP.

(i) The class $\boldsymbol{Q}$ is the collection of matrices A for which $S(q, A) \neq \emptyset$ for every q;

(ii) $\boldsymbol{Q_0}$ is the collection of matrices A such that $S(q, A) \neq \emptyset$ whenever $F(q, A) \neq \emptyset$;

(iii) $\boldsymbol{P}$ ($\boldsymbol{P_0}$) is the collection of all matrices whose principal minors are positive (nonnegative);

(iv) $\boldsymbol{R_0}$ is the collection of matrices of A such that LCP$(0, A)$ has a unique solution; and

(v) the class of *connected matrices* consists of all matrices A such that $S(q, A)$ is connected for every q.

In [3], Cao and Ferris introduced the class of connected matrices and showed that Lemke's pivotal method, from the perspective of piecewise linear homotopies and normal maps, can process LCPs in which the matrices are connected $\boldsymbol{Q_0}$-matrices. Later, Jones and Gowda [8] showed that $\boldsymbol{P_0} \cap \boldsymbol{R_0}$-matrices (equivalently, $\boldsymbol{P_0} \cap \mathbf{Q}$-matrices) are connected matrices and raised the question whether $\boldsymbol{P_0} \cap \boldsymbol{Q_0}$-matrices are also connected. Recently Murthy, Parthasarathy and Sriparna [10] have given an affirmative answer in the case of nonnegative matrices. In this article we strengthen this result further. In particular, we show, as a consequence of the connectedness result in the first part, that for any $\boldsymbol{P_0} \cap \boldsymbol{Q}$-matrix A, any connected matrix D and matrices B and C of appropriate orders, the block matrix $M = \begin{bmatrix} A & B \\ C & D \end{bmatrix}$ is connected provided C is a null (i.e., zero) matrix. We will also describe related connectedness results.

A brief outline of this paper is as follows. In the next section we shall prove the connectedness of solution sets of parametrized equations. The LCP results will be covered in Section 3.

2 SOLUTION SETS OF PARAMETRIZED EQUATIONS

For our terminology and notation, we closely follow a beautiful monograph on "Topological spaces" by Claude Berge [1]. We urge the readers to refer to [1, 4] for unexplained definitions and other details.

Let X and Y be two topological spaces. We say that a multivalued mapping $\Gamma : X \to Y$ is *upper semicontinuous* if for each $x_0 \in X$ and each open set G containing $\Gamma(x_0)$, there exists a neighborhood $N(x_0)$ of x_0 in X such that $\Gamma(x) \subseteq G$ for every $x \in N(x_0)$.

We first state the following (perhaps well known) elementary result and present a proof for completeness.

Theorem 8.1 *Let X and Y be two topological spaces and $\Gamma : X \to Y$ be an upper semicontinuous multivalued mapping. If $\Gamma(x)$ is connected for each x in X, then for any connected subset E of X, the set $\Gamma(E) := \cup_{x \in E} \Gamma(x)$ is also connected.*

Recall that a set M in a topological space Y is said to be disconnected if there exist nonempty sets C and D in Y such that

$$C \cup D = M \quad \text{and} \quad \overline{C} \cap D = \overline{D} \cap C = \emptyset$$

where $\overline{C}$ stands for closure of C, etc. Equivalently, M is disconnected if M can be written as a disjoint union of two nonempty sets which are open in M. Finally, we say that M is connected if it is not disconnected.

Proof Suppose $\Gamma(E)$ is disconnected. Then $\Gamma(E) = C \cup D$ for some nonempty subsets C and D with $\overline{C} \cap D = C \cap \overline{D} = \emptyset$. Since $\Gamma(x)$ is connected for every $x \in E$, either $\Gamma(x) \subseteq C$ or $\Gamma(x) \subseteq D$. Let

$$A = \{x \in E : \Gamma(x) \subseteq C\} \quad \text{and} \quad B = \{x \in E : \Gamma(x) \subseteq D\}.$$

Clearly $A \neq \emptyset$, $B \neq \emptyset$ and $E = A \cup B$. We now show that A and B are open in E.
Let $p \in A$. Then $\Gamma(p) \subseteq C \subseteq V$, where $V = (\overline{D})'$ is the complement of the set $\overline{D}$. Note that V is open in Y. Since Γ is upper semicontinuous at p, there exists a neighborhood U of p such that for every $x \in U$, $\Gamma(x) \subseteq V$. That is, for every $x \in E \cap U$, $\Gamma(x) \subseteq V$ and hence $\Gamma(x) \subseteq C$. Hence $E \cap U \subseteq A$ which means that A is open in E. Similarly, B is also open in E. Since $E = A \cup B$, it follows that E is disconnected. This contradicts our assumption that E is connected. Hence $\Gamma(E)$ is connected. □

To state the main result of this section, we need the following definition.

Definition 8.2 ([7]) *A continuous function g with domain X and range $g(X)$ in $\boldsymbol{R}^n$ is weakly univalent if there is a sequence of continuous one-to-one functions on X converging to g uniformly on bounded subsets of X.*

Theorem 8.3 *Let $\Omega \subseteq \boldsymbol{R}^n$ be an open set and Y be a topological space. Let*
$f : \Omega \times Y \to \boldsymbol{R}^n$ be a continuous function. Suppose for every $y \in Y$,

(i) the function $x \mapsto f(x, y)$ is weakly univalent and

(ii) the set $S_y = \{x : f(x, y) = 0\}$ is nonempty and compact.

Then, for any connected set $E \subseteq Y$, the sets

$$\cup_{y \in E} S_y = \{x : f(x, y) = 0,\ y \in E\}$$

and

$$\{(x, y) : f(x, y) = 0,\ y \in E\}$$

are connected.

The proof of the above theorem consists in combining Theorem 8.1 with the following two results.

Theorem 8.4 (Gowda-Sznajder [7]) *Let $g : \Omega \to R^n$ be weakly univalent on an open set Ω in R^n. If $\{x \in \Omega : g(x) = 0\}$ is compact, then it is connected.*

Theorem 8.5 (Ravindran-Gowda [11]) *Let $g : \Omega \to R^n$ be weakly univalent on an open set Ω in R^n. If $\{x \in \Omega : g(x) = 0\}$ is nonempty and compact, then for each small $\varepsilon > 0$, there exists a $\delta > 0$ such that*

$$\emptyset \neq \{x \in \Omega : h(x) = 0\} \subseteq \{x \in \Omega : g(x) = 0\} + \varepsilon B$$

for all h weakly univalent on Ω with

$$\sup_{\overline{V}} ||h(x) - g(x)|| < \delta$$

where $V := \{x \in \Omega : g(x) = 0\} + \varepsilon B$ and B is the open unit ball in R^n.

Proof Let Γ be the multivalued mapping from $Y \to \Omega$ defined by $\Gamma(y) = S_y$ where $S_y = \{x \in \Omega : f(x, y) = 0\}$. By our assumption, S_y is compact for each $y \in Y$. By the Gowda-Sznajder result, we see the connectedness of $\Gamma(y)$. The upper semicontinuity property of Γ at any $y^* \in Y$ follows from the Ravindran-Gowda result. Invoking Theorem 8.1, we get the first part of the conclusion. The second part is seen by applying Theorem 8.1 to the upper semicontinuous mapping $y \mapsto S_y \times \{y\}$. This completes the proof. □

3 LCP CONNECTEDNESS RESULTS

Recall that a matrix $M \in R^{n \times n}$ is connected if for each $q \in R^n$, the solution set $S(q, M)$ of LCP(q, M) is connected. In [8], Jones and Gowda proved the following.

Theorem 8.6 (Jones-Gowda) *(a) Suppose M is a $\mathbf{P_0}$-matrix. If the solution set $S(q, M)$ of LCP(q, M) (for some q) is nonempty and bounded, then it is connected.*

(b) Every $\boldsymbol{P_0} \cap \boldsymbol{R_0}$-matrix is connected.

In view of the equality $\boldsymbol{P_0} \cap \boldsymbol{R_0} = \boldsymbol{P_0} \cap \boldsymbol{Q}$ (Theorem 3.9.22, [4]), Item (b) above reads: Every $\boldsymbol{P_0} \cap \boldsymbol{Q}$-matrix is connected. In [8], Jones and Gowda raised the question whether every $\boldsymbol{P_0} \cap \boldsymbol{Q_0}$-matrix is also connected. While we do not have a complete answer to this question, in what follows, we show that certain matrices in (and in some cases, outside) $\boldsymbol{P_0} \cap \boldsymbol{Q_0}$ are connected.

Before presenting our new results, we list some well known subclasses of $\boldsymbol{P_0}$ (apart from $\boldsymbol{P_0} \cap \boldsymbol{Q}$) [4] that consists of connected matrices:

(1) **Column sufficient matrices.** These are matrices satisfying the condition

$$x * (Mx) \leq 0 \Rightarrow x * (Mx) = 0$$

where $x * (Mx)$ denotes the componentwise product of vectors x and Mx. An equivalent formulation (Theorem 3.5.8, [4]) is: For all q, $S(q, M)$ is convex. Clearly, column sufficient matrices are connected.

(2) **Sufficient matrices.** A matrix M is sufficient if M and M^T are column sufficient. Sufficient matrices are connected and contained in $\boldsymbol{P_0} \cap \boldsymbol{Q_0}$ (Section 5, [4]).

(3) **Positive semidefinite matrices.** These are matrices satisfying the condition

$$x^t(Mx) \geq 0 \quad \text{for all} \quad x.$$

It is well known that positive semidefinite matrices are sufficient.

(4) **$\mathbf{P_1}$-matrices.** These are $\mathbf{P_0}$-matrices in which exactly one principal minor is zero. It is known (see Corollary 4.1.11, [4]) that $\mathbf{P_1}$-matrices are sufficient and hence belong to $\boldsymbol{P_0} \cap \boldsymbol{Q_0}$.

(5) **P-matrices.** While the definition says that every principal minor of such a matrix is positive, an equivalent formulation is: For every q, $S(q, M)$ is a singleton set. Trivially, **P**-matrices are sufficient and connected.

Remark 8.7 *We make the following observation regarding (column) sufficient matrices. Let*

$$M = \begin{bmatrix} 0 & 0 & b^t \\ 0 & 0 & b^t \\ c & c & D \end{bmatrix}$$

be in $R^{(k+2)\times(k+2)}$ with $b, c \in R^k$, and $D \in R^{k\times k}$. If the submatrix N obtained by deleting the first row and first column in M is (column) sufficient, then M is (column) sufficient and hence connected. (To see this, suppose N is column sufficient. If $x * (Mx) \leq 0$ with $x^t = (x_1, x_2, u) \in R \times R \times R^k$, then $y * (Ny) \leq 0$ where $y = (x_1 + x_2, u)^t$. It follows that $y * (Ny) = 0$ and hence $x * (Mx) = 0$. When N is sufficient, we work with M and M^T.)

In what follows, we present new LCP connectedness results. We first begin with an application of Theorem 8.3.

Theorem 8.8 *Consider a partitioned matrix of the form*

$$M = \begin{bmatrix} A & B \\ 0 & D \end{bmatrix}$$

where $A \in R^{m\times m}$ is a $\boldsymbol{P_0} \cap \boldsymbol{Q}$-matrix, $D \in R^{n\times n}$ is a connected matrix, and B is arbitrary. Then M is connected.

Proof In view of the Jones-Gowda result mentioned above, we may assume that both m and n are positive. Let $q \in \boldsymbol{R}^{(m+n)}$ with $S(q, M) \neq \emptyset$. We write $q = (r, s)^t$ with $r \in \boldsymbol{R}^m$ and $s \in \boldsymbol{R}^n$. Using the symbol '$\wedge$' for the componentwise minimum, we see that

$$\begin{aligned} S(q, M) &= \{(x, y)^t : x \wedge (Ax + By + r) = 0 \text{ and } y \wedge (Dy + s) = 0\} \\ &= \{(x, y)^t : x \wedge (Ax + By + r) = 0 \text{ and } y \in S(s, D)\} \\ &= \{(x, y)^t : f(x, y) = 0,\ y \in S(s, D)\} \end{aligned}$$

where $f(x, y) := x \wedge (Ax + By + r)$. Since A is a $\boldsymbol{P_0}$-matrix, for each y, the function $f(\cdot, y)$ is weakly univalent (see the proof of Theorem 7 in [7]). Since A is also a $\boldsymbol{Q}$-matrix, the set $S(By + r, A) \neq \emptyset$ for $y \in S(s, D)$. Now A, being a $\boldsymbol{P_0} \cap \boldsymbol{Q}$-matrix, is an $\boldsymbol{R_0}$-matrix [4]. This gives the compactness of $S(By + r, A)$. Thus for each $y \in S(s, D)$, $S_y = \{x : f(x, y) = 0\}$ is nonempty and compact. Since by assumption, $S(s, D)$ is connected, we can invoke Theorem 8.3 to get the desired result.

□

We now consider two examples.

Example 8.9 *Let*

$$M = \begin{bmatrix} 0 & 1 & 1 \\ 0 & 0 & 1 \\ 0 & 0 & 0 \end{bmatrix} \quad \textit{and} \quad q^* = \begin{bmatrix} -1 \\ 0 \\ 0 \end{bmatrix}.$$

Stone, see [8], has shown that $S(q^*, M)$ is not connected. In the partitioned form, $C = [0, 0]$ and $D = 0$. The matrix M (hence A) is $\boldsymbol{P_0}$. So this example shows that the conclusion of Theorem 8.8 may not hold if the condition $A \in \boldsymbol{P_0} \cap \boldsymbol{Q}$ is weakened to $A \in \boldsymbol{P_0}$.

Example 8.10 *Let*

$$M = \begin{bmatrix} A & b^t \\ c & D \end{bmatrix}$$

where

$$A = [1],\ b = c = (-1, 0, 0)^t,\ \textit{and}\ D = \begin{bmatrix} 1 & 1 & 1 \\ 0 & 0 & 1 \\ 0 & 0 & 0 \end{bmatrix}.$$

Note that A is a $\boldsymbol{P}$-matrix, D is connected and nonnegative, and c is nonzero. We consider the principal pivotal transform (Section 2.3, [4]) H of M with respect to A. Then

$$H = \begin{bmatrix} 1 & 1 & 0 & 0 \\ -1 & 0 & 1 & 1 \\ 0 & 0 & 0 & 1 \\ 0 & 0 & 0 & 0 \end{bmatrix}.$$

(Note that H is a slightly modified version of Stone's example, see [8]). For $q = (0, -1, 0, 0)^t$, we can check that $S(q, H)$ is a disconnected set. Since connectedness is preserved under principal pivotal transformations, we see that M is also not connected. This example shows that the conclusion of Theorem 8.8 may not hold if the zero block in M (of Theorem 8.8) is replaced by a nonzero matrix.

We now deduce a result of Murthy, Parthasarathy, and Sriparna [10] as a consequence of Theorem 8.8.

Corollary 8.11 *Let $M \in \boldsymbol{R}^{k \times k}$ be a nonnegative $\boldsymbol{P_0} \cap \boldsymbol{Q_0}$-matrix. Then M is connected.*

Proof It is well known that a nonnegative matrix with positive diagonal is a $\boldsymbol{Q}$-matrix. If M has a positive diagonal, then it is a $\boldsymbol{P_0} \cap \boldsymbol{Q}$-matrix and hence connected by the Jones-Gowda result mentioned before. If M has a zero diagonal entry, say $m_{ii} = 0$, then from Theorem 2.5 of [9], we see that the ith row of M is zero. Consequently, when M has some zero diagonal entries, we can partition M as

$$M = \begin{bmatrix} A & B \\ 0 & 0 \end{bmatrix},$$

where A (if non-vacuous) is a $\boldsymbol{P_0} \cap \boldsymbol{Q}$-matrix. We can now invoke Theorem 8.8 to get the desired result. □

A slightly stronger version of the above corollary can be obtained in the following way. We first recall the definition of a *chain* and a related result from [2].

Definition 8.12 *Suppose $A \in \boldsymbol{R}^{n \times n}$. We say that A has a chain if there exist distinct indices i_1, i_2, ..., i_k such that*

(i) $a_{i_1 i_2}$ is the only positive entry in $A_{i_1 \cdot}$,

(ii) for each $j \in \{2, 3, \ldots, k-1\}$, $a_{i_j i_{j+1}}$ is the only positive (or the only negative) entry in row $A_{i_j \cdot}$,

(iii) $A_{i_k \cdot} \geq 0$.

Proposition 8.13 *Suppose $A \in \boldsymbol{R}^{n \times n} \cap \boldsymbol{Q_0}$. Then A cannot have any chain.*

Proof Suppose A has a chain $i_1, i_2, \ldots, i_k$. Let m be the largest index in $\{1, 2, \ldots, k-1\}$ with $a_{i_m i_{m+1}} > 0$, so that either $m = k-1$ or $m < k-1$ in which case, $a_{i_{m+1} i_{m+2}}, a_{i_{m+2} i_{m+3}}, \ldots, a_{i_{k-1} i_k}$ are all negative. Define $q \in \boldsymbol{R}^n$ with $q_{i_m} = -1$, $q_j = \lambda$ for $j \neq i_m$. It is easy to check that for all large (positive) values of λ, LCP(q, A) has a feasible solution but no complementary solution. This contradicts that $A \in \boldsymbol{Q_0}$. Thus A cannot have any chain. □

Remark 8.14 *The proof of the above proposition given in [2] is incorrect and needs a minor modification. For the matrix*

$$A = \begin{bmatrix} -1 & 1 & -1 \\ -1 & -1 & 1 \\ 0 & 0 & 0 \end{bmatrix}$$

and the vector $q = (-1, \ -\lambda, \ 1)^t$, according to the proof given in [2], LCP(q, A) should have feasible solutions for all sufficiently large positive λ. This is clearly not true. However, as in the new proof above, if we take $q = (\lambda, \ -1, \ \lambda)^t$ where λ is a large positive number, then LCP(q, A) has a feasible solution but no complementary solution. Therefore, A is not a $\boldsymbol{Q_0}$-matrix.

Corollary 8.15 *Let $M = \begin{bmatrix} A & B \\ C & D \end{bmatrix} \in \boldsymbol{R}^{n \times n}$ be a $\boldsymbol{P_0} \cap \boldsymbol{Q_0}$-matrix. Assume that A is a $\boldsymbol{Q}$-matrix and that C and D are nonnegative matrices. Then M is a connected matrix.*

Proof Let G be the largest principal submatrix of M which is a $\boldsymbol{Q}$-matrix and which contains A as a principal submatrix. We may write without loss of generality,

$$M = \begin{bmatrix} G & J \\ H & L \end{bmatrix}.$$

Note that H and L are nonnegative. Suppose, if possible, H has a nonzero row, say the first row $H_{1.}$. Then

$$N := \begin{bmatrix} G & J_{.1} \\ H_{1.} & l_{11} \end{bmatrix}$$

is a $\boldsymbol{Q_0}$-matrix by repeated applications of Corollary 2.3 in [9]. Since G is a $\boldsymbol{Q}$-matrix and $H_{1.}$ is nonzero and nonnegative, it is easily seen that N is an **S**-matrix. Now N, being a $\boldsymbol{Q_0} \cap \mathbf{S}$-matrix, is a $\boldsymbol{Q}$-matrix (see Page 146, [4]). But this contradicts the definition of G. It follows that $H = 0$. If L has a nonzero diagonal entry, then we can derive a similar contradiction. Thus the diagonal of L is zero. If L has a (positive) off-diagonal entry, we consider a principal submatrix of M of the form

$$\overline{N} := \begin{bmatrix} G & R \\ 0 & S \end{bmatrix}$$

where S is a nonzero 2×2 principal submatrix of L. Once again, by repeated applications of Corollary 2.3 in [9], we see that $\overline{N}$ is a $\boldsymbol{Q_0}$-matrix. However, by considering the last two rows of $\overline{N}$, we can produce a chain in $\overline{N}$ contradicting Proposition 8.13. Thus, $L = 0$. (Alternatively, we can show that $L = 0$ by noting that $L \in \boldsymbol{Q_0}$ (since $M \in \boldsymbol{Q_0}$ and $H = 0$) and nonnegative, and by applying Theorem 2.5 in [9].) The result now follows from Theorem 8.8. □

Theorem 8.16 *Let $M \in \boldsymbol{R}^{n \times n} \cap \boldsymbol{P_0}$ and π, $\delta \in \boldsymbol{R}^n$ be two strictly positive vectors with $\delta^t M = 0$ and $M\pi = 0$. Further, suppose rank of M is $n-1$. Then $M \in \boldsymbol{P_0} \cap \boldsymbol{Q_0}$ and connected.*

Proof It follows from Theorem 8.6, [6] that M is a $\boldsymbol{Q_0}$-matrix. Consider a q with $S(q, M)$ nonempty. If $\overline{x} \in S(q, M)$, then $M\overline{x} + q \geq 0$ so that

$$\delta^t q = \delta^t M\overline{x} + \delta^t q = \delta^t (M\overline{x} + q) \geq 0.$$

We consider two cases.
Case (i): Suppose $\delta^t q > 0$.

If we show $S(q, M)$ is a bounded set, connectedness of $S(q, M)$ follows from the Jones-Gowda result mentioned earlier. Suppose $S(q, M)$ is unbounded; let $\{x^k\}$ be an unbounded sequence in $S(q, M)$ and (without loss of generality) $y^k := x^k/||x^k|| \to y^0$. From the inequalities

$$x^k \geq 0 \quad \text{and} \quad Mx^k + q \geq 0,$$

we see that y^0 is a nonnegative vector with unit norm and $My^0 \geq 0$. Since δ is a strictly positive vector with $\delta^t M = 0$, we have $My^0 = 0$. Since rank of M is $n - 1$, y^0 is a multiple of π and hence must also be strictly positive. Thus, x^k is a strictly positive vector for large k. Then by complementarity, for all k sufficiently large, we have $Mx^k + q = 0$ and so $\delta^t (Mx^k + q) = 0$. However,

$$\delta^t (Mx^k + q) = \delta^t Mx^k + \delta^t q = \delta^t q > 0$$

is a contradiction.

Case (*ii*): Suppose $\delta^t q = 0$.

Let $x \in S(q, M)$. Then

$$Mx + q = w \geq 0 \text{ with } x^t w = 0.$$

Now, from $\delta^t M = 0$ and $\delta^t q = 0$, we get

$$\delta^t w = \delta^t(Mx + q) = \delta^t Mx + \delta^t q = 0.$$

Since δ is a strictly positive vector, $w = 0$. In other words, $S(q, M) = \{x \geq 0 : Mx + q = 0\}$. This set is convex and hence connected. This completes the proof. □

Remark 8.17 *In the proof of Theorem 8.16, we showed that $\delta^t q \geq 0$ whenever $S(q, M)$ is nonempty. Letting $K(M)$ denote the set of all q's for which $S(q, M)$ is nonempty, we conclude that $K(M) \subseteq \{q \in \boldsymbol{R}^n : \delta^t q \geq 0\}$. It turns out that the reverse inclusion holds so that*

$$K(M) = \{q \in \boldsymbol{R}^n : \delta^t q \geq 0\}.$$

To see this, fix a q with $\delta^t q \geq 0$ and suppose $q \notin K(M)$. Since M is a $\mathbf{Q_0}$-matrix, $K(M) = pos(I, -M)$ (from Proposition 3.2.1, [4]). Now from a separation theorem (see Corollary 2.7.3, [4]) applied to $q \notin pos(I, -M)$, we get a nonzero vector $d \in R^n$ such that

$$d^t q < 0,\ d \geq 0, \text{ and } \ d^t M \leq 0.$$

From $d^t M \leq 0$, $\pi > 0$ and $M\pi = 0$, we get $d^t M = 0$. Since M is of rank $n-1$ and $\delta^t M = 0$, we see that $d = \lambda\delta$ for some scalar λ. Since both d and δ are nonzero nonnegative vectors, λ is positive. But then we cannot have $\delta^t q \geq 0$ and $d^t q < 0$ simultaneously, leading to a contradiction. Hence $\delta^t q \geq 0$ implies that $q \in K(M)$, proving the required equality. We note here that Cottle and Stone [5] prove the above equality under the (stronger) assumption that $M \in \mathbf{P_1} \setminus \mathbf{Q}$ with $\det(M) = 0$. The above argument (similar to the one in the final remark of [5]) was suggested by a referee.

Here is an example of a matrix satisfying the hypothesis of the above theorem.

Example 8.18 *Let*

$$M = \begin{bmatrix} 1 & 1 & -1 & -1 \\ 1 & 1 & -2 & 0 \\ 1 & 2 & 1 & -4 \\ -3 & -4 & 2 & 5 \end{bmatrix}.$$

Note $e^t M = (0, 0, 0, 0)$ and $Me = (0, 0, 0, 0)^t$ where $e = (1, 1, 1, 1)^t$. Also, rank of M is 3.

4 CONCLUDING REMARKS

In this paper, we have described a connectedness result for the solution set defined by parametrized equations. We have proved that certain subclasses of $\boldsymbol{P_0} \cap \boldsymbol{Q_0}$ contain connected matrices. Prompted by these results, we (re)state the question raised by Jones and Gowda in the form of a

Conjecture: *Every* $\boldsymbol{P_0} \cap \boldsymbol{Q_0}$*-matrix is connected.*

Acknowledgments

Part of this work was carried out when the third author visited the department of Mathematics and Statistics, University of Maryland, Baltimore County during May-June 1999. He would like to thank the department for its kind hospitality and Dr. B. Sriparna for some useful comments. Our special thanks are due to the referees for their detailed suggestions and for pointing out a flaw in the previous proof of Proposition 8.13 (which was based on [2]). The example cited in Remark 8.14 is due to one of the referees.

References

[1] C. Berge, Topological Spaces, Oliver and Boyd, First English Edition, Edinburgh and London, 1963.

[2] A.K. Biswas and G.S.R. Murthy, "A chain condition for $\boldsymbol{Q_0}$-matrices," in Game Theoretical Applications To Economics And Operations Research, Kluwer Academic Publishers, Netherlands, pp. 149-152, 1997.

[3] M. Cao and M.C. Ferris, "$\boldsymbol{P_c}$-matrices and the linear complementarity problem," Linear Algebra and its Applications, vol. 246, pp. 231-249, 1996.

[4] R.W. Cottle, J.S. Pang and R.E. Stone, The Linear Complementarity Problem, Academic Press, New York, 1992.

[5] R.W. Cottle and R.E. Stone, "On the uniqueness of solutions to linear complementarity problems," Mathematical Programming, vol. 27, pp. 191-213, 1983.

[6] N. Eagambaram and S.R. Mohan, "On some classes of linear complementarity problems with matrices of order n and rank $(n-1)$," Mathematics of Operations Research, vol. 15, pp. 243-257, 1990.

[7] M.S. Gowda and R. Sznajder, "Weak univalence and connectedness of inverse images of continuous functions," Mathematics of Operatations Research, vol. 24, pp. 255-261, 1999.

[8] C. Jones and M.S. Gowda, "On the connectedness of solution sets of linear complementarity problems," Linear Algebra and its Applications, vol. 272, pp. 33-44, 1998.

[9] G.S.R. Murthy and T. Parthasarathy, "Some properties of fully semimonotone $\boldsymbol{Q_0}$-matrices," SIAM Journal on Matrix Analysis and Applications, vol. 16(4), pp. 1268-1286, 1995.

[10] G.S.R. Murthy, T. Parthasarathy and B. Sriparna, "On the solution sets of linear complementarity problems," SIAM Journal on Matrix Analysis and Applications, to appear.

[11] G. Ravindran and M.S. Gowda, "Regularization of P_0-functions in box variational inequality problems," Research Report, Department of Mathematics and Statistics, University of Maryland Baltimore County, Baltimore, Maryland 21250, October 9, 1997.

Chapter 9

AN ACTIVE SET-TYPE NEWTON METHOD FOR CONSTRAINED NONLINEAR SYSTEMS

Christian Kanzow
Institute of Applied Mathematics, University of Hamburg, Bundesstrasse 55, 20146 Hamburg, Germany
kanzow@math.uni-hamburg.de

Abstract We consider the problem of finding a solution of a nonlinear system of equations subject to some box constraints. To this end, we introduce a new active set-type Newton method. This method is shown to be globally convergent in the sense that every accumulation point is a stationary point of a corresponding box constrained optimization problem. Moreover, the method is locally superlinearly or quadratically convergent under a suitable regularity condition. Furthermore the method generates feasible iterates and has to solve only one linear system of equations at each iteration. Due to our active set strategy, this linear system is of reduced dimension. Some preliminary numerical results are included.

Keywords: Nonlinear equations, box constraints, Newton's method, active set strategy, projected gradient, global convergence, quadratic convergence

1 INTRODUCTION

The problem we address in this paper is to find a solution of the constrained nonlinear system

$$F(x) = 0, \quad x \in [l, u], \tag{9.1}$$

where $F : [l, u] \to \boldsymbol{R}^n$ is a given function which is assumed to be continuously differentiable in an open set containing the box $[l, u]$, and where

$$l = (l_1, \dots, l_n)^T, \quad u = (u_1, \dots, u_n)^T$$

M.C. Ferris et al. (eds.), Complementarity: Applications, Algorithms and Extensions, 179–200.

with $l_i \in \boldsymbol{R} \cup \{-\infty\}$ and $u_i \in \boldsymbol{R} \cup \{+\infty\}$ for $i = 1, \ldots, n$ are given lower and upper bounds satisfying $l_i < u_i$ for all $i = 1, \ldots, n$.

Our motivation for studying this type of problem comes from different areas:

1. There are a couple of different mathematical programming problems like Karush-Kuhn-Tucker systems or complementarity problems which immediately result into a problem of the kind (9.1), see [23] for some more details.

2. The component functions F_i of the mapping F in (9.1) are not always defined on the whole space $\boldsymbol{R}^n$, so we are urged to restrict the domain of F, and this can usually be done by putting suitable bounds on some or all of the variables.

3. In a couple of applications, one already has an idea to which area a solution of an unconstrained nonlinear system $F(x) = 0$ belongs to; in this case, it might be very helpful to have the possibility to describe this area by some lower and upper bounds on the variables.

In view of these comments, we think that a constrained nonlinear system of the type (9.1) is a very important problem. Of course, it is possible to solve this problem by reformulating it either as a box constrained optimization problem or as a mixed complementarity problem and then applying some standard software to this reformulated problem. However, such an approach does not seem to take into account the particular structure of the constrained nonlinear system (9.1) and, therefore, does not seem to be an adequate way for solving it.

Hence we believe that it is very important to design algorithms for problem (9.1) which exploit the special properties of this system. Surprisingly, it seems that there is not a rich literature on this topic. In fact, we are only aware of the papers [16, 23, 22, 21]. In [16], Kozakevich, Martínez and Santos describe an inexact Gauss-Newton-type method for constrained nonlinear systems of equations, while Wang, Monteiro and Pang design a potential reduction-type interior-point method in [23]. Both references allow more general constraints than box constraints, but are completely different from our method. Somewhat closer to our approach is the trust-region framework used by Ulbrich [22] as well as the Newton-type method presented by Sun [21] on a recent conference.

However, the method we are going to present here is the only one using an active-set strategy. As a consequence, we only have to deal with linear systems of reduced dimension. This can be a significant advantage for large-scale problems since reduced problems can be solved much more efficiently than full dimensional systems.

Note that, in contrast to some of the above mentioned papers, we only deal with nonlinear systems having box constraints here. This is partially due to our motivation (the problems discussed there can usually be solved by introducing bounds on the variables) and partially due to the way we will try to solve the constrained nonlinear system (9.1) (e.g., we will compute projections onto the feasible set which is a trivial task for box constraints but becomes a much more difficult task for more general constraints). However, we stress that it is easily possible to deal also with, e.g., general inequality constraints. To this end, one just has to add a nonnegative slack variable in order to get a problem of the form (9.1) with one more equation and one more variable.

The algorithm which we will develop in this paper has a number of desirable properties: First, it generates feasible iterates only. This is highly important in view of the motivation we gave directly after the statement of problem (9.1). Second, it solves at each iteration only one linear system, and this linear system is actually of reduced dimension. Finally, the algorithm is globally and, under certain assumptions, locally superlinearly/quadratically convergent.

Basically, the idea of our method is to solve the unconstrained nonlinear system $F(x) = 0$ by using Newton's method. We then use some recent ideas from the identification of active constraints in order to take into account the box constraints. This, more or less, gives us a feasible and locally fast convergent method. In order to make it globally convergent, we will combine it in a simple way with a projected gradient method.

The organization of this paper is as follows: Section 2 gives a detailed statement of our algorithm and contains further motivation for the design of this method. In Section 3, we then investigate the global and local convergence properties of our algorithm. Some numerical results are given in Section 4, and we conclude this paper with some final remarks in Section 5.

Notation: If $x \in \boldsymbol{R}^n$ is any given vector with components x_i and $\mathcal{K} \subseteq \{1, \ldots, n\}$ is any given subset, we write $x_{\mathcal{K}}$ for the subvector having components $x_i, i \in \mathcal{K}$. Similarly, if $A \in \boldsymbol{R}^{n \times n}$ is a given matrix with entries a_{ij} and $\mathcal{K}, \mathcal{L} \subseteq \{1, \ldots, n\}$ are two subsets, we denote by $A_{\mathcal{KL}}$ the $|\mathcal{K}| \times |\mathcal{L}|$ submatrix with elements $a_{ij}, i \in \mathcal{K}, j \in \mathcal{L}$. The Jacobian of our mapping $F : [l, u] \to \boldsymbol{R}^n$ is denoted by $F'(x)$; F_i is the ith component function of F, and $\nabla F_i(x)$ is the gradient of F_i evaluated at the point x. Note that, throughout this manuscript, we view all gradient vectors as column vectors. Hence, the rows of the Jacobian $F'(x)$ are the transposed gradients of the component functions. Furthermore, we denote by $\|\cdot\|$ the Euclidean vector norm and its corresponding matrix norm. Finally, the projection of a vector $y \in \mathbb{R}^n$ on the feasible set $[l, u]$ is denoted by y_+.

2 ALGORITHM

In this section, we give a detailed description of our algorithm for the solution of the constrained nonlinear system (9.1). Basically, our algorithm is a Newton-type method for the solution of the nonlinear system

$$F(x) = 0$$

which incorporates an active-set strategy in order to stay feasible with respect to the box $[l, u]$ and which measures progress by minimizing the related optimization problem

$$\min f(x) \quad \text{s.t.} \quad x \in [l, u], \tag{9.2}$$

where, throughout this paper, $f : [l, u] \to \mathbb{R}$ is defined by

$$f(x) := \frac{1}{2} F(x)^T F(x) = \frac{1}{2} \|F(x)\|^2.$$

The precise statement of our algorithm is as follows.

Algorithm 9.1 *(Active Set-type Newton Method)*

(S.0) (Initialization)
Choose $x^0 \in [l, u], \beta, \sigma \in (0, 1), \gamma \in (0, 1), s > 0, \delta > 0$ *sufficiently small (see (9.5) below),* $c > 0$*, and set* $k := 0$.

(S.1) (Termination Criterion)
If x^k *is a stationary point of (9.2): STOP.*

(S.2) (Active Set Strategy)
Define

$$\delta_k := \min\{\delta, c\sqrt{\|F(x^k)\|}\}$$

and

$$\begin{aligned} \mathcal{A}_k &:= \{i \mid x_i^k - l_i \le \delta_k \text{ or } u_i - x_i^k \le \delta_k\}, \\ \mathcal{I}_k &:= \{1, \dots, n\} \setminus \mathcal{A}_k. \end{aligned}$$

(S.3) (Search Direction Calculation)
Try to compute a vector $d^k \in \mathbb{R}^n$ *in the following way: For* $i \in \mathcal{A}_k$*, set*

$$d_i^k := \begin{cases} l_i - x_i^k & \text{if } x_i^k - l_i \le \delta_k, \\ u_i - x_i^k & \text{if } u_i - x_i^k \le \delta_k, \end{cases}$$

then solve the linear system

$$F'(x^k)_{\mathcal{I}_k\mathcal{I}_k} d_{\mathcal{I}_k} = -F(x^k)_{\mathcal{I}_k} - F'(x^k)_{\mathcal{I}_k\mathcal{A}_k} d^k_{\mathcal{A}_k} \tag{9.3}$$

in order to get the components d_i^k for $i \in \mathcal{I}_k$.

(S.4) (Feasibility Preserving Safeguard)
Compute

$$\begin{aligned} \bar{\tau}_k &:= \sup\{\tau > 0 \,|\, x_i^k + \tau d_i^k \in [l_i, u_i] \ \forall i \in \mathcal{I}_k\}, \\ \tau_k &:= \min\{1, \bar{\tau}_k\}. \end{aligned}$$

(S.5) (Computation of New Iterate)
If the computation of d^k in Step (S.3) was successful and d^k satisfies the descent condition

$$f(x^k + \tau_k d^k) \le \gamma f(x^k), \tag{9.4}$$

set $x^{k+1} := x^k + \tau_k d^k$; otherwise use a projected gradient step, i.e., set $x^{k+1} := x^k(t_k)$, where $x^k(t) := [x^k - t\nabla f(x^k)]_+$ and $t_k =$ $\max\{s\beta^\ell \,|\, \ell = 0, 1, 2, \dots\}$ *such that*

$$f(x^k(t_k)) \le f(x^k) - \sigma \nabla f(x^k)^T (x^k - x^k(t_k)).$$

(S.6) (Update)
Set $k \leftarrow k + 1$, and go to (S.1).

Throughout this paper, we assume that the constant $\delta > 0$ is chosen sufficiently small such that

$$\delta < \frac{1}{2} \min_{i=1,\dots,n} |u_i - l_i|. \tag{9.5}$$

This, in particular, implies that we cannot have $x_i^k - l_i \le \delta_k$ and $u_i - x_i^k \le \delta_k$ for the same index $i \in \mathcal{A}_k$, so there is no ambiguity in the definition of the components d_i^k for $i \in \mathcal{A}_k$ in Step (S.3) of Algorithm 9.1.

Let us now give some further comments on each step of Algorithm 9.1 in order to get a better understanding of the method.

Apart from some parameters which have an obvious meaning, we only assume in Step (S.0) that we have a feasible starting point x^0. Of course, finding such a starting point is a trivial task due to our simple constraints. Step (S.1) contains a standard termination criterion (which will be modified in our actual implementation).

In Step (S.2), we define a set $\mathcal{A}_k$ which is viewed as a kind of active set: It contains all those indices which, at a current iterate x^k, are close to the boundary (but not necessarily equal to one of the bounds). Later, we will see that the set $\mathcal{A}_k$ is eventually equal to the set of active constraints at a solution x^* of the constrained nonlinear system (9.1) if this solution satisfies an appropriate regularity condition. For some further motivation regarding the definition of $\mathcal{A}_k$, we refer the interested reader to the paper [7].

In Step (S.3) of Algorithm 9.1, we then try to compute a Newton-type search direction d^k. For the components d_i^k with $i \in \mathcal{A}_k$, we use a simple formula whose aim is, basically, to push the corresponding components of our iterates to come closer to the boundary. This makes sense since, due to our current information, we believe that these components will be active at a solution. In order to understand the formula for the computation of the components d_i^k for $i \in \mathcal{I}_k$, note that, after a possible permutation of the rows and columns, we can rewrite the standard (unconstrained) Newton equation $F'(x^k)d = -F(x^k)$ as

$$\begin{pmatrix} F'(x^k)_{\mathcal{I}_k\mathcal{I}_k} & F'(x^k)_{\mathcal{I}_k\mathcal{A}_k} \\ F'(x^k)_{\mathcal{A}_k\mathcal{I}_k} & F'(x^k)_{\mathcal{A}_k\mathcal{A}_k} \end{pmatrix} \begin{pmatrix} d_{\mathcal{I}_k} \\ d_{\mathcal{A}_k} \end{pmatrix} = - \begin{pmatrix} F(x^k)_{\mathcal{I}_k} \\ F(x^k)_{\mathcal{A}_k} \end{pmatrix}. \tag{9.6}$$

Now it is easy to see that the linear system (9.3) is nothing else but the first block row in (9.6) (recall that we already have an expression for $d_{\mathcal{A}_k}^k$). Note that our Newton-type search direction is somewhat different from those suggested in some projected Newton methods for constrained optimization problems in, e.g., [1, 17]. In particular, there are two main differences if compared with Bertsekas' projected Newton method from [1] (also described in the recent book [15]): On the one hand, our method is based on a different active set strategy, and on the other hand, the second term on the right-hand side of the linear system (9.3) does not occur in Bertsekas' method. These two changes will allow us to prove local fast convergence of our method also to degenerate solutions (in contrast to [1]).

In Step (S.4), we compute the largest possible step $\tau \in [0, 1]$ such that the point $x^k + \tau d^k$ would still be feasible. Note that there is a simple closed-form expression for the value of $\tau = \tau_k$, namely

$$\tau_k = \min\{1, \bar{\tau}_k\}$$

with

$$\bar{\tau}_k := \min \left\{ \min_{\{i \in \mathcal{I}_k \,|\, d_i^k < 0\}} \frac{l_i - x_i^k}{d_i^k}, \min_{\{i \in \mathcal{I}_k \,|\, d_i^k > 0\}} \frac{u_i - x_i^k}{d_i^k} \right\}.$$

Here, we assume that $\bar{\tau}_k = +\infty$ if the minimum is taken over the empty set.

Step (S.5) of Algorithm 9.1 then tries to accept the point $x^k + \tau_k d^k$ as the new iterate x^{k+1} if this gives a sufficient decrease in the objective function f in the sense that condition (9.4) is satisfied. From a practical point of view, this test is somewhat restrictive and will be replaced by some kind of arc search, i.e., in practice we will try to find a stepsize $t_k \in [0, 1]$ which minimizes the value of f along the projected arc $[x^k + td^k]_+$ approximately (note that this is a piecewise linear arc). As long as the step t_k is chosen in such a way that

$$f([x^k + t_k d^k]_+) \leq f(x^k + \tau_k d^k),$$

this modification does not destroy the global and local convergence properties of Algorithm 9.1. If, however, the test (9.4) (or its modification just described) is not satisfied either because the search direction d^k is a bad direction or because we were not able to compute d^k in Step (S.3), we switch to a projected gradient step, cf. [11, 3, 2] for further details. (Note that the projections onto $[l, u]$ can be carried out trivially in our situation.)

However, the projected gradient step is there not only as a safeguard. It is also used in order to change the active set $\mathcal{A}_k$. In fact, this set can change rapidly when using a projected gradient step, i.e., $\mathcal{A}_{k+1}$ can differ from $\mathcal{A}_k$ by more than just one element in contrast to many classical active set methods in constrained optimization [10].

On the other hand, it is also possible that the set $\mathcal{A}_k$ changes when using a Newton-type step in Algorithm 9.1. This is because $\mathcal{A}_k$ contains also those indices which are close to the boundary but not necessarily on the boundary. In fact, this is a major difference to some other active set Newton-type methods for constrained optimization, see, e.g., [1, 17] and references therein.

Finally, we note that the simple descent test (9.4) is motivated by the fact that we want to solve a (constrained) system of equations so that we know a priori that zero is the optimal value of the function f. However, if we would adapt our method to the solution of box constrained optimization problems, a test like (9.4) would not be possible any more, and it is an interesting question whether, in this situation, it would be possible to use a line search globalization as in Bertsekas' method [1].

3 CONVERGENCE

In this section, we discuss the theoretical properties of Algorithm 9.1. Our first result shows that Algorithm 9.1 generates only feasible iterates. This follows relatively easily from the statements of Algorithm 9.1, hence we do not include a formal proof here.

Lemma 9.2 *Any sequence $\{x^k\}$ generated by Algorithm 9.1 belongs to the feasible set $[l, u]$.*

The global convergence properties of Algorithm 9.1 are discussed in our next result.

Theorem 9.3 *Any accumulation point of a sequence $\{x^k\}$ generated by Algorithm 9.1 is at least a stationary point of the equivalent minimization problem (9.2).*

Proof. Let x^* be an accumulation point of a sequence $\{x^k\}$ generated by Algorithm 9.1. If condition (9.4) is satisfied infinitely many times, it follows immediately that $f(x^*) = 0$, i.e., x^* is a global minimum of f and therefore, in particular, a stationary point of (9.2).

On the other hand, if condition (9.4) holds for only finitely many iterations, we eventually take projected gradient steps only, and in this case the assertion follows from, e.g., Proposition 2.3.3 (b) in Bertsekas [2]. □

Let us denote by

$$\mathcal{A}_* := \{i \,|\, x_i^* \in \{l_i, u_i\}\}$$

the set of active constraints at a given solution x^* of our constrained nonlinear system (9.1). In analogy to our notation used for the index set $\mathcal{A}_k$, we denote by $\mathcal{I}_*$ the complement of $\mathcal{A}_*$ in $\{1, \ldots, n\}$:

$$\mathcal{I}_* := \{1, \ldots, n\} \setminus \mathcal{A}_* = \{i \,|\, x_i^* \in (l_i, u_i)\}.$$

We next show that $\mathcal{A}_k$ is always a subset of $\mathcal{A}_*$ if x^k is sufficiently close to a solution x^* of the constrained nonlinear system (9.1). Later, we will show that these two sets actually coincide under a suitable regularity assumption, cf. Lemma 9.7.

Lemma 9.4 *Let x^* be a solution of the constrained nonlinear system (9.1). Then $\mathcal{A}_k \subseteq \mathcal{A}_*$ for all $x^k \in [l, u]$ sufficiently close to x^*.*

Proof. Define

$$\nu := \min\{\min\{x_i^* - l_i, u_i - x_i^*\} \,|\, i \in \mathcal{I}_*\} > 0,$$

i.e., ν is the smallest distance of the inactive components x_i^* to the boundary of the feasible set $[l, u]$. Let $x^k \in [l, u]$ be sufficiently close to x^* such that

$$\|x^k - x^*\| \leq \frac{\nu}{4}$$

and

$$c\sqrt{\|F(x^k)\|} \le \frac{\nu}{4}.$$

Choose $i \in \mathcal{A}_k$ arbitrarily. Then $x_i^k - l_i \le \delta_k$ or $u_i - x_i^k \le \delta_k$. If $x_i^k - l_i \le \delta_k$, we get from

$$|x_i^k - x_i^*| \le \|x^k - x^*\| \le \frac{\nu}{4}$$

and (using Lemma 9.2)

$$|x_i^k - l_i| = x_i^k - l_i \le \delta_k \le c\sqrt{\|F(x^k)\|} \le \frac{\nu}{4}$$

that

$$|x_i^* - l_i| \le |x_i^* - x_i^k| + |x_i^k - l_i| \le \frac{\nu}{2},$$

i.e., we have $i \in \mathcal{A}_*$ in view of the very definition of the constant ν.

If, on the other hand, we have $u_i - x_i^k \le \delta_k$, we can show in a similar way that

$$|x_i^* - u_i| \le \frac{\nu}{2},$$

so that $i \in \mathcal{A}_*$ follows also in this case. Hence we have $\mathcal{A}_k \subseteq \mathcal{A}_*$ for all $x^k \in [l, u]$ sufficiently close to x^*. □

As a simple consequence of Lemma 9.4, we obtain the following corollary.

Corollary 9.5 *Let x^* be a solution of the constrained nonlinear system (9.1). Then $d^k_{\mathcal{A}_k} = x^*_{\mathcal{A}_k} - x^k_{\mathcal{A}_k}$ for all $x^k \in [l, u]$ sufficiently close to x^*.*

Proof. Assume that $x^k \in [l, u]$ is close enough to x^* such that Lemma 9.4 holds, i.e., assume that $\mathcal{A}_k$ is a subset of $\mathcal{A}_*$.

Take any index $i \in \mathcal{A}_k$. Then $i \in \mathcal{A}_*$ so that $x_i^* = l_i$ or $x_i^* = u_i$. In fact, if x^k is sufficiently close to x^*, it follows from our choice of δ that $x_i^* = l_i$ if $x_i^k - l_i \le \delta_k$ and $x_i^* = u_i$ if $u_i - x_i^k \le \delta_k$. Hence we have

$$d_i^k = \begin{cases} l_i - x_i^k = x_i^* - x_i^k & \text{if } x_i^k - l_i \le \delta_k, \\ u_i - x_i^k = x_i^* - x_i^k & \text{if } u_i - x_i^k \le \delta_k. \end{cases}$$

This completes the proof. □

We next introduce our central assumption which will be used in order to show local fast convergence of Algorithm 9.1.

Assumption 9.6 *Let x^* be a solution of the constrained nonlinear system (9.1). We call x^** regular *if both the Jacobian $F'(x^*)$ and the submatrix $F'(x^*)_{\mathcal{I}_*\mathcal{I}_*}$ are nonsingular.*

Note that, in particular, $\mathcal{A}_*$ is the empty set if all lower and upper bounds are infinite, so in this case the assumption that x^* is a regular solution reduces to the standard nonsingularity condition for the Jacobian $F'(x^*)$. Moreover, it follows from Assumption 9.6 and well-known results for the solution of nonlinear systems of equations (see, e.g., [4]) that there is a constant $\alpha > 0$ such that

$$\|F(x)\| \geq \alpha \|x - x^*\| \tag{9.7}$$

for all x sufficiently close to x^*. In particular, this implies that x^* is a locally isolated solution of the standard nonlinear system $F(x) = 0$ and, therefore, also of our constrained nonlinear system (9.1). These facts will be used later.

As another consequence of Assumption 9.6, we now prove that our approximation $\mathcal{A}_k$ of the active set $\mathcal{A}_*$ eventually coincides with $\mathcal{A}_*$.

Lemma 9.7 *Let x^* be a regular solution of the constrained nonlinear system (9.1). Then $\mathcal{A}_k = \mathcal{A}_*$ for all $x^k \in [l, u]$ sufficiently close to x^*.*

Proof. We first note that (9.7) implies

$$|x_i^k - x_i^*| \leq \|F(x^k)\|/\alpha \tag{9.8}$$

for some constant $\alpha > 0$, all $i \in \{1, \dots, n\}$, and all $x^k \in [l, u]$ sufficiently close to x^*.

In order to prove our statement, it suffices to show that $\mathcal{A}_* \subseteq \mathcal{A}_k$ since the other inclusion has already been shown in Lemma 9.4. So let $i \in \mathcal{A}_*$ be any fixed index. Then $x_i^* = l_i$ or $x_i^* = u_i$.

Assume first that $x_i^* = l_i$. Then

$$x_i^k - l_i = |x_i^k - x_i^*| \leq \|F(x^k)\|/\alpha$$

because of (9.8) and Lemma 9.2. Since $\|F(x^k)\| \to 0$ for $x^k \to x^*$ and $\delta_k = O(\sqrt{\|F(x^k)\|})$ in view of the definition of δ_k, we have

$$\|F(x^k)\|/\alpha \leq \delta_k$$

for all $x^k \in [l, u]$ sufficiently close to x^*. Therefore, we obtain

$$x_i^k - l_i \leq \delta_k,$$

i.e., $i \in \mathcal{A}_k$.

If, on the other hand, we have $x_i^* = u_i$, we can show in an analogous way that

$$u_i - x_i^k \leq \delta_k.$$

This proves that $\mathcal{A}_* \subseteq \mathcal{A}_k$, so that both sets eventually coincide by Lemma 9.4. □

We stress that the proof of the previous result clearly shows why we took the square root of $\|F(x^k)\|$ in the definition of δ_k, cf. Step (S.2) of Algorithm 9.1. In fact, all results would still be true if we would take an arbitrary power $\|F(x^k)\|^p$ with $p \in (0,1)$.

The next result is a relatively simple consequence of Lemma 9.7.

Lemma 9.8 *Let x^* be a regular solution of the constrained nonlinear system (9.1). Then there exist constants $\kappa_1 > 0$ and $\kappa_2 > 0$ such that*

(a) the submatrices $F'(x^k)_{\mathcal{I}_k\mathcal{I}_k}$ are nonsingular with

$$\|F'(x^k)^{-1}_{\mathcal{I}_k\mathcal{I}_k}\| \le \kappa_1,$$

and

(b) the linear systems (9.3) have a unique solution satisfying the inequality

$$\|d^k_{\mathcal{I}_k}\| \le \kappa_2 \|F(x^k)\|$$

for all $x^k \in [l,u]$ sufficiently close to x^*.

Proof. Part (a) follows immediately from Lemma 9.7 and a standard perturbation result (see [4]) for linear systems of equations. In particular, the linear systems (9.3) have a unique solution $d^k_{\mathcal{I}_k}$.

In order to verify the second statement, we first note that Corollary 9.5 and (9.7) imply that

$$\|d^k_{\mathcal{A}_k}\| = \|x^k_{\mathcal{A}_k} - x^*_{\mathcal{A}_k}\| \le \|x^k - x^*\| \le \|F(x^k)\|/\alpha$$

for some constant $\alpha > 0$ and all x^k close enough to x^*. Moreover, since F' is continuous, there is a constant $\kappa > 0$ such that

$$\|F'(x^k)_{\mathcal{I}_k\mathcal{A}_k}\| \le \kappa$$

for all x^k close to x^*. Combining these inequalities with the first statement, we finally obtain from (9.3):

$$\begin{aligned}
\|d^k_{\mathcal{I}_k}\| &\le \|F'(x^k)^{-1}_{\mathcal{I}_k\mathcal{I}_k}\| \left(\|F(x^k)_{\mathcal{I}_k}\| + \|F'(x^k)_{\mathcal{I}_k\mathcal{A}_k}\| \, \|d^k_{\mathcal{A}_k}\| \right) \\
&\le \kappa_1 \left(\|F(x^k)\| + \frac{\kappa}{\alpha} \|F(x^k)\| \right) \\
&= \kappa_2 \|F(x^k)\|
\end{aligned}$$

for all $x^k \in [l, u]$ sufficiently close to x^*, where $\kappa_2 := \kappa_1 + \kappa/\alpha$. □

We next want to show that the entire sequence $\{x^k\}$ generated by Algorithm 9.1 converges to a solution x^* of the constrained nonlinear system if this solution is regular and is at least an accumulation point of the sequence $\{x^k\}$. To this end, we need the following result by Moré and Sorensen [19] (in fact, the result stated here is a refinement of the one given in [19], however, a closer look at the proof given in [19] shows that actually this refined result holds, cf. [13]).

Proposition 9.9 *Assume that $x^* \in \boldsymbol{R}^n$ is an isolated accumulation point of an arbitrary sequence $\{x^k\} \subseteq \boldsymbol{R}^n$ (not necessarily generated by Algorithm 9.1) such that, for every subsequence $\{x^k\}_K$ converging to x^*, there is an infinite subset $\tilde{K} \subseteq K$ with $\{\|x^{k+1} - x^k\|\}_{\tilde{K}} \to 0$. Then the whole sequence $\{x^k\}$ converges to x^*.*

Verifying the condition in Proposition 9.9, we are now able to state the following convergence result.

Proposition 9.10 *Let x^* be a regular solution of the constrained nonlinear system (9.1), and assume that x^* is an accumulation point of a sequence $\{x^k\}$ generated by Algorithm 9.1. Then the entire sequence $\{x^k\}$ converges to x^*.*

Proof. As noted after Assumption 9.6, a regular solution is an isolated solution of the constrained nonlinear system (9.1). Since the sequence $\{f(x^k)\}$ is monotonically decreasing, it can easily be seen that this implies that x^* is an isolated accumulation point of the sequence $\{x^k\}$.

Next, let $\{x^k\}_K$ be a subsequence converging to x^*. Then, by continuity, we have

$$\{\|F(x^k)\|\}_K \to \|F(x^*)\| = 0. \tag{9.9}$$

We now consider two cases:

Case 1: There is an infinite subset $\tilde{K} \subseteq K$ such that Algorithm 9.1 takes projected gradient steps for all $k \in \tilde{K}$.

In the same way as in the proof of Theorem 4.5 in [9], we then obtain from Lemma 9.2 and the fact that the projection operator is nonexpansive:

$$\begin{aligned}
\|x^{k+1} - x^k\| &= \|x^k(t_k) - x^k\| \\
&= \|[x^k - t_k \nabla f(x^k)]_+ - x^k\| \\
&= \|[x^k - t_k \nabla f(x^k)]_+ - [x^k]_+\| \\
&\leq \|t_k \nabla f(x^k)\| \\
&\leq s\|\nabla f(x^k)\| \\
&\to_{\tilde{K}} 0
\end{aligned}$$

since x^* solves the constrained nonlinear system (9.1) so that x^* is a global minimizer and hence an unconstrained stationary point of f (recall that F and therefore also f is defined and continuously differentiable in an open set containing the box $[l, u]$).

Case 2: There are only finitely many indices $k \in K$ for which Algorithm 9.1 takes a projected gradient step.

Then we can assume without loss of generality that the search direction d^k is always the one computed in Step (S.3) of Algorithm 9.1. Since

$$\|d^k_{\mathcal{I}_k}\| \leq \kappa_2 \|F(x^k)\|$$

for all k sufficiently large by Lemma 9.8, we have at least

$$\{\|d^k_{\mathcal{I}_k}\|\}_K \to 0$$

because of (9.9). If, on the other hand, we have an index $i \in \mathcal{A}_k$, then $i \in \mathcal{A}_*$ by Lemma 9.4, so that we either have

$$d^k_i = l_i - x^k_i \to l_i - x^*_i = 0 \qquad (\text{if } x^*_i = l_i)$$

or

$$d^k_i = u_i - x^k_i \to u_i - x^*_i = 0 \qquad (\text{if } x^*_i = u_i).$$

This shows that

$$\{\|d^k\|\}_K \to 0$$

which, in turn, implies that

$$\{\|x^{k+1} - x^k\|\}_K \to 0.$$

Summarizing both cases, we see that

$$\{\|x^{k+1} - x^k\|\}_{\tilde{K}} \to 0$$

for the infinite subset $\tilde{K} \subseteq K$ defined in Case 1. Hence our statement follows immediately from Proposition 9.9. □

The following result shows that the feasibility preserving safeguard in Step (S.4) of Algorithm 9.1 becomes inactive eventually, i.e., a full step stays within the feasible set $[l, u]$ if x^k is close enough to a regular solution.

Lemma 9.11 *Let x^* be a regular solution of the constrained nonlinear system (9.1). Then $\tau_k = 1$ for all $x^k \in [l, u]$ sufficiently close to x^*.*

Proof. In view of Lemma 9.8, there exists a constant $\kappa_2 > 0$ such that

$$\|d^k_{\mathcal{I}_k}\| \leq \kappa_2 \|F(x^k)\| \tag{9.10}$$

for all $x^k \in [l, u]$ close enough to x^*. Since we have $x^k_i - l_i > \delta_k$ and $u_i - x^k_i > \delta_k$ for $i \in \mathcal{I}_k$, we consider two cases.

Case 1: $x^k_i - l_i > \delta_k$.
Since $\|F(x^k)\| \to 0$ for $x^k \to x^*$, we obtain from (9.10) and the definition of δ_k:

$$\begin{aligned} x^k_i + d^k_i &> l_i + \delta_k + d^k_i \\ &\geq l_i + \delta_k - \kappa_2 \|F(x^k)\| \\ &= l_i + \min\{\delta, c\sqrt{\|F(x^k)\|}\} - \kappa_2 \|F(x^k)\| \\ &= l_i + c\sqrt{\|F(x^k)\|} - \kappa_2 \|F(x^k)\| \\ &= l_i + \left(c/\sqrt{\|F(x^k)\|} - \kappa_2\right) \|F(x^k)\| \\ &\geq l_i \end{aligned}$$

for all $x^k \in [l, u]$ sufficiently close to x^*.

Case 2: $u_i - x^k_i > \delta_k$.
Similar to the previous case, we can show that

$$x^k_i + d^k_i < u_i$$

for all $x^k \in [l, u]$ in a small neighborhood of x^*.

Combining these two cases together with the definition of $\bar{\tau}_k$, it follows that $\bar{\tau}_k \geq 1$ and therefore $\tau_k = 1$ for all $x^k \in [l, u]$ sufficiently close to x^*. □

In order to prove local fast convergence, we will also employ the following result which is a specialization of a more general theorem in [13] (see also [8]).

Proposition 9.12 *Let $G : [l, u] \to \mathbf{R}^n$ be continuously differentiable in an open set containing $[l, u]$, $x^* \in [l, u]$ with $G(x^*) = 0$ and $G'(x^*)$ nonsingular be given, and assume that there are two sequences (not necessarily generated by Algorithm 9.1) $\{x^k\} \subseteq \mathbf{R}^n$ and $\{d^k\} \subseteq \mathbf{R}^n$ with $\{x^k\} \to x^*$ and $\|x^k + d^k - x^*\| = o(\|x^k - x^*\|)$. Then $\|G(x^k + d^k)\| = o(\|G(x^k)\|)$.*

In order to verify the assumptions of Proposition 9.12, we next show that the search directions computed in Step (S.3) of Algorithm 9.1 are "superlinearly convergent search directions".

Lemma 9.13 *Let x^* be a regular solution of the constrained nonlinear system (9.1), and let $\{x^k\}$ be any sequence (not necessarily generated by Algorithm 9.1) converging to x^*. Then*

$$\|x^k + d^k - x^*\| = o(\|x^k - x^*\|),$$

where d^k denotes the search direction calculated in Step (S.3) of Algorithm 9.1. Furthermore, if F' is locally Lipschitzian around x^*, then

$$\|x^k + d^k - x^*\| = O(\|x^k - x^*\|^2).$$

Proof. Since $x^k \to x^*$ by assumption, it follows from Lemma 9.8 that the search directions d^k as computed in Step (S.3) of Algorithm 9.1 actually exist for all k sufficiently large. Moreover, Lemma 9.7 guarantees that we have $\mathcal{A}_k = \mathcal{A}_*$ for all k large enough. In the rest of this proof, we therefore assume implicitly that k is sufficiently large so that the above two statements hold.

First consider an index $i \in \mathcal{A}_*$. Then $x_i^* = l_i$ or $x_i^* = u_i$. If $x_i^* = l_i$, then $d_i^k = l_i - x_i^k$, and it follows that

$$|x_i^k + d_i^k - x_i^*| = |x_i^k + l_i - x_i^k - l_i| = 0 = o(\|x^k - x^*\|). \tag{9.11}$$

Similarly, if $x_i^* = u_i$, then $d_i^k = u_i - x_i^k$, and we obtain

$$|x_i^k + d_i^k - x_i^*| = |x_i^k + u_i - x_i^k - u_i| = 0 = o(\|x^k - x^*\|). \tag{9.12}$$

Next consider indices $i \in \mathcal{I}_*$. Using the notation $\mathcal{N} := \{1, \ldots, n\}$, we obtain from the linear system (9.3), Corollary 9.5 and the fact that $\mathcal{I}_k = \mathcal{I}_*$:

$$\begin{aligned}
& F'(x^k)_{\mathcal{I}_*\mathcal{I}_*}(x^k_{\mathcal{I}_*} + d^k_{\mathcal{I}_*} - x^*_{\mathcal{I}_*}) \\
= \; & F'(x^k)_{\mathcal{I}_*\mathcal{I}_*} d^k_{\mathcal{I}_*} + F'(x^k)_{\mathcal{I}_*\mathcal{I}_*}(x^k_{\mathcal{I}_*} - x^*_{\mathcal{I}_*}) \\
= \; & -F(x^k)_{\mathcal{I}_*} - F'(x^k)_{\mathcal{I}_*\mathcal{A}_*} d^k_{\mathcal{A}_*} + F'(x^k)_{\mathcal{I}_*\mathcal{I}_*}(x^k_{\mathcal{I}_*} - x^*_{\mathcal{I}_*}) \\
= \; & -F(x^k)_{\mathcal{I}_*} + F(x^*)_{\mathcal{I}_*} + F'(x^k)_{\mathcal{I}_*\mathcal{A}_*}(x^k_{\mathcal{A}_*} - x^*_{\mathcal{A}_*}) \\
& + F'(x^k)_{\mathcal{I}_*\mathcal{I}_*}(x^k_{\mathcal{I}_*} - x^*_{\mathcal{I}_*}) \\
= \; & -F(x^k)_{\mathcal{I}_*} + F(x^*)_{\mathcal{I}_*} + F'(x^k)_{\mathcal{I}_*\mathcal{N}}(x^k - x^*),
\end{aligned} \tag{9.13}$$

since the solution x^* satisfies $F(x^*) = 0$. Since F is continuously differentiable, we have

$$|F_i(x^k) - F_i(x^*) - \nabla F_i(x^k)^T(x^k - x^*)| = o(\|x^k - x^*\|)$$

for all $i \in \{1, \ldots, n\}$. Hence we obtain

$$\|F_{\mathcal{I}_*}(x^k) - F_{\mathcal{I}_*}(x^*) - F'(x^k)_{\mathcal{I}_*\mathcal{N}}(x^k - x^*)\| = o(\|x^k - x^*\|),$$

so it follows from (9.13) and Lemma 9.8 that

$$\|x^k_{\mathcal{I}_*} + d^k_{\mathcal{I}_*} - x^*_{\mathcal{I}_*}\| = o(\|x^k - x^*\|). \tag{9.14}$$

Using (9.11), (9.12) and (9.14), we have

$$\|x^k + d^k - x^*\| = o(\|x^k - x^*\|).$$

In a similar way, one can show that

$$\|x^k + d^k - x^*\| = O(\|x^k - x^*\|^2)$$

holds if F' is locally Lipschitzian around x^*. □

Summarizing the previous results, we are now in the position to prove our main local convergence theorem.

Theorem 9.14 *Let x^* be a regular solution of the constrained nonlinear system (9.1), and assume that x^* is an accumulation point of a sequence $\{x^k\}$ generated by Algorithm 9.1. Then the following statements hold:*

(a) The entire sequence $\{x^k\}$ converges to x^*.

(b) Eventually, the algorithm takes only Newton steps, i.e., no projected gradient steps are carried out for k sufficiently large.

(c) The rate of convergence is Q-superlinear.

(d) The rate of convergence is Q-quadratic if F' is locally Lipschitzian around x^*.

Proof. Statement (a) follows directly from Proposition 9.10.

Furthermore, Lemma 9.8 implies that the search directions d^k do at least exist in Step (S.3) of Algorithm 9.1 for all k sufficiently large, and Proposition 9.12 together with Lemma 9.13 show that

$$\|F(x^k + d^k)\| = o(\|F(x^k)\|).$$

Hence

$$f(x^k + d^k) = o(f(x^k))$$

so that the descent condition (9.4) is eventually satisfied with $\tau_k = 1$. This proves the second statement.

Statements (c) and (d) are now immediate consequences of Lemma 9.13 and the fact that $x^{k+1} = x^k + d^k$ for all sufficiently large k (recall that $\tau_k = 1$ for k large enough by Lemma 9.11). □

4 SOME NUMERICAL EXAMPLES

We implemented Algorithm 9.1 in MATLAB using the stopping criterion

$$f(x^k) \le \varepsilon$$

and the parameters

$$\beta = 0.5,\ \sigma = 10^{-4},\ \gamma = 0.995,\ \delta = 10^{-4},\ s = c = 1,\ \varepsilon = 10^{-9}.$$

In addition to the formal description of Algorithm 9.1, we also included a simple backtracking in the descent test (9.4), i.e., if the "full" step $t_k = \tau_k$ does not satisfy criterion (9.4), we first check whether $t_k = \beta^\ell \tau_k$ satisfies (9.4) with $\ell = 0, 1, \dots, 5$ before we switch to a projected gradient step. It is not difficult to see that this modification does not destroy the theoretical convergence properties of Algorithm 9.1.

We tested our method on the following two examples.

Example 9.15 *(Chemical Equilibrium)*
This example is taken from [5] and describes a chemical equilibrium problem with propane. Its dimension is $n = 10$, and the mapping F is defined by

$$F(x) := \begin{pmatrix} x_1 + x_4 - 3 \\ 2x_1 + x_2 + x_4 + x_7 + x_8 + x_9 + 2x_{10} - \lambda \\ 2x_2 + 2x_5 + x_6 + x_7 - 8 \\ 2x_3 + x_9 - 4\lambda \\ x_1 x_5 - 0.193 x_2 x_4 \\ x_6^2 x_1 - 0.67444 \cdot 10^{-5} x_2 x_4 s \\ x_7^2 x_4 - 0.1189 \cdot 10^{-4} x_1 x_2 s \\ x_8 x_4 - 0.1799 \cdot 10^{-4} x_1 s \\ x_9^2 x_4^2 - 0.4644 \cdot 10^{-7} x_1^2 x_3 s \\ x_{10} x_4^2 - 0.3846 \cdot 10^{-4} x_1^2 s \end{pmatrix},$$

where $\lambda \ge 3$ is a parameter and s denotes the sum

$$s := \sum_{i=1}^{10} x_i.$$

Since the variables x_i correspond to chemical concentrations, we require all variables to be nonnegative, i.e., we set $l_i = 0$ and $u_i = +\infty$ for all $i = 1, \dots, 10$ in this example.

Example 9.16 *(Chandrasekhar's H-Equation)*
This example is a discretized version of Chandrasekhar's H-equation, see [14] and references therein for some further details. The dimension n is variable (depending on the discretization) and the mapping F is defined componentwise by

$$F_i(x) := x_i - \frac{1}{1 - \frac{c}{2n}\sum_{j=1}^{n} \frac{\mu_i x_j}{\mu_i + \mu_j}} \quad (i = 1, \dots, n),$$

where $c \in (0,1)$ denotes a constant and

$$\mu_i := (i - 0.5)/n \quad \forall i = 1, \dots, n.$$

We add lower bounds $l_i = 0$ to this problem since otherwise the solution has no physical meaning.

Table 9.1 Numerical results for Example 9.15 with $\lambda = 3.1$

k	$f(x^k)$	t_k	m_k	type
0	1.18e+02	0.125	10	PG
1	4.98e+01	0.320	0	N
2	2.22e+01	0.367	1	N
3	8.54e+00	0.695	1	N
4	7.84e-01	0.957	2	N
5	2.52e-03	0.976	3	N
6	2.11e-03	0.500	5	PG
7	1.06e-03	0.031	5	PG
8	9.98e-04	0.198	0	N
9	6.45e-04	1.000	1	N
10	3.87e-07	1.000	3	N
11	2.08e-08	1.000	3	N
12	1.54e-09	1.000	3	N
13	3.42e-10			

Application of Algorithm 9.1 to Example 9.15 with $\lambda = 3.1$ and starting point $x^0 = (0, \dots, 0)^T$ gives the result in Table 1, where we present the following data:

k:	iteration number
$f(x^k)$:	function value at the kth iterate
t_k:	stepsize at the kth iteration
m_k:	number of approximately active indices, i.e., $m_k = \|\mathcal{A}_k\|$
type:	type of step (N for Newton, PG for projected gradient)

Our method found the solution

$$x^* \approx (1.26\cdot10^{-2}, 8.53\cdot10^{-2}, 6.20, 2.99, 3.91, 0, 2.07\cdot10^{-3}, 0, 5.91\cdot10^{-5}, 0)^T$$

without any problems in 13 iterations. For this example, it is interesting to note that three components of the solution vector x^* seem to be exactly zero, and that this is also indicated correctly by our active set strategy ($m_k = 3$ in the last three iterations) although some of the other components of the solution vector are rather small.

Our next experiment is also based on Example 9.15. We start with $\lambda = 3.0$ and $x^0 = (0, \dots, 0)^T$. Having solved this problem, we increase the value of λ by 0.1 and use the solution of the previous problem as a starting point for this new problem. We proceed in this way until λ reaches 4.0. The results are given in Table 2, where, for each value of λ, we report the number of projected gradient steps, the number of Newton steps, and the final value of f. Again, the method does not seem to have any troubles with this example.

Table 9.2 Numerical results for Example 9.15 with different λ

λ	# PG	# N	$f(x^k)$
3.0	13	13	4.37e-10
3.1	5	1	6.70e-10
3.2	2	0	2.64e-11
3.3	2	0	4.80e-11
3.4	2	0	9.13e-11
3.5	2	0	1.53e-10
3.6	2	0	2.36e-10
3.7	2	0	3.46e-10
3.8	2	0	4.88e-10
3.9	2	0	6.68e-10
4.0	2	0	8.94e-10

Finally, we applied Algorithm 9.1 to Example 9.16 with $n = 100, c = 0.9$ and starting point $x^0 = (0, \dots, 0)^T$. The corresponding results are given in Table 3. The rows in Table 3 have the same meaning as the rows in Table 1. Note that the value of m_k is always zero (except for the starting point). This is not surprising since all components of the solution vector x^* found for this example belong to the interval $(1, 2)$.

Table 9.3 Numerical results for Example 9.16

k	$f(x^k)$	t_k	m_k	type
0	5.00e+01	1.0	100	PG
1	5.23e+00	1.0	0	N
2	6.31e-02	1.0	0	N
3	3.77e-02	0.5	0	N
4	1.37e-02	0.5	0	N
5	1.91e-05	0.5	0	N
6	5.69e-06	1.0	0	N
7	1.30e-06	0.5	0	N
8	3.78e-08	0.5	0	N
9	1.44e-09	0.5	0	N
10	2.84e-11			

5 CONCLUDING REMARKS

We presented an active set-type Newton method for the solution of constrained nonlinear systems. This method generates only feasible iterates and has some strong theoretical convergence properties. Some preliminary numerical results were quite promising.

A modification of this method (using a somewhat different globalization strategy and generating strictly feasible iterates) has recently been investigated in [12], where it is used in order to solve reformulations of mixed complementarity problems as box constrained nonlinear systems of equations. The manuscript [12] also contains some numerical results which indicate that Newton-type methods combined with an active set strategy as described in this paper seem to work quite well in practice.

Acknowledgments

The author would like to thank a referee for some very detailed comments. In particular, the current version of Assumption 9.6, which is weaker than in the original manuscript, was suggested by the referee.

References

[1] Bertsekas, D.P. (1982). Projected Newton methods for optimization problems with simple constraints. *SIAM J. Control Optim.*, 20: 221–246.

[2] Bertsekas, D.P. (1995). *Nonlinear Programming.* Athena Scientific, Belmont, MA.

[3] Calamai, P.H. and Moré, J.J. (1987). Projected gradient methods for linearly constrained problems. *Math. Programming*, 39: 93–116.

[4] Dennis, Jr., J.E. and Schnabel, R.B. (1983). *Numerical Methods for Unconstrained Optimization and Nonlinear Equations.* Prentice-Hall, Englewood Cliffs, NJ.

[5] Deuflhard, P. and Hohmann, A. (1995). *Numerical Analysis. A First Course in Scientific Computation.* Verlag de Gruyter, Berlin, New York.

[6] Facchinei, F., Fischer, A. and Kanzow. C. (1997). A semismooth Newton method for variational inequalities: The case of box constraints. In Ferris, M.C. and Pang, J.-S., editors, *Complementarity and Variational Problems: State of the Art*, pages 76–90. SIAM, Philadelphia, PA.

[7] Facchinei, F., Fischer, A. and Kanzow, C. (1999). On the accurate identification of active constraints. *SIAM J. Optim.*, 9: 14–32.

[8] Facchinei, F. and Soares, J. (1997). A new merit function for nonlinear complementarity problems and a related algorithm. *SIAM J. Optim.* 7: 225–247.

[9] Ferris, M.C., Kanzow, C. and Munson, T.S. (1999). Feasible descent algorithms for mixed complementarity problems. *Math. Programming*, 86: 475–497.

[10] Fletcher, R. (1987). *Practical Methods of Optimization.* John Wiley & Sons, New York, NY.

[11] Gafni, E.M. and Bertsekas, D.P. (1982). Convergence of a gradient projection method. Report LIDS-P-1201, Laboratory for Information and Decision Systems, Massachusetts Institute of Technology, MA.

[12] Kanzow, C. (1999) Strictly feasible equation-based methods for mixed complementarity problems. Preprint 145, Institute of Applied Mathematics, University of Hamburg, Hamburg, Germany.

[13] Kanzow, C. and Qi, H.-D. (1999). A QP-free constrained Newton-type method for variational inequality problems. *Math. Programming*, 85: 81–106.

[14] Kelley, C.T. (1995). *Iterative Methods for Linear and Nonlinear Equations.* SIAM, Philadelphia, PA.

[15] Kelley, C.T. (1999): *Iterative Methods for Optimization.* SIAM, Philadelphia, PA.

[16] Kozakevich, D.N., Martínez, J.M. and Santos, S.A. Solving nonlinear systems of equations with simple constraints. *Computational and Applied Mathematics*, to appear.

[17] Lin, C.-J. and Moré, J.J. (1999). Newton's method for large bound-constrained optimization problems. *SIAM J. Optim.*, 9: 1100–1127.

[18] Liu, J. (1995). Strong stability in variational inequalities. *SIAM J. Control Optim.*, 33: 725–749.

[19] Moré, J.J. and Sorensen, D.C. (1983). Computing a trust region step. *SIAM J. Sci. Stat. Comp.*, 4: 553–572.

[20] Robinson, S.M. (1980). Strongly regular generalized equations. *Math. Oper. Res.*, 5: 43–62.

[21] Sun, D. (1998). Projected Newton-type methods for constrained nonsmooth equations. Talk presented at the "Workshop on Nonsmooth and Smoothing Methods" in Hong Kong, China.

[22] Ulbrich, M. Non-monotone trust-region methods for bound-constrained semismooth equations with applications to nonlinear mixed complementarity problems. *SIAM J. Optim.*, to appear.

[23] Wang, T., Monteiro, R.D.C., and Pang. J.-S. (1996). An interior point potential reduction method for constrained equations. *Math. Programming*, 74: 159–195.

Chapter 10

MATHEMATICAL PROGRAMMING IN ENGINEERING MECHANICS: SOME CURRENT PROBLEMS

G. Maier
Department of Structural Engineering, Technical University (Politecnico) of Milan, Piazza Leonardo da Vinci 32, 20133 Milan, Italy

G. Bolzon
Department of Structural Engineering, Technical University (Politecnico) of Milan, Piazza Leonardo da Vinci 32, 20133 Milan, Italy

F. Tin-Loi[†]
School of Civil and Environmental Engineering, University of New South Wales, Sydney 2052, Australia

Abstract The application of mathematical programming methods in a variety of practically motivated engineering mechanics problems provides a fertile field for interdisciplinary interaction between the mathematical programming and engineering communities. This paper briefly outlines several topical problems in engineering mechanics involving the use of mathematical programming techniques. The intention is to attract the attention of mathematical programming experts to some of the still open questions in the intersection of the two fields.

[*]This paper is dedicated to Professor Richard Cottle on the occasion of his 65th birthday.
[†]Corresponding author.

M.C. Ferris et al. (eds.), Complementarity: Applications, Algorithms and Extensions, 201–231.

1 INTRODUCTION

The first fruitful encounter between engineering mechanics and mathematical programming (MP) occurred perhaps in the Fifties, when the assessment of the safety factor with respect to plastic collapse, based on either the static or the kinematic limit analysis theorem, for elastic perfectly plastic beams and frames was recognized to be a problem in linear programming (LP).

In the Sixties, analogous recognition emerged for shakedown analysis which can be interpreted as a generalization of limit analysis to engineering situations characterized by "variable repeated" loading, i.e. by external actions varying in time with an infinity of unknown time histories within a known domain in the space of the load-governing variables. Simultaneously, the analogy between the duality theory of LP and the theory of limit analysis and shakedown analysis was noticed and computationally exploited, while nonlinear programming (NLP) became the mathematical tool for limit and shakedown analyses when nonlinear constitutive yield loci were adopted. Techniques suitable for generating and minimizing upper bounds on post-shakedown history-dependent quantities in piecewise linear plasticity were also established. These techniques implied merely the solution of appropriate LP and quadratic programming (QP) problems. In the Seventies, convex QP with its associated notions of duality, and the equivalent (via the Karush-Kuhn-Tucker optimality conditions) linear complementarity problem (LCP) were recognized as being convenient mathematical models for piecewise linear elastic-plastic step-by-step structural analysis up to incipient failure.

Meanwhile, the fast growth of structural optimization (encompassing limit, elastic and elastic-plastic optimal design) was fostered by parallel developments in MP.

"Marriages à la mode" between concepts and methods of engineering plasticity and those of MP frequently occurred in those years, and led to the NATO conference on "Engineering Plasticity by Mathematical Programming" held at the University of Waterloo, Canada in 1977, with substantial contributions by prominent MP researchers, *in primis* Professor R. Cottle. For the interested reader, the proceedings volume of that conference [1], together with two later survey articles [2, 3], may still represent a source of information on the present subject up to the middle Eighties. More recent sources of reference, specifically on some engineering and economic applications of complementarity problems, are the review paper by Ferris and Pang [4] and the proceedings of the first International Conference on Complementarity Problems held at The Johns Hopkins University in 1995 [5].

This paper is not intended to provide a conspectus of recent developments in the several areas where engineering mechanics may benefit from inputs from MP. Its purpose is to concisely describe a representative set of current research results and open problems apt to promote possible further interactions and cross fertilizations between research workers in the engineering mechanics and MP communities. Consistent with this objective, each one of the following Sections contains: a brief description of the mechanical and engineering motivation of the problem considered; a formulation of the mathematical problem preceded by a concise description of the mechanical concepts and methodology leading to it; an outline of the procedure adopted by the writers and their co-workers for numerical solution; and finally remarks primarily intended to point out MP-related open questions and prospects for future developments arising from the computational experiences gained so far.

The topical list is as follows: assessment of the strength (by a kinematic approach leading to an equality constrained minimization of a nonsmooth function) of ductile heterogeneous materials with a periodic heterogeneity texture subjected to proportionally increasing average (or "macroscopic") stresses (Section 2); a representative problem of bounding post-shakedown quantities formulated as an optimization problem (Section 3); evaluation by LP of the safety factor with respect to incremental collapse or alternating plasticity in poroplasticity, i.e. for two-phase (liquid-fluid) fully saturated systems subjected to fluctuating external actions, such as masonry dams in the presence of diffuse crack damage (Section 4); simulations, via linear (LCP) and nonlinear (NCP) complementarity problems, of fracture processes concerning structures made of quasibrittle (concrete-like) materials (Section 5); and identification of fracture parameters of quasibrittle materials using the mathematical models of Section 5 as a minimization of a suitably defined discrepancy between experimentally measured and computed quantities under complementarity constraints (Section 6).

2 LIMIT AND SHAKEDOWN ANALYSES OF DUCTILE PERIODIC COMPOSITE MATERIALS

2.1 MODEL

The reference problem dealt with in this Section can be described as follows. Consider a heterogeneous medium endowed with periodic microstructure, typically a ductile metal-matrix fibre-reinforced composite. Many engineering applications of such material require the determination of its "strength". Specifically, in the six-dimensional space of average (or "macroscopic") stress components gathered in vector Σ, it is required to determine the domain Ω ("elastic domain") which defines the carrying capacity of the material, in the sense that $\Sigma \in \Omega$ can be sustained, $\Sigma \notin \Omega$ cannot and $\Sigma \in \partial\Omega$ (denoting by $\partial\Omega$ the boundary of Ω) is consistent with the manifestation of a plastic collapse mechanism say $\dot{E}$ (denoting by E the vectorial representation of the average strain tensor with an "engineering definition" of shear strains and dots representing derivatives with respect to an ordinative, not necessarily chronological, time variable).

The basic hypotheses of classical plasticity are adopted, namely linear kinematics ("small deformations"), and associative nonhardening flow rules. In particular, the isotropic von Mises constitutive model is attributed to both matrix and fibre in terms of local stresses σ and strains ε. Perfect adhesion (no debonding) is assumed to apply at the matrix-fibre interfaces.

The periodicity of the material texture implies the following: (i) a "representative volume" V with boundary Γ can be defined as the minimum portion of the solid which contains all information on the local geometrical and physical properties, so that the whole solid can be conceived as generated by multiple translations of V, translations which dictate the periods in the material texture; (ii) in the presence of uniform fields Σ and E throughout the heterogeneous solid, the fields of local strains ε, displacements u and stresses σ exhibit periodicity properties, see e.g. [6] and references therein. Thus,

$$\varepsilon = E + \tilde{\varepsilon}, \qquad \tilde{\varepsilon} \text{ periodic on } \Gamma, \qquad \frac{1}{V}\int_V \tilde{\varepsilon}\, dV = 0, \tag{10.1}$$

$$u = xE + \tilde{u}, \qquad \tilde{u} \text{ periodic on } \Gamma, \tag{10.2}$$

$$\tilde{t} = n\sigma, \qquad \text{antiperiodic on } \Gamma, \tag{10.3}$$

where x and n are matrices containing as entries in suitable locations the x space coordinates and outward unit normals to Γ, respectively.

For assigned macroscopic stresses Σ, the limit factor s which amplifies Σ up to plastic collapse of the heterogeneous medium can be evaluated on the basis of the kinematic theorem of limit analysis (see e.g. [1]) applied to representative volume V, namely

$$s = \min_{\dot{E},\dot{\tilde{u}}} \left\{ \dot{\pi} \equiv \frac{1}{V}\int_V \dot{D}(x\dot{E} + \dot{\tilde{u}})dV \right\}, \tag{10.4a}$$

$$\text{subj. to: } \Sigma^T \left(\frac{1}{V}\int_V \dot{\varepsilon}(x\dot{E} + \dot{\tilde{u}})dV \right) = 1, \quad \dot{\tilde{u}} \text{ periodic on } \Gamma. \tag{10.4b}$$

In the formulation given by (10.4a)-(10.4b), the objective functional $\dot{\pi}$ represents the total dissipation rate. The (local) dissipation density $\dot{D}$ is uniquely identified by its argument, namely by the velocity fields through the linear geometric compatibility operator leading from $\dot{u}$ to $\dot{\varepsilon}$ and through the "maximum work" principle implied by the plastic constitutive laws (associative with convex elastic domain in accordance with Drucker's postulate of material stability). The constraint (10.4b) expresses the normalization of the external power which entails the identification of the total dissipated power $\dot{\pi}$ with the "kinematically admissible" load factor to minimize.

Let the following provisions be adopted to transform problem (10.4a)-(10.4b): (i) the plastic constitution is specialized to von Mises model; (ii) the displacement field is modeled by conventional finite elements (say 8 node "bricks" in 3D) and integrations are performed numerically by a Gauss procedure; (iii) the kinematic boundary conditions of displacement periodicity in constraint (10.4b) are enforced on the vector $\dot{U}$ of all nodal displacement rates of the finite element aggregates, leading to the reduced vector $\dot{\tilde{U}}^*$ of independent optimization variables.

Thus, problem (10.4a)-(10.4b) in functional space is specialized and discretized to the following problem in finite-dimensional vector space:

$$s = \min_{\dot{E},\dot{\tilde{U}}^*} \left\{ \sqrt{\frac{2}{3}}\frac{1}{V} \sum_{i \in \mathcal{I}} \sigma_o^{r(i)} \rho_i \left|J\right|_i \sqrt{\dot{E}^T D \dot{E} + 2\dot{E}^T D B \dot{\tilde{U}}^* + \dot{\tilde{U}}^{*T} R \dot{\tilde{U}}^*} \right\}, \tag{10.5a}$$

$$\text{subj. to: } \Sigma^T \dot{E} = 1, \tag{10.5b}$$

$$Y^T(\dot{E} + B_i^* \dot{\tilde{U}}^*) = 0, \quad i = 1, \ldots, n_G. \tag{10.5c}$$

The new symbols in (10.5a)-(10.5c) have the following meanings: index i runs over the set $\mathcal{I}$ of all n_G Gauss integration points; $\sigma_o^{r(i)}$ denotes the yield limit of the r-th material component at point i; ρ_i is a Gauss weight; $\left|J\right|_i$ represents the Jacobian determinant of the master-to-slave element mapping; D is the matrix of constants arising from the von Mises model when the relevant dissipation density is made explicit; B denotes the compatibility matrix so that $\varepsilon = BU$ in the chosen model; $R = B^T D B$; $Y^T = \{ 1 \;\; 1 \;\; 1 \;\; 0 \;\; 0 \;\; 0 \}$ is the operator which extracts the volumetric strain from the 6-vector of strain components. The constraint (10.5c) expresses the plastic incompressibility entailed by von Mises material model.

2.2 NUMERICAL SOLUTION

Problem (10.5a)-(10.5c) consists of the minimization of a convex non-smooth function under linear equality constraints. In [6] numerical solution has been achieved by means of the iterative algorithm outlined below and derived, with slight changes, from that originally devised in [7]:

(1) Deal with constraint (10.5b) by the Lagrangian method.

(2) Enforce constraint (10.5c) by penalization, i.e. adding to the objective function the product of a penalty factor α and of the sum, with respect of index i, of the squares of the LHS of (10.5c).

(3) Express the stationarity of the Lagrangian function by setting to zero its derivatives with respect to the variables E, $\tilde{U}$ and the Lagrange multipliers λ. This leads to the following system of nonlinear equations in the unknowns $\dot{\tilde{U}}^*$, $\dot{E}$, λ (Σ being the average stresses with the role of input basic "loads"):

$$\sqrt{\frac{2}{3}}\frac{1}{V}\sum_{i\in\mathcal{I}} \sigma_o^{r(i)} \rho_i \left|J\right|_i \frac{R_i^* \dot{\tilde{U}}^* + B_i^{*T} D \dot{E}}{H^i} + \alpha \sum_{i\in\mathcal{I}} \rho_i \left|J\right|_i (B_i^{*T} C B_i^* \dot{\tilde{U}}^* + B_i^{*T} C \dot{E}) = 0, \tag{10.6a}$$

$$\sqrt{\frac{2}{3}}\frac{1}{V}\sum_{i\in\mathcal{I}} \sigma_o^{r(i)} \rho_i \left|J\right|_i \frac{D B_i^* \dot{\tilde{U}}^* + D \dot{E}}{H^i} + \alpha \sum_{i\in\mathcal{I}} \rho_i \left|J\right|_i (C B_i^* \dot{\tilde{U}}^* + C \dot{E}) = \lambda \Sigma, \tag{10.6b}$$

$$\Sigma^T \dot{E} = 1, \tag{10.6c}$$

having set

$$H_i = \sqrt{\dot{E}^T D \dot{E} + 2 \dot{E}^T D B_i^* \dot{\tilde{U}}^* + \dot{\tilde{U}}^{*T} R_i^* \dot{\tilde{U}}^*}, \qquad C = YY^T. \tag{10.7}$$

Clearly, if H_i (which measures the energy dissipation rate at the Gauss point i) were known, (10.6a)-(10.6c) would become linear.

(4) In order to initialize an iterative procedure, assume for example that $H_i = 1$ for any i (implying that plastic yielding is assumed to occur everywhere) and solve the thus linearized equations (10.6a)-(10.6c).

(5) At iteration $k+1$, if $H_i^k > \gamma$ (γ being a suitable tolerance) set $H_i^{k+1} = H_i^k$; otherwise set $H_i^{k+1} = \beta$ (where $\beta \ll 1$). Then, solve the thus linearized equations (10.6a)-(10.6c).

(6) Go to (5) until some convergence criterion is satisfied.

2.3 REMARKS

(a) Shakedown analysis under fluctuating loading can be interpreted as a generalization of limit analysis under proportional loading. Such generalization does not alter the essential qualitative features of the resulting mathematical problems to solve. In fact, the above outlined kinematic formulation and algorithm were applied to both limit and shakedown analyses. Two kinds of engineering situations have been studied: ductile periodic composites, referred to in what precedes (Section 2.2); pressurized pipelines and pressure vessels with slots (as a result of damage due to corrosion) in their walls.

(b) After a suitable choice, on an empirical basis, of the penalty parameter α and of the tolerances β and γ for each class of industrial problems mentioned above at (a), the algorithm outlined in Section 2.2 turned out to solve efficiently limit and shakedown analysis problems for a number of different geometries with a few thousands of variables. The cost effectiveness ascertained by comparisons with other approaches (MP algorithms based on the static limit theorems, evolutive elastoplastic analysis by a commercial code) and the accuracy checked by comparisons with exact analytical solutions (Prandtl indentation problem in plane strains, see e.g. [8]) lead us to believe that the present limit analysis procedure is a satisfactory computational tool for engineering practice. However, at present, the following questions of a mathematical nature appear to be still open. Can convergence be rigorously proven? Are there *a priori* criteria to choose the penalty parameter α and the tolerances β and γ? What is the influence of these parameters and of the initialization vector on the convergence and its rate of the iteration sequence?

(c) The combined effects of plastic incompressibility (10.5c) and of modeling the displacement field in the sense of the traditional finite element (FE) method, are known to possibly induce "locking phenomena" [9]. These basically consist of excessive kinematic restrictions which hinder the development of a subset of the compatible stressless plastic mechanism admitted by the original continuum and, hence, fictitiously increase, sometimes dramatically, the safety factors. A popular anti-locking remedy in FE work is "reduced integration". An alternative provision adopted herein is the "soft" enforcement of constraint (10.5c) by penalization. In fact, by suitable choice of factor α, local violations of plastic incompressibility turn out to be reasonably small and, simultaneously, the factor s is very accurate when compared to analytical solutions.

(d) On the theoretical side, the formulation adopted herein, stemming from the kinematic theorem and amounting to the minimization of a nonsmooth convex function under linear equality constraints, can be proven to be equivalent to the minimization of a smooth objective function under inequality constraints. The latter can in fact be formulated from the dual problem involving the maximization (using the static limit analysis theorem) of the load factor over the convex feasible domain defined by the yield inequalities and the (approximate) equilibrium equations. Such an equivalence deserves investigation both from the mechanical standpoint as for the role of FE field modeling in space, and from the MP standpoint as for solution methods arising from the static formulation and its dual. Also, the possible advantages of using concepts of nonsmooth mathematics (such as subdifferentials) appear worthy of investigation.

3 BOUNDS ON POST-SHAKEDOWN QUANTITIES

3.1 MODEL

There is no doubt that for ductile structures subjected to variable repeated loads, a shakedown analysis is an elegant and cost-effective approach to the safety evaluation. Current theory, basically an offshoot of the seminal works of Bleich and Melan in the late Thirties and of Koiter and Symonds in the Fifties, is now fairly well-established and can provide various procedures suitable for computing the safety factor with respect to the occurrence of shakedown or the lack of shakedown (inadaptation) for a structure under various conditions, including dynamic effects, nonassociativity of constitutive laws and advanced material models.

However, it is useful, indeed often mandatory as in the case of a theoretically unbounded shakedown limit, to supplement a shakedown analysis with upper bounds on various quantities of interest at certain designated points on the structure. Again, the theoretical results for achieving this aim are well-developed. A recent paper in this area [10] summarizes the development of bounding theorems and derives results of wide applicability. In particular, the so-called "fundamental inequality" [11] is extended to provide a general bounding theorem which can supply theoretical bounding results on quantities of interest (such as meaningful displacements) for a wide range of conditions, involving dynamic loading, nonassociative flow rules and nonlinear hardening under suitable restriction of material stability. The search for the best bound necessarily involves an optimization problem.

In the following, we focus on the key ingredients of the discretized mathematical model leading in particular to a typical example of such a class of optimization problems which, in our view, still requires further research from the algorithmic standpoint. The ultimate aim is to compare and validate the generated bounds with experimental data, as was done, for example, in the early Eighties for grillages [12]. Paucity of space precludes detailed derivations and a presentation in its full generality; the interested reader is referred to [10] for details.

The governing relations for a broad class of discrete FE models, in the small deformation regime, can be described by the following relations, expressing, respectively, geometric compatibility (under the hypotheses of fixed constraints), equilibrium and (symbolically) a path-dependent constitutive law:

$$\varepsilon(t) = Bu(t), \tag{10.8}$$

$$B^T \sigma(t) = f(t), \tag{10.9}$$

$$\sigma = \sigma(\varepsilon(\tau)), \qquad \tau \leq t, \tag{10.10}$$

where u, ε and σ are vectors describing the unknown displacements, generalized strains and generalized stresses, respectively; f is a vector of known quasistatic externally applied loads which vary in a repeated manner with time t; and B is a compatibility matrix of full column rank.

Here, a multifield modeling in space is implicitly referred to, so that the scalar product (and its energy meaning) of conjugate variables is preserved in passing from local to element formulations and essential features of operators (such as symmetry and sign-definiteness) are the same in the continuum and in its discrete model consisting of a FE aggregate. These circumstances make the FE model a description of the system physically legitimate as the continuum formulation and a basis for theoretical developments. Such nonconventional space discretization, called "natural" or one in "Prager's generalized variables" was investigated for example in [13] A familiar illustrative special case is represented, for 3D solids, by 4-node tetrahedral FE with linear interpolation for displacements and constant local strains and stresses: the "natural" generalized variables vectors ε and σ, referred to in the above equations (10.8)-(10.10), are 6 edge elongations and 6 self-balanced edge forces, respectively, in one-to-one correspondence with the element local strain and stress tensors, respectively [15].

The use of generalized variables enables us to describe the constitutive laws at the element level as follows:

$$\varepsilon = \varepsilon^e + \varepsilon^p, \tag{10.11}$$

$$\sigma = \frac{\partial U}{\partial \varepsilon^e}, \qquad \chi = \frac{\partial W}{\partial \eta}, \tag{10.12}$$

$$\dot{\varepsilon}^p = \frac{\partial \phi^T}{\partial \sigma}\dot{\lambda}, \qquad \dot{\eta} = -\frac{\partial \phi^T}{\partial \chi}\dot{\lambda}, \tag{10.13}$$

$$\phi(\sigma, \chi) \leq 0, \qquad \dot{\lambda} \geq 0, \qquad \phi^T \dot{\lambda} = 0. \tag{10.14}$$

Equation (10.11) expresses assumed additivity of elastic and plastic generalized strains. Relations (10.12) postulate the existence of two potential functions U (external recoverable elastic strain energy) and W (internal inelastic strain energy) with the latter described through the vector η of kinematic variables conjugate with a vector χ of static internal variables. Normality (i.e. plastic flow is perpendicular to the yield function) is described by (10.13). Finally, (10.14) defines the yield condition ϕ and enforces its complementarity relationship with the vector of plastic multipliers $\dot{\lambda}$. A simple example, considered below for brevity, of the potential functions in (10.12) is $U = \frac{1}{2}\varepsilon^{eT} E \varepsilon^e$ and $W = \frac{1}{2}\eta^T H \eta$ describing a common and well-known case of linear elasticity (through the positive definite matrix E of elastic stiffnesses) and linear hardening (through matrix H).

We now turn to the specific issue of bounds under repeated loading conditions. Without going into details (see [10, 11]), a general upper bound result can be derived from consideration of a fictitious loading process involving, in our case, the superposition of a purely elastic stress state $\sigma^E(t)$ with a self-stress state ρ, constant in time, as follows:

$$\tilde{\sigma}(t) = \sigma^E(t) + \rho, \tag{10.15}$$

$$\sigma^E(t) = EBK^{-1}f(t), \tag{10.16}$$

$$B\rho = 0. \tag{10.17}$$

Through the introduction of so-called "perturbation parameters", various results concerning bounds can be specifically generated by means of general theorems presented in [10]. For the sake of simplicity of exposition, hopefully without undue loss of generality, we consider the specific and usually desirable example of attempting to bound plastic strains ε^p up to an instant $\bar{t}$. We further assume that no plastic deformations are present at $t = 0$ and that fictitious initial conditions are the same as actual conditions. After defining vector r as one which collects perturbation coefficients, the desired bound is provided by

$$r^T \varepsilon^p(\bar{t}) \leq L_e + L_s, \tag{10.18}$$

$$\phi(\tilde{\sigma}(t) + r, \tilde{\chi}) \leq 0, \qquad 0 \leq t \leq \bar{t}, \tag{10.19}$$

where, for the assumed linear elastic, linear hardening case

$$L_e = \frac{1}{2}\rho^T E^{-1} \rho, \qquad L_s = \frac{1}{2}\tilde{\eta}^T H \tilde{\eta}, \qquad \tilde{\chi} = H\tilde{\eta}, \tag{10.20}$$

If we wish to obtain the best upper bound on the j-th generalized plastic strain, $\varepsilon^p_j(\bar{t})$ can be selected from the vector $\varepsilon^p(\bar{t})$ by setting $r^T = (0, \ldots, r_j = \omega, \ldots, 0)$ and solving the following optimization problem:

$$\varepsilon^p_j \leq (\varepsilon^p_j)^{opt} = \min_{\omega,\rho,\tilde{\eta}} \left\{ \frac{1}{\omega}(L_e + L_s) \right\}, \tag{10.21a}$$

$$\text{subj. to: } \omega > 0, \quad \phi(\sigma^E(t) + \rho + r, H\tilde{\eta}) \leq 0, \quad 0 \leq t \leq \bar{t}. \tag{10.21b}$$

The objective function is convex and quadratic in all variables but ω, the constraints are convex since yield functions in vector ϕ are convex too.

3.2 REMARKS

(a) It must be stressed, as already alluded to, that we have deliberately described a simple case. More general results provided in [10] cover such cases as nonassociativity (through the notion of a "reduced domain"), nonlinear hardening and dynamics.

(b) Numerical solution of optimization problems such as given by (10.21a)-(10.21b) is still an open problem. More precisely, there is a need for the development of efficient and robust optimization tools aimed at providing the tightest bounds.

(c) A specific instance of this problem was studied some 17 years ago by Kaneko [14]: reference was to truss-like structures, any element of which entails a single (instead of 6, as for the afore-mentioned 4-node tetrahedra) generalized strain (bar elongation) and stress (bar force), and involves linear yield inequalities (two for each truss member).We summarize herein some key findings in order to highlight some of the problems associated with solving (10.21a)-(10.21b). The problem considered in [14] can be stated explicitly as follows (we retain the symbols used in that paper for easy reference):

$$\min_{\omega,z,y} \left\{ \frac{1}{\omega}(\tfrac{1}{2}z^T Hz + K_o) \right\}, \tag{10.22a}$$

$$\text{subj. to: } Hz - N^T y + k - \omega d \geq 0, \tag{10.22b}$$

$$C^T y = f, \tag{10.22c}$$

$$\omega > 0, \qquad z \geq 0. \tag{10.22d}$$

The variables of problem (10.22a)-(10.22d) are: scalar ω, n-vector z, and m-vector y. The data are: $n \times n$ symmetric positive semidefinite matrix H, nonnegative scalar K_o, $m \times n$ matrix N, $m \times t$ matrix C, n-vectors k and d, and t-vector f. In the first instance, the feasibility of problem (10.22a)-(10.22d) can be determined by solving a linear program. Assuming that the problem is feasible, Kaneko [14] proposed two methods for its solution.

Since the objective function is convex and differentiable, and the constraints are all linear, a standard NLP algorithm can, in principle, be applied. However, an NLP approach requires the replacement of constraint $\omega > 0$ by $\omega \geq \omega'$ where ω' is a small positive number. If ω' is too small, possible numerical instability may occur whereas if ω' is too large, the true optimum may be overlooked. Indeed, as reported in [14], numerical difficulties were experienced with a series of small size test problems.

The second approach and the one which is the more promising, according to Kaneko [14], is to tackle the problem as a QP parametric in ω. For every fixed $\omega \in [\underline{\omega}, \infty)$, where $\underline{\omega}$ is the minimum of $\omega \geq 0$ such that (ω, z, y) satisfies (10.22b) and (10.22c), let

$$s(\omega) = \min_{z,y} \left\{ \frac{1}{\omega}(\tfrac{1}{2}z^T Hz + K_o) \right\}, \tag{10.23a}$$

$$\text{subj. to: } Hz - N^T y + k - \omega d \geq 0, \tag{10.23b}$$

$$C^T y = f, \tag{10.23c}$$

$$z \geq 0. \tag{10.23d}$$

Then, (10.23a)-(10.23d) can be solved as the one-dimensional line search problem

$$\min \{ s(\omega) : \omega \in [\underline{\omega}, \infty), \quad \omega \neq 0 \}. \tag{10.24}$$

There are two main advantages in this approach: (i) not having to artificially choose ω' and (ii) the solution involves QPs (of finite termination) instead of an NLP. Kaneko [14] suggests the use of Cottle's algorithm to solve (10.23a)-(10.23d) as an LCP. He obtained satisfactory results with this approach, albeit on some limited small size problems only. Clearly, large scale computing (involving some thousands of variables) is a frequent requirement in real-life engineering situations and still represents, also in the present context of bounds, a challenge for the near future.

4 SHAKEDOWN ANALYSIS IN POROPLASTICITY

4.1 MODEL

The integrity assessment of large dams (existing or in the design process) is one of the most fascinating and timely issues in civil engineering. In fact, besides its obvious social and economic importance, this technological issue has a scientific appeal since it stimulates applications of recent results and further research in various branches of present day mechanics, such as quasibrittle fracture, computational techniques, mechanics of materials, size effects, seismic dynamics and coupled multifield problems.

One of the current research topics related to dam engineering (albeit also to petroleum and environmental engineering issues, such as subsidence and hydraulic fracture) concerns poroplasticity and, more specifically, direct non-evolutive methods for limit-state analysis in poroplasticity. This topic will be briefly considered below, since it represents a natural though novel extension of the mechanical concepts dealt with in Sections 2 and 3, and sets further requirements in terms of large scale computing in MP. As for the first results in this direction, details can be found in [16]; one should refer to [17] for the underlying mechanics of multiphase media and coupled problems tackled by FE procedures.

Consider a masonry or concrete dam (or a rock geological formation beneath it) affected by diffuse cracks permeated with water. For overall analyses, the interpretation of the material behavior as a two-phase (porous solid skeleton filled by liquid) fully saturated poroplastic medium is currently widely accepted. It can be described mathematically by the following constitutive model:

$$\varepsilon = \varepsilon^e + \varepsilon^p, \qquad \zeta = \zeta^e + \zeta^p, \tag{10.25}$$

$$\varepsilon^e = C\sigma + Pp, \tag{10.26}$$

$$\zeta^e = P^T\sigma + Qp, \tag{10.27}$$

$$\phi = N_\sigma \sigma + N_p p - Y_o^o - H\lambda, \tag{10.28}$$

$$\dot{\varepsilon}^p = N_\sigma^T \dot{\lambda}, \qquad \dot{\zeta}^p = N_p^T \dot{\lambda}, \tag{10.29}$$

$$\phi \leq 0, \qquad \dot{\lambda} \geq 0, \qquad \phi^T \dot{\lambda} = 0. \tag{10.30}$$

As in Section 3, a FE multifield space discretization in "natural" generalized variables is referred to, and the above relations concern elemen-twise the whole assembled FE aggregate.

The traditional additivity hypothesis (10.25) now applies not only to the strains ε of the solid skeleton, but also to the fluid content variable ζ which measures the amount of liquid per unit volume of the fluid. In fact, an irreversible ("plastic") addend ζ^p is envisaged in order to allow for inelastic re-arrangements of the solid microscale, which can cause "permanent" capture or release of pore fluid. The reversible ("elastic") addends of both quantities are linearly related by (10.26) and (10.27) to their conjugate counterparts, namely total stresses σ and pressures p, respectively. These equations express Biot classical (coupled) poroelastic constitution, see e.g. [17], where C and Q (both symmetric and positive definite, the former collecting the elastic compliances) and P are matrices depending on material parameters and also on the adopted FE model. The yield functions are formulated by (10.28) as linear (or linearized, as an approximation) in σ, p (the fluid pressure vector) and in the plastic multipliers λ through a hardening matrix H which is symmetric and positive semidefinite as a consequence of assumed material stability. "Associativity" and "consistency" of the flow rules are expressed by (10.29) and (10.30), respectively. Associativity (or "normality rule") requires that each rate vector of the irreversible kinematic variables (contained in $\dot{\varepsilon}^p$ and $\dot{\zeta}^p$) is proportional to the gradient of the relevant yield function ϕ with respect to the corresponding static variables (contained in σ and p). These gradients are assembled in matrices N. It is worth noting that, if there is no fluid phase (i.e. if ζ and p are dropped), (10.25)-(10.30) become a special case of (10.11)-(10.14), namely a piecewise linearized version (with $\eta = \lambda$).

The FE discretized field equations of compatibility (10.8) and equilibrium (10.9) still hold, now interpreting u as encompassing all nodal displacements and f both body (gravity) forces on the mixture and given tractions. They must be supplemented by the mass conservation law and filtration (Darcy) law, which, combined, can be expressed in the form

$$\dot{\zeta} = -Vp + F. \tag{10.31}$$

In (10.31), matrix V (symmetric and positive definite) reflects the permeability properties of the material and vector F contains boundary fluxes and gravity forces on the fluid alone.

Now, let us: (i) substitute ε^e, ζ^e and σ using (10.25), (10.26) and (10.31); (ii) impose the boundary conditions, namely tractions and displacements (set equal to zero) on two complementary parts of the boundary, pressure and fluxes (set equal to zero) on the other two complementary portions of the boundary. These algebraic manipulations, see [16], lead to:

$$Ku - Lp = \mu c(t) + R\varepsilon^p, \tag{10.32}$$

$$L\dot{u} + S\dot{p} + Vp = \mu d(t) + P\dot{\varepsilon}^p - \dot{\zeta}^p, \tag{10.33}$$

where c and d are vectors of data, and scalar μ is the load factor which is assumed to amplify all external actions.

The linear ordinary differential equations (10.32) and (10.33) and the piecewise linear "inelastic" constitutive relations (10.28)-(10.30), together with the initial condition $p(0) = p_o$, govern the time evolution of the space-discretized poroplastic system under loading at level μ.

4.2 SHAKEDOWN THEOREMS AND LINEAR PROGRAMMING

Similar to the engineering situations referred to in the preceding Sections, it is practically and conceptually important to assess, also in the present poroplastic context, the load carrying capacity "directly", i.e. by avoiding expensive repeated (by trial-and-error for various μ) solutions of the above-formulated nonlinear initial-value problem.

Within the class of models described in Section 4.1, the carrying capacity is measured by the scalar s (safety factor) such that: for $\mu < s$ the system shakes down; for $\mu > s$ it does not, in the sense that (like in plasticity) the cumulative dissipated energy is unbounded in time, a circumstance which characterizes incremental collapse, alternating plasticity and, as a special case for non-fluctuating loads, plastic collapse.

With respect to traditional plasticity, poroplasticity implies additional fields (pressure p and fluid content ζ), fluid-structure coupling, through (10.32) and (10.33), and time dependence, through (10.33). In such context, a novel shakedown theory [16] was recently developed; if specialized to the piecewise linear associative model of Section 4.1, under the hypothesis of periodic external actions, this theory reduces the search for the safety factor s to the following pair of dual LP problems:

$$s = \max\{\mu\}, \tag{10.34a}$$
$$\text{subj. to: } \mu M - A\lambda \leqslant Y_o, \qquad \mu \geqslant 0, \qquad \lambda \geqslant 0, \tag{10.34b}$$

$$s = \min\{Y_o^T \lambda\}, \tag{10.35a}$$
$$\text{subj. to: } M^T\lambda \geqslant 1, \qquad A\lambda \leqslant 0, \qquad \lambda \geqslant 0, \tag{10.35b}$$

having set

$$A \equiv H - N_\sigma Z N_\sigma^T, \qquad Z \equiv EBK^{-1}B^T E - E, \tag{10.36}$$
$$M \equiv \max\{N_\sigma \sigma^E(t) + N_p p^E(t)\}. \tag{10.37}$$

In (10.37), vectors σ^E and p^E represent, respectively, the total stress and pressure response of the system to the loads in a fictitious, linear, purely poroelastic regime and, if the loads are periodic, can be economically computed in closed form, disregarding the transient regime due to initial conditions. In (10.36), matrix $-Z$ and, hence A, are symmetric positive semidefinite with low rank.

4.3 REMARKS

(a) The underlying theory [16] confers mechanical meaning to the above LP problems. In fact, (10.34a)-(10.34b) and (10.35a)-(10.35b) embody generalizations of the static and kinematic theorems of classical plasticity, respectively. The optimal vector associated with the common optimal values, in the primal LP (10.34a)-(10.34b), defines one of the sets of inelastic strains likely to generate "stabilizing" self-stresses at shakedown; in the dual LP (10.35a)-(10.35b), it represents one of the possible compatible collapse mechanisms at inadaptation just beyond s.

(b) The limited numerical experience achieved so far in poroplastic shakedown analysis highlights two main circumstances: (i) the computing effort required by direct procedures via (10.34a)-(10.34b) and/or (10.35a)-(10.35b) is smaller at least by one order of magnitude than that needed for repeated time-stepping solutions by a commercial nonlinear FE code; (ii) solutions of the dual LP (10.35a)-(10.35b) are generally more expensive than those of the primal (10.34a)-(10.34b); (iii) in passing from a few dozens of variables in academic tests to several hundreds for more realistic applications, only one (CPLEX) out of five adopted LP computing codes consistently provided reliable solutions.

(c) There are alternative LP formulations (e.g., when $H = 0$, in σ only or in the "redundancies" only) which can be derived from the new shakedown theorems in poroplasticity and might alleviate the computational difficulties above-mentioned in (iii). Some current research concerns are: NLP approaches (i.e. abandoning the cumbersome piecewise linear approximation of yield functions) and their computational merits with respect to the LP formulations outlined herein; extension to poroplasticity of bounding techniques similar to those in Section 3 and their application in terms of large scale computing; and, finally, extensions to more sophisticated material models such as partially saturated damageable two-phase media.

5 QUASIBRITTLE FRACTURE ANALYSIS

5.1 MODEL

Several materials of engineering relevance (e.g. concrete, mortar and some rocks, ceramics and polymers) are called "quasibrittle". As one of their main characteristic features, the following idealization of crack propagation process is widely regarded as realistic for overall analysis purposes: cohesive tractions continue to be transmitted across the two faces of developing crack surfaces ("process zone") up to some critical value of the relative displacements, while in the domain outside the fracture surface, linear elastic behavior can still be assumed. These tractions and the corresponding relative displacements are linked by a constitutive law ("interface law") which exhibits "softening", i.e. tractions decay with the increase of opening/sliding relative displacements.

Let us consider, for instance, the widely used so-called 3-point bending (3PB) test, or the 2-notch tensile (2NT) test (see e.g. [18, 19] for schematic diagrams of these tests). These are special cases of mode I (i.e. opening mode) fracture, for which the geometry and the location of the potential fracture surface, say Γ, are *a priori* known and the latter coincides with the symmetry axis of each specimen. For the computer simulation of such fracture processes, the distribution, along axis x, of the opening displacements $w(x)$ across Γ, and of the tractions $t(x)$ has to be determined as a function of some input scalar parameter α. This problem can be given the following compact and computationally effective formulation which exploits its peculiar features [20, 21].

The linearity of the material behavior outside Γ allows us to express the tractions $t(x)$ by superposition as the sum of two contributions, namely

$$t(x) = \alpha t^E(x) + \int_\Gamma z(x,\xi)w(\xi)d\Gamma, \qquad x,\xi \in \Gamma, \tag{10.38}$$

where $t^E(x)$ denotes the distribution of tractions generated by the external "load" in a fictitious elastic regime, i.e. in the absence of displacement discontinuities, while the Green's (influence) function $z(x,\xi)$, which reflects the geometric and elastic properties of the considered specimen, provides the effects of the opening displacements in the otherwise unloaded body.

A further relationship between $t(x)$ and $w(x)$ is provided by the interface law.

For progressive fracture phenomena, i.e. when local closure of cracks can be ruled out *a priori*, the interface law can be described by the following (so-called "holonomic" or reversible) analytical representation in a complementarity format:

$$\varphi_w(x) = f(w(x)) - f(\lambda(x) + w_c) - t(x) \geq 0, \quad w(x) \geq 0,$$
$$\varphi_w(x)w(x) = 0, \tag{10.39a}$$
$$\varphi_\lambda(x) = w_c - w(x) + \lambda(x) \geq 0, \quad \lambda(x) \geq 0,$$
$$\varphi_\lambda(x)\lambda(x) = 0. \tag{10.39b}$$

where f is a given function such that $f(0) = t_c$ and $f(w_c) = 0$, where t_c and w_c represent the material tensile strength and the critical opening displacement, respectively, while $\lambda(x)$ denotes an auxiliary variable field [22, 23]. Linear, piecewise linear and nonlinear expressions have been proposed for function f to represent specific material behavior (see e.g. [23] and the references contained therein). The above set of relationships allows us to describe the behavior of all points along the interface Γ, belonging either to the still undamaged material ($w = \lambda = 0$, $t < t_c$), or to the "process zone" ($0 \leq w \leq w_c$, $\lambda = 0$, $t = f(w)$), or to the true crack ($w > w_c$, $t = 0$). If desired, the tractions can be eliminated in the process by combining the interface relations (10.39a)-(10.39b) with the integral equation (10.38).

An LCP or an NCP can then be obtained as a final formulation of the problem. It is noteworthy that this holonomic formulation is particularly suited for the parameter identification problem which will be presented in Section 6.

An LCP always arises when the nonholonomic (irreversible) material behavior in the process has to be accounted for as required, for instance, for a reliable simulation of the 2NT test [19]. In this case, the irreversible behavior at one of the points belonging to the process zone can be described as follows:

$$\dot{\varphi}(x) = h_w(x)\dot{w}(x) - \dot{t}(x) \geq 0, \quad \dot{w}(x) \geq 0,$$
$$\dot{\varphi}(x)\dot{w}(x) = 0, \tag{10.40}$$

where $h_w(x)$ represents the (negative) slope of the $f(w)$ curve at the considered point in the current configuration and dots mark rates (i.e. derivatives with respect to ordinative "time").

Along Γ, but outside the process zone, the material exhibits a reversible behavior characterized by the conditions: $\dot{t} = 0$ on the true crack (where $w > w_c$), and $\dot{w} = 0$ (and $t < t_c$) in the undamaged zone.

To obtain the desired formulation, the nonholonomic interface relations (10.40) have to be coupled with the corresponding rate form of the integral equation (10.38) and then supplemented by an appropriate time-integration tip-advancement procedure capable of following the crack propagation evolution in the specimen.

For computational purposes, the above problems, either in total terms or in rate form, are usually discretized through FE or boundary element (BE) modeling. In these space-discretized formulations, the concept and methodology of generalized variables outlined in Section 3 can be fruitfully applied.

Discretization transforms the integral relationship (10.38) into a system of linear algebraic equations relating vector T, governing the modeled traction field, to the opening displacement vector W through a matrix Z. This matrix may or may not, depending on the adopted discretization technique, preserve the original properties of the integral operator it approximates (namely, symmetry and negative-definiteness or semi-definiteness). In particular, FE and symmetric Galerkin BE methods preserve these features while standard BE collocation techniques do not [25, 26, 27]. Discretization can also entail the introduction of variables (more numerous in FE approaches than in BE ones which do not involve domain variables) other than W and T, depending on the chosen techniques. However, static condensation of any variable entering the discretization, other than T and W, allows the discrete counterpart of relation (10.38) to be always expressed as

$$T = \alpha T^E + ZW. \tag{10.41}$$

The description of the interface behavior as well has to be reduced to a finite-dimensional space through space modeling. In particular, the discrete counterparts of the interface rate relations (10.41), namely

$$\dot{\Phi} = H_w \dot{W} - \dot{T} \geq 0, \qquad \dot{W} \geq 0, \qquad \dot{\Phi}^T \dot{W} = 0 \tag{10.42}$$

are characterized by the (hardening/softening) matrix H_w which reflects the distribution of the moduli h_w. Hence, matrix H_w may be sign-indefinite or negative definite. Evolutive analyses can then be reduced to a sequence of LCPs or NCPs in finite increments by means of some time integration scheme.

The discrete version of the holonomic relations (10.39a)-(10.39b) acquires the following format of an LCP or NCP directly in total variables:

$$\Phi_w = F(W) - F(\Lambda + w_c e) - T \geq 0, \qquad W \geq 0, \qquad \Phi_w^T W = 0, \tag{10.43a}$$

$$\Phi_\lambda = w_c I - W + \Lambda \geq 0, \qquad \Lambda \geq 0, \qquad \Phi_\lambda^T \Lambda = 0. \tag{10.43b}$$

Here, symbol F denotes one set of properly defined vector-valued functions which reflects the characteristics of the continuum function $f(w)$ [23], Λ represents the vector which governs the space-modeled auxiliary variables λ, while e represents a vector of ones.

5.2 NUMERICAL SOLUTION

LCPs and NCPs arising within the quasibrittle fracture context as shown in Section 5.1, may admit one solution, no solution or multiple solutions, thus reflecting some crucial aspects of the mechanical problem. For instance, for the holonomic formulation, lack of solution would indicate that the load carrying capacity of the structure is below the applied load level. A multiplicity of solutions corresponds, in rates, to a phenomenon of response path branching or bifurcation; in holonomic single step analyses to a variety of equilibrium configurations for the same applied load — both frequent occurrences in the presence of softening.

In computer simulations it would be desirable to capture all solutions, as each one has a particular physical meaning. Unfortunately, this task still represents a formidable challenge for large size problems.

For LCPs of order up to about 150, we have successfully used the enumerative approach proposed by Judice and Mitra [28]. As expected, due to the intrinsic combinatorial nature of the method, computing time exponentially grows with the problem dimension and becomes prohibitive for practical, large size engineering problems [21].

For large size LCPs and for NCPs, apparently no existing method guarantees to capture all solutions or to show that none exists with a reasonable computing effort. However, we have had considerable success with the PATH code [29]. Based on some *a priori* knowledge of the "shape" of the crack pattern, all solutions could be captured in some benchmark exercises [21, 22] by using different starting vectors to initialize the solver. In particular, we have used vectors which have been filled by zero and nonzero entries, as suggested by the expected opening and closing crack patterns. This heuristic, however, still requires fairly extensive testing on various other problems.

5.3 REMARKS

(a) From a computational point of view, quasibrittle fracture analyses resting on holonomic descriptions are more demanding since the complementarity problem involves all nodes on the potential crack itinerary. Further, auxiliary variables Λ are often also introduced, besides W, leading to large size problems. The problem dimension is of great concern when branch-and-bounds procedures are to be used for finding all alternative equilibrium paths of a mechanical problem formulated as an LCP. Equilibrium branching is usually more appropriately dealt with by incremental procedures for which much smaller LCPs are involved. Moreover, there is a definite need to develop robust and efficient procedures guaranteed to yield all solutions for practical, often large size engineering problems.

(b) Current work, which has yielded as yet unreported encouraging results, aims at using the framework provided by the modeling system GAMS [30], for which the PATH solver is available [31]. Particular advantages, from the application point of view, are that (i) it can efficiently manage a large volume of data, (ii) it allows the problem to be formulated in its most natural mixed complementarity problem (MCP) format, rather than requiring an explicit elimination of variables to give an LCP or NCP, (iii) it allows the user to concentrate on the model rather than the solution methodology, and (iv) it provides automatic differentiation facilities which are particularly advantageous for nonlinear interface models.

6 IDENTIFICATION OF FRACTURE PARAMETERS

6.1 MODEL

Engineering-oriented mathematical models are designed to simulate real mechanical responses more and more accurately, and, in particular, are characterized by the rapidly increasing complexity of constitutive laws apt to reflect the essential features of structural material behavior. These sophisticated laws naturally embody a large number of material parameters which have to be calibrated on the basis of experimental data.

Interface models have been introduced for analyzing quasibrittle fracture processes (see Section 5). Also in this context, the behavior of any material is quantitatively described by a set of parameters, some of which (e.g. the tensile strength t_c) have a direct physical meaning and can be experimentally measured directly; others (e.g. the critical opening displacement w_c) can hardly be measured; others still may have no direct physical meaning.

A powerful, albeit difficult, methodology for identifying all such material parameters entails the solution of an inverse problem. In essence, this methodology requires one: (i) to carry out proper experiments on material samples and to record meaningful quantities (e.g. reaction forces, displacement components, etc.); (ii) to formulate a suitably accurate mathematical model of the experiment for computing (implicitly, as functions of the parameters) the quantities corresponding to the measured ones; (iii) to select a function which defines the discrepancy between measured and computed quantities; and (iv) to minimize this error function using the parameters as optimization variables.

Experimental data relevant to quasibrittle fracture processes can be acquired on the basis of standard 3PB tests which are routinely and economically performed also in industrial environments. The 3PB test is usually run under displacement control and the reaction force Q^m is measured for each recorded displacement. Information from the load-displacement curve can then be used directly to determine the material fracture energy G_f which represents the area under the traction t versus crack opening w curve. In some cases, further experimental data such as the displacements U^m of some reference points (arranged in a grid-like pattern) on the specimen surface can be obtained [18, 23], usually by optical methods (e.g. interferometry).

The inverse identification problem can be formulated as follows. It is assumed that a test such as the above-mentioned one has been carried out. Reactions Q^m and displacements U^m have been measured (marked by the superscript m). Let Q^c and U^c denote the reaction and displacement values, respectively, computed (as indicated by the superscript c) for the same corresponding loadings (imposed displacements), for instance by means of the structural model materialized in the complementarity problem (10.41), (10.43a)-(10.43b). A natural measure of the discrepancy (or error ω) between measured and theoretical quantities is given by the square of the Euclidean norm, weighted appropriately if desired, of their difference. Since Q^c and U^c can be expressed [18] simply as

$$Q^c = \alpha Q^E + RW, \qquad U^c = \alpha U^E + MW, \tag{10.44}$$

where R and M are the influence vector and matrix, respectively, of the displacement discontinuities W, then the error ω can be expressed as follows:

$$\omega = (U^c - U^m)^T D_U (U^c - U^m)^T + D_Q (Q^c - Q^m)^2, \tag{10.45}$$

where D_U and D_Q represent a positive definite symmetric weighting matrix and a positive weighting factor, respectively, used to confer more "weight" to the less uncertain experimental data.

The unknown fracture parameters $(t_c, w_c, \dots)$ pertaining to the cohesive model of Section 5 can then be evaluated by minimizing ω subject to the complementarity constraints (10.43a)-(10.43b). After some simple algebraic manipulations, this optimization problem becomes

$$\min_{W,\Lambda,t_c,w_c,\dots} \left\{\omega \equiv W^T C W + d^T W + a\right\}, \tag{10.46a}$$

$$\text{subj. to: } \Phi_w(W, \Lambda, t_c, w_c, \dots) \geq 0, \qquad W \geq 0, \qquad \Phi_w^T W = 0, \tag{10.46b}$$

$$\Phi_\lambda(W, \Lambda, w_c) \geq 0, \qquad \Lambda \geq 0, \qquad \Phi_\lambda^T \Lambda = 0, \tag{10.46c}$$

where C, d and a are appropriate constants [18, 23]. Further constraints representing upper and lower bounds on the parameters can also be introduced on the basis of *a priori* "expert knowledge" or engineering judgement.

Optimization problem (10.46a)-(10.46c) is clearly a generally nonconvex mathematical program with equilibrium constraints (MPEC) [32]. Ideally, we seek its global optimum.

6.2 NUMERICAL SOLUTION

Our computational experiences with solution of the fracture identification problem given by (10.46a)-(10.46c) are still limited. To date, two different algorithms have been employed, resting both on "pseudo experimental" (artificially generated) data and on actual experimental results concerning a quasibrittle polymeric composite (epoxy matrix with silicon microspherical hollow inclusions) [18, 23]. A linear (one-branch) or piecewise linear (two-branch) cohesive crack model was adopted, for which two and four parameters, respectively, need to be identified. The experimental 3PB test was computer simulated using 1840 4-node finite elements with 10 nodes on the crack interface Γ. A maximum of 48 loading situations (corresponding to different applied displacement levels α) can be included, so that maximum dimensions are: 480-vector W and 1440-vector Λ [18].

In the first instance, a classical penalty method was employed. In particular, this involves adding the penalized complementarity term $\rho(\Phi_w^T W + \Phi_\lambda^T \Lambda)$ to the objective ω, and solving a series of NLPs for increasing values of the penalty factor ρ (> 0) until the complementarity gap is reduced to a preset tolerance. The computations were carried out within MATLAB; both the optimization toolbox routine "constr" and the NAG routine "e04uef" have been tried for solving the NLP subproblems. Exact derivatives were supplied to help the solution process. Convergence was generally good but in the presence of imprecise data (perturbed pseudo experimental data or truly experimental values), different sets of parameters were identified depending on the number and load levels in the identification. This is expected, not only because of "noise" in the data, but also because, as is well-known, there is no guarantee that the solution obtained with the penalty procedure is even a local minimum of the original MPEC. For a simple example regarding this last point, the reader is referred to p. 105 of Luo *et al.* [32], where for any finite value of the penalty parameter, the optimal solution to the penalized problem is never feasible to the MPEC.

The second approach tried was a simple application of a genetic algorithm. Such algorithms have been shown to be effective in numerous kinds of optimization problems [33, 44] and are particularly appealing to the engineering community in view of the simplicity of the underlying concepts. In the present application, the genetic procedure was started with a randomly generated population, each member of which represents a set of the n material parameters to be identified (in our case $n = 4$), inside the n-dimensional box defined by upper and lower bounds which are chosen on the basis of *a priori* conjectures. For each member ("individual") of the population, a solution to the analysis problem given by MCP (10.41), (10.43a)-(10.43b) was first obtained using PATH [29]. Displacements U^c and the reaction force Q^c were then computed through (10.44) and, hence, the error ("fitness") ω through (10.45). According to a typical genetic strategy, a subset of better fit individuals is then selected from the initial population and from subsequent generations to produce a new population by such devices as "cross-over" and "mutation". The process of reproduction is continued until further reduction of the objective function is no longer observed in the fittest member. In our numerical tests, reasonable success has been achieved, and computing times required by the genetic algorithm combined with PATH were found to be comparable to those of the penalty approach.

The present problem of fracture parameters identification is mathematically almost identical to one problem investigated in the early Eighties [34, 35, 36] which required the identification of yield limits and hardening moduli (included in piecewise linear constitutive laws), presuming the availability of some measured information on the displacement response of an elastoplastic structure subject to a given quasistatic loading condition. As a preliminary to assessing and designing solution algorithms for the fracture problem (10.46a)-(10.46c), we have recently revisited that work. In fact, one of the test examples considers the meaningful (from the computational viewpoint) case of a hypothetical elastoplastic beam on elastoplastic foundation [36] structure which is ideally suited for testing MPEC algorithms.

Maier *et al.* [35] considered the case of unknown yield limits only, so that the complementarity constraints turned out to be linear. In essence, they proposed an enumerative scheme in their attempt to obtain a global minimum of the nonconvex optimization problem. Nappi [36] later extended the formulation to identify hardening parameters as well, however, under the assumption of perfect measurements. The combination of unknown yield limits and unknown hardening parameters leads, as expected, to nonlinear complementarity constraints. Nappi's numerical approach consists of minimizing the sum of error and complementarity terms. Since zero error (perfect measurements) was assumed, it is obviously possible to ascertain whether a global (zero) optimum has been reached. However, the assumption of a penalty factor of one does not guarantee that the complementarity gap can be reduced to zero in all instances.

We summarize briefly some recent preliminary findings in the following; exact as well as perturbed data (to simulate measurement error) were used in all cases. Jiang *et al.* [37] demonstrated the viability of a piecewise sequential quadratic programming (SQP) approach [38] to capture a local minimum of the optimization problem for the unknown yield limits only case. This decomposition-based technique has recently been extended to the MPEC with nonlinear complementarity constraints, but with less success. A simple heuristic, unfortunately without any sound theoretical support, has also been devised and tested by Tin-Loi and Ferris [39] for both linear and nonlinear complementarity constraints. The algorithm involves reformulating the MPEC as a parametric nonlinear program, in which we attempt to drive the complementarity gap, treated as the objective function, to zero. Whilst promising, the computational results obtained were not as good as those obtained by using a standard penalty approach (i.e. adding the penalized constraints to the objective term) or by simply treating the MPEC iteratively as a series of NLPs [40] with the complementarity conditions appropriately smoothed out (e.g. by using a smoothing of the Fischer function originally proposed by Kanzow [41]). Detailed descriptions of these approaches in the specific structural identification context, together with some comparative numerical performance of these two algorithms, are given in [42]. Motivated by computational efficiency, Jiang and Ralph [43] have extended the NLP smoothing scheme to a class of smooth SQP methods. Preliminary, as yet unreported, work using the smooth SQP scheme on the beam on elastic foundation problem has been encouraging. Finally, a robust and efficient approach, in view of current numerical tests being carried out at present, appears to be the penalty interior point algorithm (PIPA) developed by Luo *et al.* [32].

6.3 REMARKS

(a) The reliability of the identification process rests on both the confidence level attributed to the experimental data and on the appropriate design of the inverse procedure. Reliability obviously increases with the amount of data used in the inverse procedure. In view of the consequent increased size of the problems and of the inherent difficulties in solving the underlying MPEC, there is a need to develop robust and computationally efficient procedures for solving the identification problem. In this respect, the possibility of using nondeterministic techniques, such as Kalman-Bucy filtering, as used in some early work on parameter identification by Bittanti *et al.* [45] deserves investigation.

(b) The use of a modeling framework such as GAMS also deserves consideration, especially in view of such developments as the recent introduction of the "MPEC" model type into the GAMS language and of appropriate MATLAB interfaces to GAMS [46]. In fact, current ongoing work with the experimental data set using GAMS with the NLP solver CONOPT [47] indicates several orders of magnitudes increase in computing efficiency for the penalty approach as compared to a MATLAB implementation. It is interesting to note, however, that a MATLAB version of PIPA, albeit coded to take advantage of the special mathematical structures involved in the quasibrittle fracture identification problem, proved extremely robust and fast.

(c) Various aspects targeted for future research include: (i) development and comparison of further algorithms for solving the identification problem; (ii) extension of the described mode I fracture model to one of mixed mode; in preparation for this task, we have already carried out successfully, using both a penalty approach (with CONOPT as the NLP solver) and PIPA, some preliminary identification of nonlinear softening laws; (iii) investigation of regularization techniques aimed at reducing the inherent ill-posedness of the identification problem; and (iv) devising methods to efficiently attain the challenging goal of finding a global minimum to the inverse problem.

7 CONCLUSIONS

In this paper, a brief survey of five topical engineering mechanics problems, formulated as various MP problems, has been presented. As indicated in the introduction, our aim is not to provide a review of the extensive interaction of engineering mechanics and MP, but to foster an exchange of ideas regarding these specific areas. We have not attempted to give precise mathematical descriptions of any of these various problems, but rather we have tried to impart some idea of the types of open questions that exist in the hope of motivating input from MP experts.

Acknowledgments

Support of the Australian Research Council in the form of an ARC grant is gratefully acknowledged. We also thank the following friends and colleagues who have contributed in many ways to various aspects of the work described: Sergio Bittanti, Michael Ferris, Franco Giannessi, Houyuan Jiang, Jong-Shi Pang, Danny Ralph.

References

[1] M.Z. Cohn and G. Maier (eds.), *Engineering Plasticity by Mathematical Programming*, NATO Advanced Study Institute, University of Waterloo 1977, Pergamon Press, New York, 1979.

[2] G. Maier and J. Munro, Mathematical programming applications to engineering plastic analysis, *Applied Mechanics Reviews* **35**, 1982, 1631–1643.

[3] G. Maier and D. Lloyd Smith, Update to "Mathematical programming applications to engineering plastic analysis", *ASME Applied Mechanics Update*, 1985, 377–383.

[4] M.C. Ferris and J.S. Pang, Engineering and economic applications of complementarity problems, *SIAM Review* **39**, 1997, 669–713.

[5] M.C. Ferris and J.S. Pang (eds.), *Complementarity and Variational Problems: State of the Art*, SIAM, Philadelphia, Pennsylvania, 1997.

[6] V. Carvelli, *Limit and Shakedown Analysis of Three-Dimensional Structures and Periodic Heterogeneous Materials*, Ph.D. Thesis, Politecnico di Milano, 1999.

[7] Y.G. Zhang and M.W. Lu, An algorithm for plastic limit analysis, *Computer Methods in Applied Mechanics and Engineering* **126**, 1995, 333–341.

[8] J. Lubliner, *Plasticity Theory*, Macmillan Publishing Company, 1990.

[9] A. Capsoni and L. Corradi, A finite element formulation of the rigid-plastic limit analysis problem, *International Journal for Numerical Methods in Engineering* **40**, 1997, 2063–2086.

[10] A. Corigliano, G. Maier and S. Pycko, Dynamic shakedown analysis and bounds for elastoplastic structures with nonassociative, internal variable constitutive laws, *International Journal of Solids and Structures* **32**, 1995, 3145–3166.

[11] C. Polizzotto, A unified treatment of shakedown theory and related bounding techniques, *Solid Mechanics Archives* **7**, 1982, 19–75.

[12] F. Tin-Loi, Deflection bounding at shakedown, *Proceedings ASCE, Journal of Structural Division* **106**, 1980, 1209–1215.

[13] C. Comi, G. Maier and U. Perego, Generalized variable finite element modeling and extremum theorems in stepwise holonomic elastoplasticity with internal variables, *Computer Methods in Applied Mechanics and Engineering* **96**, 1992, 213–237.

[14] I. Kaneko, On some recent engineering applications of complementarity problems, *Mathematical Programming Study* **17**, 1982, 111–125.

[15] G. Maier, A matrix structural theory of piecewise linear plasticity with interacting yield planes, *Meccanica* **5**, 1970, 55–66.

[16] G. Cocchetti and G. Maier, Static shakedown theorems in piecewise linearized poroplasticity, *Archive of Applied Mechanics* **68**, 1998, 651–661.

[17] R.W. Lewis and B.A. Schrefler, *The Finite Element Method in the Deformation and Consolidation of Porous Media*, John Wiley & Sons, Chichester, UK, 1998.

[18] G. Bolzon, D. Ghilotti and G. Maier, Parameter identification of the cohesive crack model, in *Material Identification Using Mixed Numerical Experimental Methods* (H. Sol and C.W.J. Oomens, eds.), Dordrecht, Kluwer Academic Publishers, 1997, pp. 213–222.

[19] G. Bolzon, G. Maier and F. Tin-Loi, On multiplicity of solutions in quasi-brittle fracture computations, *Computational Mechanics* **19**, 1997, 511–516.

[20] G. Bolzon, G. Maier and G. Novati, Some aspects of quasi-brittle fracture analysis as a linear complementarity problem, in *Fracture and Damage in Quasibrittle Structures* (Z.P. Bažant, Z. Bittnar, M. Jirásek and J. Mazars, eds.), E&FN Spon, London, 1994, pp. 159–174.

[21] G. Bolzon, G. Maier and F. Tin-Loi, F, Holonomic and nonholonomic simulations of quasi-brittle fracture: a comparative study of mathematical programming approaches, in *Fracture Mechanics of Concrete Structures* (F.H. Wittmann, ed.), Aedificatio Publishers, Freiburg, 1995, pp. 885–898.

[22] F. Tin-Loi and M.C. Ferris, Holonomic analysis of quasibrittle fracture with nonlinear softening, in *Advances in Fracture Research* (B.L. Karihaloo, Y.W. Mai, M.I. Ripley and R.O. Ritchie, eds.), Pergamon Press, 1997, pp. 2183–2190.

[23] G. Bolzon and G. Maier, Identification of cohesive crack models for concrete on the basis of three-point-bending tests, in *Computational Modelling of Concrete Structures* (R. de Borst, N. Bicanic, H. Mang and G. Meschke, eds.), Balkema, Rotterdam, 1998, pp. 301–310

[24] R.W. Cottle, J.S. Pang and R.E. Stone, *The Linear Complementarity Problem*, Academic Press, 1992.

[25] Z. Cen and G. Maier, Bifurcations and instabilities in fracture of cohesive-softening structures: a boundary element analysis, *Fatigue and Fracture of Engineering Materials and Structures* **15**, 1992, 911–928.

[26] G. Maier, G. Novati and Z. Cen, Symmetric Galerkin boundary element method for quasi-brittle fracture and frictional contact problems, *Computational Mechanics* **13**, 1993, 74–89.

[27] G. Bolzon, Hybrid finite element approach to quasi-brittle fracture, *Computers and Structures* **60**, 1996, 733–741.

[28] J.J. Judice and G. Mitra, An enumerative method for the solution of linear complementarity problems, *European Journal of Operations Research* **36**, 1988, 122–128.

[29] S.P. Dirkse and M.C. Ferris, The PATH solver: a non-monotone stabilization scheme for mixed complementarity problems, *Optimization Methods & Software* **5**, 1995, 123–156.

[30] A. Brooke, D. Kendrick, A. Meeraus and R. Raman, *GAMS: A User's Guide*, Gams Development Corporation, Washington, DC 20007, 1998.

[31] M.C. Ferris and T.S. Munson, Complementarity problems in GAMS and the PATH solver, *Journal of Economic Dynamics and Control* **24**, 2000, 165–188.

[32] Z.Q. Luo, J.S. Pang and D. Ralph, *Mathematical Programs with Equilibrium Constraints*, Cambridge University Press, 1996.

[33] D.E. Goldberg, *Genetic Algorithms in Search, Optimization and Machine Learning*, Addison-Wesley, 1989.

[34] G. Maier, Inverse problem in engineering plasticity: a quadratic programming approach, *Academia Nazionale dei Lincei*, Serie VIII, vol LXX, 1981, 203–209.

[35] G. Maier, F. Giannessi and A. Nappi, Indirect identification of yield limits by mathematical programming, *Engineering Structures* **4**, 1982, 86–98.

[36] A. Nappi, System identification for yield limits and hardening moduli in discrete elastic-plastic structures by nonlinear programming, *Applied Mathematical Modelling* **6**, 1982, 441–448.

[37] H. Jiang, D. Ralph and F. Tin-Loi, Identification of yield limits as a mathematical program with equilibrium constraints, in *Proceedings, 15th Australasian Conference on the Mechanics of Structures and Materials (ACMSM15)* (R.H. Grzebieta, R. Al-Mahaidi and J.L. Wilson, eds.), Balkema, 1997, pp. 399–404.

[38] D. Ralph, A piecewise sequential quadratic programming method for mathematical programs with linear complementarity constraints, in *Proceedings, 7th Conference on Computational Techniques and Applications (CTAC95)* (A. Easton and R. May, eds.), World Scientific Press, 1995, pp. 663–668.

[39] F. Tin-Loi and M.C. Ferris, A simple mathematical programming method to a structural identification problem, in *Proceedings, 7th International Conference on Computing in Civil and Building Engineering (ICCCBE-VII)* (C.K. Choi, C.B. Yun and H.G. Kwak, eds.), Techno-Press, 1997, pp. 511–518.

[40] F. Facchinei, H. Jiang and L. Qi, A smoothing method for mathematical programs with equilibrium constraints, *Mathematical Programming* **85**, 1999, 107–134.

[41] C. Kanzow, Some noninterior continuation methods for linear complementarity problems, *SIAM Journal on Matrix Analysis and Applications* **17**, 1996, 851–868.

[42] M.C. Ferris and F. Tin-Loi, Nonlinear programming approach for a class of inverse problems in elastoplasticity, *Structural Engineering and Mechanics* **6**, 1998, 857–870.

[43] H. Jiang and D. Ralph, Smooth SQP methods for mathematical programs with nonlinear complementarity constraints, Research Report, Department of Mathematics and Statistics, The University of Melbourne, 1998.

[44] L.D. Davis, *Handbook of Genetic Algorithms*, Van Nostrand Reinhold, New York, 1991.

[45] S. Bittanti, G. Maier and Nappi, Inverse problems in structural elastoplasticity: a Kalman filter approach, in *Plasticity Today* (A. Sawczuk and G. Bianchi, eds.), Elsevier Applied Science Publishers, London, 1993, pp. 311–329.

[46] S.P. Dirkse and M.C. Ferris, Modeling and solution environments for MPEC: GAMS & MATLAB, in *Reformulation: Nonsmooth, Piecewise Smooth, Semismooth and Smoothing Methods* (M. Fukushima and L. Qi, eds.), Kluwer Academic Publishers, 1999, pp. 127–148.

[47] A. Drud, CONOPT – a large-scale GRG code, *ORSA Journal on Computing* **6**, 1994, 207–216.

Chapter 11

DATA DISCRIMINATION VIA NONLINEAR GENERALIZED SUPPORT VECTOR MACHINES

O. L. Mangasarian

Computer Sciences Department, University of Wisconsin, 1210 West Dayton Street, Madison, WI 53706

olvi@cs.wisc.edu

David R. Musicant

Computer Sciences Department, University of Wisconsin, 1210 West Dayton Street, Madison, WI 53706

musicant@cs.wisc.edu

Abstract The main purpose of this paper is to show that new formulations of support vector machines can generate nonlinear separating surfaces which can discriminate between elements of a given set better than a linear surface. The principal approach used is that of generalized support vector machines (GSVMs) [21] which employ possibly indefinite kernels. The GSVM training procedure is carried out by either a simple successive overrelaxation (SOR) [22] iterative method or by linear programming. This novel combination of powerful support vector machines [28, 7] with the highly effective SOR computational algorithm [19, 20, 17], or with linear programming, allows us to use a *nonlinear* surface to discriminate between elements of a dataset that belong to one of two categories. Numerical results on a number of datasets show improved testing set correctness, by as much as a factor of two, when comparing the nonlinear GSVM surface to a linear separating surface.

1 INTRODUCTION

Support vector machines (SVMs) discriminate between the elements of a finite set in an n-dimensional real space, each point of which be-

M.C. Ferris et al. (eds.), Complementarity: Applications, Algorithms and Extensions, 233–251.

longs to one of two categories. The SVM constructs a separating plane, in the original n-dimensional space or in a higher dimensional space, that is mid-way between two parallel planes that bound the two categories [6, 28, 7]. A conventional SVM in its dual formulation contains bound constraints, as well as an equality constraint that requires special treatment in iterative procedures such as successive overrelaxation (SOR). A very simple convex quadratic program with bound constraints *only* can be obtained by taking the dual of the quadratic program associated with a support vector machine that maximizes the margin (distance between bounding separating planes) with respect to *all* the parameters determining the bounding planes [22]. This can be achieved by maximizing the margin (distance between bounding separating planes) with respect to *both* orientation of the separating planes as well as their location relative to the origin. This quadratic program can be solved by successive overrelaxation for datasets with as many as 10 million points [22] to obtain a *linear* separating surface. In the present paper we show how to obtain a *nonlinear* separating surface for more complex datasets by using SOR, thus opening the possibility of solving large scale nonlinear separation problems. The nonlinear surface is induced by a possibly indefinite kernel. We also use linear programming as another approach for nonlinear kernel separation, where the finite simplex algorithm [8] is used.

We note that nonlinear kernel separation is a standard technique in the SVM literature [6, 28, 7]. The techniques that we offer in this paper differ from previous work in some important respects. SOR for support vector machines has been shown to be effective in [22] for linear kernels. In this paper, we show that SOR can be successfully used in conjunction with nonlinear kernels. The SVM literature generally requires that nonlinear kernels be of a certain type, namely kernels that satisfy Mercer's positive definiteness condition [6]. In [21], it is shown theoretically and by means of a small example that linear programming based SVMs do not require Mercer's condition. In this work, we demonstrate the effectiveness and power of this approach using different nonlinear kernels on five publicly available datasets.

In Section 2 we first state our linear discrimination problem as a conventional SVM quadratic program (11.1) and its dual (11.5). We then introduce our modification of the SVM quadratic program (11.1), which is given in problem (11.4) and its dual in (11.6). This modification is an SOR-solvable convex quadratic program with bound constraints only and a positive semidefinite kernel. The dual problems (11.5) and (11.6) for a linear separator are generalized to a kernel-induced nonlinear separator in (11.8) and (11.11). Another version of our generalized nonlinear support vector machine is also given in (11.12) below, in which the kernel is completely arbitrary but the problem is convex and hence also SOR-solvable. We note that in our previous work [22] only a *linear kernel* (polynomial kernel of degree one) was implemented, whereas various *nonlinear kernels* that may be indefinite will be used in this paper. In Section 3 we state our SOR algorithm for nonlinear kernels and establish its linear convergence. In Section 4 we give linear programming formulations of the generalized support vector machine problem, including a novel one that includes two or more arbitrary kernels. In Section 5 we give numerical results for various datasets all of which show the improvement obtained in using the generalized support vector machine with a nonlinear kernel over a linear one. Section 6 draws some conclusions and points out future directions such as parallel SOR implementations that may lead to the solution of very large discrimination problems with nonlinear separating surfaces.

A word about our notation. All vectors will be column vectors unless transposed to a row vector by a prime superscript $'$. For a vector x in the n-dimensional real space R^n, the plus function x_+ is defined as $(x_+)_i = \max\ \{0, x_i\}$, $i = 1, \ldots, n$. The scalar (inner) product of two vectors x and y in the n-dimensional real space R^n will be denoted by $x'y$. For an $m \times n$ matrix A, A_i will denote the ith row of A. The identity matrix in a real space of arbitrary dimension will be denoted by I, while a column vector of ones of arbitrary dimension will be denoted by e. We shall employ the MATLAB "dot" notation [23] to signify application of a function to all components of a matrix or a vector. For example if $A \in R^{m\times n}$, then $A^2_\bullet \in R^{m\times n}$ will denote the matrix obtained by squaring each element of A. For $A \in R^{m\times n}$ and $B \in R^{n\times \ell}$, the **kernel** $K(A, B)$ maps $R^{m\times n} \times R^{n\times \ell}$ into $R^{m\times \ell}$. In particular if x and y are column vectors in R^n then, $K(x', A')$ is a row vector in R^m, $K(x', y)$ is a real number and $K(A, A')$ is an $m \times m$ matrix. Note that for our purposes here $K(A, A')$ will be assumed to symmetric, that is $K(A, A') = K(A, A')'$. For a vector x in the n-dimensional real space R^n, the step function $x_{\bullet *}$ is defined as a vector in R^n of minus ones, zeros and plus ones, corresponding to negative, zero and positive components of x respectively. Similarly for an $m \times n$ matrix M, $M_{\bullet *}$ will denote an $m \times n$ matrix of minus ones, zeros and plus ones.

2 THE SUPPORT VECTOR MACHINE AND ITS GENERALIZATION: QUADRATIC FORMULATION

We consider the problem of discriminating between m points in the n dimensional real space R^n, represented by the $m \times n$ matrix A, according to membership of each point A_i in the classes 1 or -1 as specified by a given $m \times m$ diagonal matrix D with ones or minus ones along its diagonal. For this problem the standard linear support vector machine with a linear kernel AA' [28, 7] is given by the following for some $\nu > 0$:

$$\begin{array}{rrcl} \min\limits_{w,\gamma,y} & \nu e'y + \frac{1}{2}w'w & & \\ \text{s.t.} & D(Aw - e\gamma) + y & \geq & e \\ & y & \geq & 0. \end{array} \tag{11.1}$$

Here w is the normal to the bounding planes:

$$\begin{array}{rcrcl} x'w & - & \gamma & = & +1 \\ x'w & - & \gamma & = & -1. \end{array} \tag{11.2}$$

The first plane above bounds the class 1 points and the second plane bounds the class -1 points, if the two classes are linearly separable and $y = 0$. This is illustrated in Figure 11.1.

If the classes are linearly inseparable then the two planes bound the two classes with a "soft margin" determined by a slack variable $y \geq 0$, that is:

$$\begin{array}{rcrcrcl} x'w & - & \gamma & + & y_i & \geq & +1, \ \text{for } x = A_i \text{ and } D_{ii} = +1, \\ x'w & - & \gamma & - & y_i & \leq & -1, \ \text{for } x = A_i \text{ and } D_{ii} = -1. \end{array} \tag{11.3}$$

The one-norm of the slack variable y is minimized with weight ν in (11.1). The quadratic term in (11.1), which is twice the reciprocal of the square of the 2-norm distance $\frac{2}{\|w\|_2}$ between the two planes of (11.2) in the n-dimensional space of $w \in R^n$ for a *fixed* γ, maximizes that distance. In our approach here, which is similar to that of [2, 14, 13], we measure the distance between the planes in the $(n+1)$-dimensional space of $[w; \gamma] \in R^{n+1}$ which is $\frac{2}{\|[w;\gamma]\|_2}$. Using this measure of distance instead results in our modification of the SVM problem as follows:

$$\begin{array}{rrcl} \min\limits_{w,\gamma,y} & \nu e'y + \frac{1}{2}(w'w + \gamma^2) & & \\ \text{s.t.} & D(Aw - e\gamma) + y & \geq & e \\ & y & \geq & 0. \end{array} \tag{11.4}$$

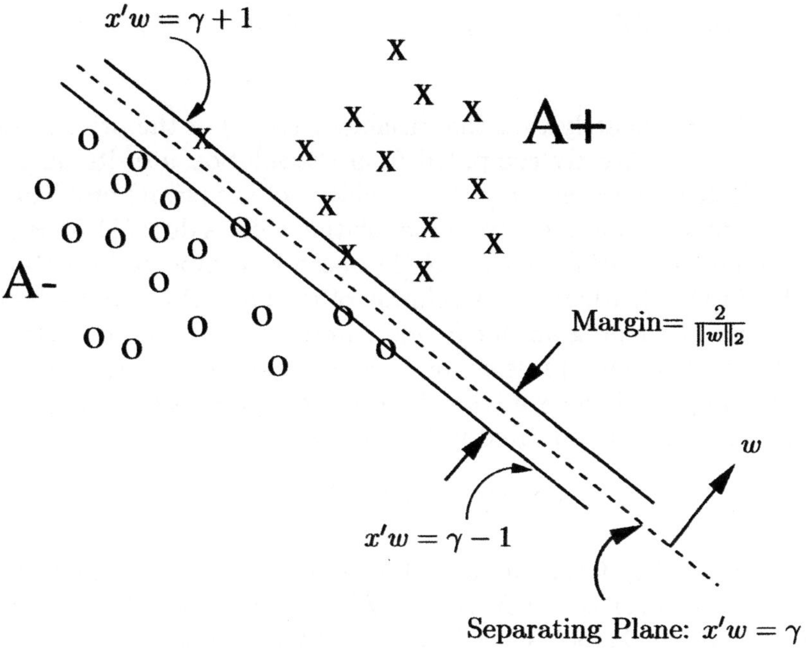

Figure 11.1 The Linearly Separable Case: The bounding planes of equation (11.2) with margin $\frac{2}{\|w\|_2}$

The Wolfe duals [18, Section 8.2] to the quadratic programs (11.1) and (11.4) are as follows.

$$\begin{array}{c} \max_{u} \; -\frac{1}{2}u'DAA'Du + e'u, \text{ s.t. } e'Du = 0, \; 0 \le u \le \nu e \\ (w = A'Du). \end{array} \tag{11.5}$$

$$\begin{array}{c} \max_{u} \; -\frac{1}{2}u'DAA'Du - \frac{1}{2}u'Dee'Du + e'u, \text{ s.t. } 0 \le u \le \nu e \\ (w = A'Du, \; \gamma = -e'Du, \; y = (e - D(Aw - e\gamma))_+). \end{array} \tag{11.6}$$

We note immediately that the variables (w, γ, y) of the primal problem (11.4) can be directly computed from the solution u of its dual (11.6) as indicated. However, only the variable w of the primal problem (11.1) can be directly computed from the solution u of its dual (11.5) as shown. The remaining variables (γ, y) of (11.1) can be computed by setting $w = A'Du$ in (11.1), where u is a solution of its dual (11.5), and solving the resulting linear program for (γ, y). Alternatively, γ can be determined by minimizing the expression for $e'y = e'(e - D(Aw - e\gamma))_+$ as a function of the single variable γ after w has been expressed as a function of the dual solution u as indicated in (11.5), that is:

$$\min_{\gamma \in R} e'(e - D(AA'Du - e\gamma))_+). \tag{11.7}$$

We note that formulations (11.5) and (11.6) can be extended to a general nonlinear kernel $K(A, A') : R^{m \times n} \times R^{n \times m} \to R^{m \times m}$ by replacing AA' by the kernel $K(A, A')$ and the SOR approach can similarly be extended to a general nonlinear kernel [21]. Thus the conventional support vector machine with a positive semidefinite kernel $K(A, A')$ [28, 7] is given by the following dual convex quadratic program for some $\nu > 0$ which is obtained from (11.5) by replacing AA' by $K(A, A')$:

$$\begin{array}{lrcl} \min_{u \in R^m} & \multicolumn{3}{l}{\frac{1}{2}u'DK(A, A')Du - e'u} \\ \text{s.t.} & e'Du & = & 0 \\ & 0 \le u & \le & \nu e. \end{array} \tag{11.8}$$

A solution u of this quadratic program leads to the nonlinear separating surface

$$K(x', A')Du = \gamma, \tag{11.9}$$

where γ is defined as in (11.7) above by a solution of:

$$\min_{\gamma \in R} e'(e - D(K(A, A')Du - e\gamma))_+. \qquad (11.10)$$

Note that the nonlinear surface (11.9) in R^n obviates the need for computing w, since w is defined only in the higher dimensional space in which the nonlinear surface (11.9) is mapped into a plane.

By a similar approach we obtain the following quadratic dual with bound constraints only and a *nonlinear* kernel $K(A, A')$ by replacing AA' in (11.6) by $K(A, A')$:

$$\begin{array}{c} \min_{u} \frac{1}{2} u'D[K(A, A') + ee']Du - e'u, \text{ s.t. } 0 \leq u \leq \nu e \\ (\gamma = -e'Du), \end{array} \qquad (11.11)$$

with an explicit expression for γ and the same separating surface (11.9). The formulation (11.11) allows a direct use of the SOR algorithm to solve very large problems.

We mention in passing another possible change in (11.8) allows the use of possibly indefinite kernels. One particular formulation motivated by [21, Equation (10)] is the following one which, as in (11.11), incorporates the equality $e'Du = 0$ into the objective function:

$$\begin{array}{c} \min_{u} \frac{1}{2} u'D[K(A, A')K(A, A')' + ee']Du - e'u, \text{ s.t. } 0 \leq u \leq \nu e \\ (\gamma = -e'Du), \end{array}$$

$$(11.12)$$

with a separating surface [21, Equations (1) & (10)] different from (11.9):

$$K(x', A')K(A, A')'Du = \gamma. \qquad (11.13)$$

Note that the kernel $K(A, A')$ in the formulation (11.12) is completely arbitrary and need not satisfy any positive semidefiniteness condition in order for the objective function of (11.12) to be convex. This makes the separating surface (11.13) quite general.

3 SUCCESSIVE OVERRELAXATION FOR NONLINEAR GSVM

Our goal is now to present an algorithm which can be used to efficiently solve the support vector machines presented in the previous section. Traditional quadratic programming techniques do not exploit the structure of this problem, and run exceedingly slowly. Moreover, such techniques expect that all the data is available in memory. This criterion results in significant limitations in scaling these algorithms. Successive overrelaxation was used effectively in [22] for *linear* support vector machines in order to overcome these difficulties. We therefore apply it here to the *nonlinear* SVMs (11.11) and (11.12). Both of these problems can be stated as:

$$\min_u \frac{1}{2}u'Mu - e'u, \text{ s.t. } u \in S = \{u \mid 0 \leq u \leq \nu e\}, \tag{11.14}$$

with the symmetric matrix M defined as $D(K(A, A') + ee')D$ and $D(K(A, A')K(A, A')' + ee')D$ respectively for these two problems. Thus M will be positive semidefinite if we assume that $K(A, A')$ is positive semidefinite in the former case and under no assumptions in the latter case. If we decompose M as follows:

$$M = L + E + L', \tag{11.15}$$

where $L \in R^{m\times m}$ is the strictly lower triangular part of the symmetric matrix M, and $E \in R^{m\times m}$ is the positive diagonal of M, then a necessary and sufficient optimality condition for (11.14) for positive semidefinite M is the following gradient projection optimality condition [27, 17]:

$$u = (u - \omega E^{-1}(Mu - e))_{\#}, \ \omega > 0, \tag{11.16}$$

where $(\cdot)_{\#}$ denotes the 2-norm projection on the feasible region S of (11.14), that is:

$$((u)_{\#})_i = \left\{ \begin{array}{l} 0 \text{ if } u_i \leq 0 \\ u_i \text{ if } 0 < u_i < \nu \\ \nu \text{ if } u_i \geq \nu \end{array} \right\}, \ i = 1, \ldots, m. \tag{11.17}$$

Our SOR method, which is a matrix splitting method that converges linearly to a point $\bar{u}$ satisfying (11.16), consists of splitting the matrix M into the sum of two matrices as follows:

$$M = \omega^{-1}E(B + C), \text{ s.t. } B - C \text{ is positive definite.} \tag{11.18}$$

For our specific problem we take:

$$B = (I + \omega E^{-1}L), \ C = ((\omega - 1)I + \omega E^{-1}L'), \ 0 < \omega < 2. \tag{11.19}$$

This leads to the following linearly convergent [17, Equation (3.14)] matrix splitting algorithm:

$$u^{i+1} = (u^{i+1} - Bu^{i+1} - Cu^i + \omega E^{-1}e)_{\#}, \tag{11.20}$$

for which

$$B + C = \omega E^{-1}M, \; B - C = (2 - \omega)I + \omega E^{-1}(L - L'). \qquad (11.21)$$

Note that for $0 < \omega < 2$, the matrix $B + C$ is positive semidefinite and matrix $B - C$ is positive definite. The matrix splitting algorithm (11.20) results in the following easily implementable SOR algorithm once the values of B and C given in (11.19) are substituted in (11.20).

Algorithm 11.1 SOR Algorithm *Choose $\omega \in (0, 2)$. Start with any $u^0 \in R^m$. Having u^i compute u^{i+1} as follows:*

$$u^{i+1} = (u^i - \omega E^{-1}(Mu^i - e + L(u^{i+1} - u^i)))_{\#}, \qquad (11.22)$$

until $\|u^{i+1} - u^i\|$ is less than some prescribed tolerance.

Remark 11.2 *The components of u^{i+1} are computed in order of increasing component index. Thus the SOR iteration (11.22) consists of computing u_j^{i+1} using $(u_1^{i+1}, \ldots, u_{j-1}^{i+1}, u_j^i, \ldots, u_m^i)$. That is, the latest computed components of u are used in the computation of u_j^{i+1}. The strictly lower triangular matrix L in (11.22) can be thought of as a substitution operator, substituting $(u_1^{i+1}, \ldots, u_{j-1}^{i+1})$ for $(u_1^i, \ldots, u_{j-1}^i)$.*

We have immediately from [17, Proposition 3.5] the following linear convergence result.

Theorem 11.3 SOR Linear Convergence *The iterates $\{u^i\}$ of the SOR Algorithm 11.1 converge R-linearly to a solution of $\bar{u}$ of the dual problem (11.13), and the objective function values $\{f(u^i)\}$ of (11.13) converge Q-linearly to $f(\bar{u})$. That is for $i \geq \bar{i}$ for some $\bar{i}$:*

$$\begin{array}{rcl} \|u^i - \bar{u}\| & \leq & \mu\delta^i, \text{ for some } \mu > 0, \; \delta \in (0,1), \\ f(u^{i+1}) - f(\bar{u}) & \leq & \tau(f(u^i) - f(\bar{u})), \text{ for some } \tau \in (0,1). \end{array} \qquad (11.23)$$

We turn our attention to linear programming formulations of the nonlinear GSVM.

4 THE NONLINEAR GSVM AS A LINEAR PROGRAM

We briefly describe now how the nonlinear separating surface (11.9) can also be generated by a linear program. For greater detail, the reader is referred to [21].

We begin with the standard linear support vector machine (11.1). Motivated by the Wolfe dual (11.5), we make the substitution $w = A'Du$

into the primal and obtain:

$$\begin{array}{rrcl} \min\limits_{u,\gamma,y} & \nu e'y + \frac{1}{2}uDAA'Du & & \\ \text{s.t.} & D(AA'Du - e\gamma) + y & \geq & e \\ & y & \geq & 0. \end{array} \tag{11.24}$$

As in the SOR formulation, we substitute a nonlinear kernel for the AA' term. Additionally, we "simplify" the regularization term in the objective to depend only on the variable u in a linear fashion. We achieve this by introducing a variable s, which takes on the value of the absolute value of u. The substitution surrogates support vector suppression for maximizing the distance between separating bounding planes, which has a similar regularizing effect. This results in the linear program [21, Equation (11)]:

$$\begin{array}{rrcl} & \min\limits_{u,\gamma,y,s} \quad \nu e'y + e's & & \\ \text{s.t.} & D(K(A,A')Du - e\gamma) + y & \geq & e \\ & s \geq u & \geq & -s \\ & y & \geq & 0. \end{array} \tag{11.25}$$

A solution (u, γ, y, s) to this linear program for a completely arbitrary kernel $K(A, A')$ will provide a separating surface given by (11.9). We note that if the linear kernel $K(A, A') = AA'$ (i.e. polynomial kernel of degree 1) is used in the linear program above and w is introduced and set equal to $A'Du$ we obtain the high-performing 1-norm linear SVM proposed in [5] and utilized successfully in [4, 1, 3].

Since the linear programming formulation does not impose restrictions on the kernel employed, we are free to use more than one kernel in a given formulation. For example we can use **both** a linear kernel and a discontinuous neural network kernel in a single linear programming formulation. More generally if we let $K^1(A, A')$ and $K^2(A, A')$ be two arbitrary kernels, then the corresponding linear programming formulation becomes:

$$\begin{array}{rrcl} & \min\limits_{u^1,u^2,\gamma,y,s^1,s^2} \quad \nu e'y + e's^1 + e's^2 & & \\ \text{s.t.} & D(K^1(A,A')Du^1 + K^2(A,A')Du^2 - e\gamma) + y & \geq & e \\ & s^1 \geq u^1 & \geq & -s^1 \\ & s^2 \geq u^2 & \geq & -s^2 \\ & y & \geq & 0. \end{array} \tag{11.26}$$

The corresponding separating surface to a solution $(u^1, u^2, \gamma, y, s^1, s^2)$ of this linear program is given by:

$$K^1(x', A')Du^1 + K^2(x', A')Du^2 = \gamma \tag{11.27}$$

For example we can use a linear and a discontinuous neural network kernel for $K^1(A, A')$ and $K^2(A, A')$ as follows:

$$K^1(A', A') = (AA' - ee'); \; K^2(A', A') = (AA' - ee')_{\bullet *}, \qquad (11.28)$$

where, as defined in the Introduction, the subscript $\bullet *$ denotes replacement of each component of the matrix AA' by 1, 0, or -1, depending on whether it is positive, zero or negative, respectively. Note that in general this neural network kernel is not positive semidefinite and hence standard support vector techniques cannot make use of it. We also note that the linear programming formulation (11.26) can be extended to more than two kernels in an obvious manner.

5 NUMERICAL TESTING

We tested nonlinear kernels versus linear ones by using different formulations of the generalized SVM, namely the quadratic programming formulations as given in (11.11) and (11.12), and the linear programming formulation as given in (11.25). Most of our code development was in the MATLAB environment [23], though the actual solutions to the SVMs were handled externally. We coded the SOR algorithm in C, and accessed it via the MATLAB "mex" interface which allows one to execute external C code as a native MATLAB function [24]. The linear programs were solved with the MINOS solver [26]. All experiments were run on the University of Wisconsin Computer Sciences Department Ironsides cluster. This cluster of four Sun Enterprise E6000 machines consists of 16 UltraSPARC II 250 MHz processors and 2 gigabytes of RAM on each node resulting in a total of 64 processors and 8 gigabytes of RAM.

The SVMs were tested on the following datasets from the University of California at Irvine (UCI) repository [25]:

- The *liver-disorders* dataset contains 345 points, each consisting of six features. Class 1 contains 145 points, and class -1 contains 200 points.

- The *letter-recognition* dataset is used for recognizing letters of the alphabet. Traditionally, this dataset is used for multi-category classification. We used it here in a two class situation by taking a subset of those points which correspond to the letter "A," and a subset of those points which correspond to the letter "B." This resulted in a dataset of 600 points with 6 features, where each class contained 300 points.

- The *mushroom* dataset is a two class dataset which contains a number of categorical attributes. We transformed each categorical attribute into a series of binary attributes, one attribute for each distinct value. For example, this dataset contains an attribute called "cap surface," which can take one one of four categories, namely "fibrous," "grooves," "scaly," or "smooth." We represented this as four attributes. A 1 is assigned to the attribute that corresponds to the actual category, and a 0 to the rest. Thus each instance of this specific categorical attribute corresponds to one row of a 4 x 4 diagonal matrix of ones. A subset of the entire mushroom dataset was used, so as to shorten the running times of our experiments. The final dataset we used contained 22 features with 200 points in class 1 and 300 points in class -1.

- The *tic-tac-toe* dataset is a two class dataset that contains incomplete tic-tac-toe games. All those games with a possible winning move for "X" end up in category 1, and all other games end up in category -1. We have intentionally represented this problem with a poor representation scheme to show the power of non-linear kernels in overcoming this difficulty. Each tic-tac-toe game is represented by 9 attributes, where each attribute corresponds to a spot on the tic-tac-toe board. An "X" is represented by 1, an "O" is represented by 0, and a blank is represented by -1. We used a subset of this dataset with 9 attributes, 200 points in class 1, and 100 points in class -1.

We additionally used some synthetic datasets, namely the checkerboard dataset [16, 15] and also a synthetic dataset of Gaussian distribution. The Gaussian synthetic dataset, suggested by Usama Fayyad of Microsoft Research, was created by constructing 20 random centers, then generating "clouds" of points around these centers in a Gaussian fashion based on random covariance matrices. An arbitrary hyperplane was used to separate the centers into two classes. Each center was assigned to the class corresponding to the side of the hyperplane it fell on; furthermore, each randomly generated point was assigned to the same class as the center from which it came. We then used a scaling parameter to size the clouds so that the data was not linearly separable.

All experiments were conducted using tenfold cross validation, which is is a standard technique used for evaluating machine learning algorithms. We randomly partitioned each dataset into ten equal parts, and ran the algorithm ten times on 90% of this dataset. Each time, a different one of the ten segments was held out to serve as a test set. We used this test set to measure the generalization capability of the learn-

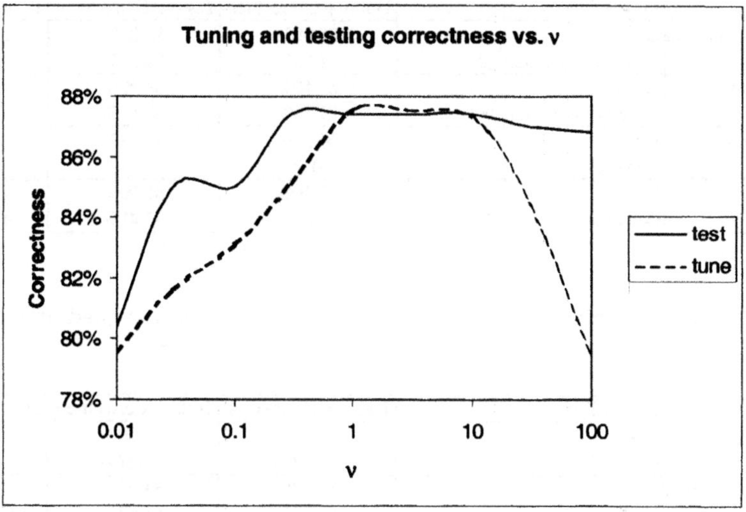

Figure 11.2 Tuning and testing set accuracy for SOR with a linear kernel on the mushroom dataset.

ing method. Kernels and parameters were chosen via the use of a *tuning* set, an artificially created testing set surrogate taken completely from the training set. We ran our experiments for a variety of parameters and kernels on the training data with the tuning set removed, and found those parameters that performed best on the tuning set. We then used these settings to retrain on the training set as a whole (tuning set included) to report our generalization results on the test set. Tuning set accuracy is usually a good predictor of test accuracy, as shown in Figure 11.2.

Table 11.1 shows results on a number of datasets, using the SOR Algorithm 11.1 with both linear and nonlinear kernels. In each of the cases shown, the chosen nonlinear kernel generalizes better than the linear kernel. This is particularly noticeable for the tic-tac-toe and Gaussian datasets. In general, the kernels we used are as follows:

Kernel 1 Polynomial $((\frac{A}{\lambda} - \rho)(\frac{A}{\lambda} - \rho)' - \mu)^d_{\bullet}$

Kernel 2 Negative Polynomial $-((\frac{A}{\lambda} - \rho)(\frac{A}{\lambda} - \rho)' - \mu)^d_{\bullet}$

The variables λ, ρ, μ are scalar parameters and d is a positive integer. When a parameter is added or subtracted from a matrix or vector, it means componentwise addition or subtraction of the parameter in conformity with MATLAB notation. The "negative squared cubic" kernel

Dataset	Kernel Type	Training Correctness	Testing Correctness	Time (secs)
Letter Recognition	Linear	85.00%	84.83%	11.1
	6th degree polynomial	92.65%	86.33%	11.0
Mushroom	Linear	87.40%	87.40%	6.6
	Negative squared cubic	89.38%	88.80%	6.7
Tic Tac Toe	Linear	60.37%	56.33%	1.4
	Quadratic	100.00%	83.33%	1.4
Gaussian	Linear	83.20%	81.70%	43.0
	Quadratic	97.68%	93.41%	43.0

Table 11.1 SOR training and testing set correctness for linear and quadratic kernels

Dataset	Kernel Type	Training Correctness	Testing Correctness	Time (secs)
Liver Disorders	Linear	71.21%	68.70%	23.5
	Quadratic	78.33%	73.37%	46.8
Letter Recognition	Linear	85.43%	84.33%	62.5
	Negative Cubic	97.15%	87.67%	553.7
Tic Tac Toe	Linear	66.67%	66.67%	17.1
	Quadratic	98.78%	90.67%	21.4
Mushroom	Linear	87.40%	87.40%	45.5
	Negative 7th degree polynomial	89.56%	88.80%	35.1
Checkerboard	Linear	51.12%	48.60%	206.8
	6th degree polynomial	99.97%	98.50%	289.0

Table 11.2 Linear programming training and testing set correctness for linear and nonlinear support vector machines

listed for the mushroom dataset is a negative polynomial kernel of degree 3, with the problem formulation as in (11.12). Other kernels use the problem formulation shown in (11.11).

Table 11.2 shows training and testing set correctness, using the linear programming formulation (11.25) with various kernels, under tenfold cross validation for the above mentioned datasets. For these experiments, we again implemented a linear kernel and various polynomial kernels. Note that these polynomial kernels are indefinite in general, and therefore do not satisfy Mercer's condition.

As in the SOR case above, the nonlinear kernels gave improved testing set results over linear kernels. The most striking results are those for the checkerboard dataset of Figure 11.3. An initial attempt using a k-nearest-neighbor algorithm yields the very coarse separation shown in Figure 11.4. We instead used a nonlinear SVM, and obtained a tenfold testing set correctness of over 97.70% for our nonlinear kernel. This is reflected in the very realistic rendering in Figure 11.5 depicting the

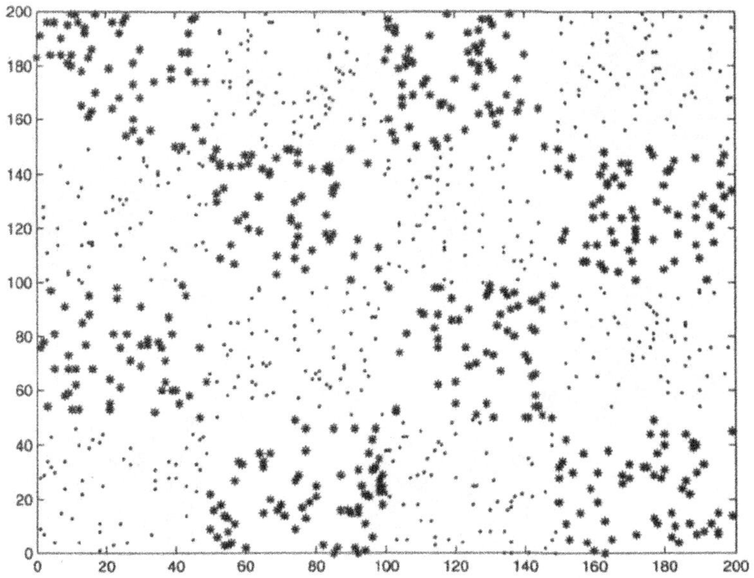

Figure 11.3 Checkerboard training dataset

resulting separation of the original dataset. Our experiments were inspired by those of Kaufman in [16] where eighth degree polynomial and radial basis kernels were used in a standard support vector machine formulation to separate the checkerboard dataset. However, our separation depicted in Figure 11.5 is considerably sharper than those of [16].

6 CONCLUSION

We have implemented a simple iterative method, successive overrelaxation, and a linear programming approach for the solution of discrimination problems using generalized support vector machines with nonlinear kernels. Both of these methods yield advantages over previous techniques for the solution of SVMs. SOR has significantly smaller memory requirements than previous techniques, and can therefore be used on larger datasets than previously considered. In fact, it can be used on some datasets that do not fit in memory [22]. The linear programming formulation yields the advantage that linear programs can be solved significantly more efficiently than quadratic programs. The methods scale up and can be parallelized by using techniques already implemented [10, 11, 9, 12]. Future work includes multicategory nonlinear discrimination via kernel methods using successive overrelaxation

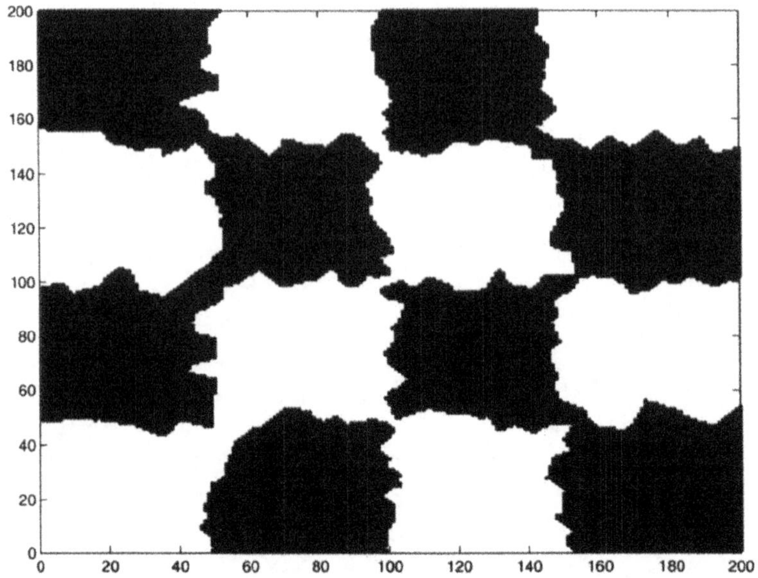

Figure 11.4 **k-nearest-neighbor separation of the checkerboard dataset (k=1, 2-norm)**

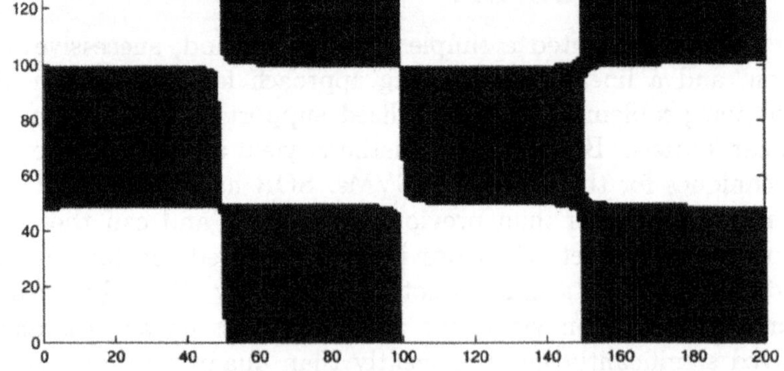

Figure 11.5 **Indefinite sixth degree polynomial kernel separation of the checkerboard dataset ($\nu = 10,000$, $\lambda = 100$, $\rho = 1$, $d = 6$, $\mu = 0.5$)**

and linear programming, as well as parallel implementation on multiprocessor machines.

Acknowledgments

We are indebted to our colleagues Usama Fayyad, Microsoft Research, for suggesting the Gaussian dataset, to Linda Kaufman, Bell Labs, for supplying us with the checkerboard dataset and to Tin Kam Ho, Bell Labs, for allowing us to make the checkerboard dataset public. This research is supported by National Science Foundation Grants CCR-9729842 and CDA-9623632, and by Air Force Office of Scientific Research Grant F49620-97-1-0326 as Mathematical Programming Technical Report 99-03, March 1999, and revised March 2000.

References

[1] K. P. Bennett, D. Hui, and L. Auslender. On support vector decision trees for database marketing. Department of Mathematical Sciences Math Report No. 98-100, Rensselaer Polytechnic Institute, Troy, NY 12180, March 1998. http://www.math.rpi.edu/~bennek/.

[2] B. E. Boser, I. M. Guyon, and V. N. Vapnik. A training algorithm for optimal margin classifiers. In D. Haussler, editor, *Proceedings of the 5th Annual ACM Workshop on Computational Learning Theory*, pages 144–152, Pittsburgh, PA, July 1992. ACM Press.

[3] P. S. Bradley and O. L. Mangasarian. Feature selection via concave minimization and support vector machines. In J. Shavlik, editor, *Machine Learning Proceedings of the Fifteenth International Conference(ICML '98)*, pages 82–90, San Francisco, California, 1998. Morgan Kaufmann. ftp://ftp.cs.wisc.edu/math-prog/tech-reports/98-03.ps.

[4] E. J. Bredensteiner. *Optimization Methods in Data Mining and Machine Learning.* PhD thesis, Department of Mathematical Sciences, Rensselaer Polytechnic Institute, Troy, NY, 1997.

[5] E. J. Bredensteiner and K. P. Bennett. Feature minimization within decision trees. *Computational Optimizations and Applications*, 10:111–126, 1998.

[6] C. J. C. Burges. A tutorial on support vector machines for pattern recognition. *Data Mining and Knowledge Discovery*, 2(2):121–167, 1998.

[7] V. Cherkassky and F. Mulier. *Learning from Data - Concepts, Theory and Methods.* John Wiley & Sons, New York, 1998.

[8] G. B. Dantzig. *Linear Programming and Extensions.* Princeton University Press, Princeton, New Jersey, 1963.

[9] R. De Leone and O. L. Mangasarian. Serial and parallel solution of large scale linear programs by augmented Lagrangian successive overrelaxation. In A. Kurzhanski, K. Neumann, and D. Pallaschke, editors, *Optimization, Parallel Processing and Applications*, pages 103–124, Berlin, 1988. Springer–Verlag. Lecture Notes in Economics and Mathematical Systems 304.

[10] R. De Leone, O. L. Mangasarian, and T.-H. Shiau. Multi-sweep asynchronous parallel successive overrelaxation for the nonsymmetric linear complementarity problem. *Annals of Operations Research*, 22:43–54, 1990.

[11] R. De Leone and M. A. Tork Roth. Massively parallel solution of quadratic programs via successive overrelaxation. *Concurrency: Practice and Experience*, 5:623–634, 1993.

[12] M. C. Ferris and O. L. Mangasarian. Parallel constraint distribution. *SIAM Journal on Optimization*, 1(4):487–500, 1991.

[13] T.-T. Friess. Support vector neural networks: The kernel adatron with bias and soft margin. Technical report, Department of Automatic Control and Systems Engineering, University of Sheffield, Sheffield, England, 1998. Revised Version: www.brunner-edv.com/friess/.

[14] T.-T. Friess, N. Cristianini, and C. Campbell. The kernel-adatron algorithm: A fast and simple learning procedure for support vector machines. In Jude Shavlik, editor, *Machine Learning Proceedings of the Fifteenth International Conference (ICML'98)*, pages 188–196, San Francisco, 1998. Morgan Kaufmann. http://svm.first.gmd.de/papers/FriCriCam98.ps.gz.

[15] Tin Kam Ho and Eugene M. Kleinberg. Building projectable classifiers of arbitrary complexity. In *Proceedings of the 13th International Conference on Pattern Recognition*, pages 880–885, Vienna, Austria, 1996. http://cm.bell-labs.com/who/tkh/pubs.html. Checker dataset at: ftp://ftp.cs.wisc.edu/math-prog/cpo-dataset/machine-learn/checker.

[16] L. Kaufman. Solving the quadratic programming problem arising in support vector classification. In Bernhard Schölkopf, Christopher J. C. Burges, and Alexander J. Smola, editors, *Advances in Kernel Methods - Support Vector Learning*, pages 147–167. MIT Press, 1999.

[17] Z.-Q. Luo and P. Tseng. Error bounds and convergence analysis of feasible descent methods: A general approach. *Annals of Operations Research*, 46:157–178, 1993.

[18] O. L. Mangasarian. *Nonlinear Programming.* McGraw–Hill, New York, 1969. Reprint: SIAM Classic in Applied Mathematics 10, 1994, Philadelphia.

[19] O. L. Mangasarian. Solution of symmetric linear complementarity problems by iterative methods. *Journal of Optimization Theory and Applications*, 22(4):465–485, August 1977.

[20] O. L. Mangasarian. On the convergence of iterates of an inexact matrix splitting algorithm for the symmetric monotone linear complementarity problem. *SIAM Journal on Optimization*, 1:114–122, 1991.

[21] O. L. Mangasarian. Generalized support vector machines. In A. Smola, P. Bartlett, B. Schölkopf, and D. Schuurmans, editors, *Advances in Large Margin Classifiers*, pages 135–146, Cambridge, MA, 2000. MIT Press. ftp://ftp.cs.wisc.edu/math-prog/tech-reports/98-14.ps.

[22] O. L. Mangasarian and David R. Musicant. Successive overrelaxation for support vector machines. *IEEE Transactions on Neural Networks*, 10:1032–1037, 1999. ftp://ftp.cs.wisc.edu/math-prog/tech-reports/98-18.ps.

[23] MATLAB. *User's Guide.* The MathWorks, Inc., Natick, MA 01760, 1992.

[24] MATLAB. *Application Program Interface Guide.* The MathWorks, Inc., Natick, MA 01760, 1997.

[25] P. M. Murphy and D. W. Aha. UCI repository of machine learning databases, 1992. www.ics.uci.edu/~mlearn/MLRepository.html.

[26] B. A. Murtagh and M. A. Saunders. MINOS 5.0 user's guide. Technical Report SOL 83.20, Stanford University, December 1983. MINOS 5.4 Release Notes, December 1992.

[27] B. T. Polyak. *Introduction to Optimization.* Optimization Software, Inc., Publications Division, New York, 1987.

[28] V. N. Vapnik. *The Nature of Statistical Learning Theory.* Springer, New York, 1995.

Chapter 12

ON CONSTRAINT QUALIFICATIONS FOR MATHEMATICAL PROGRAMS WITH MIXED COMPLEMENTARITY CONSTRAINTS

J. V. Outrata

Institute of Information Theory and Automation, Academy of Sciences of the Czech Republic, Pod vodárenskou věží 4, CZ – 182 08 Prague, Czech Republic

outrata@utia.cas.cz

Abstract The contribution concerns mathematical programs, where a mixed complementarity problem arises as a side constraint. The attention is paid above all to optimality conditions and to the respective constraint qualifications. In addition, we propose an exact penalty approach to the numerical solution of such problems.

1 INTRODUCTION

In recent years, considerable attention has been paid to mathematical programs, where a parameter-dependent (nonlinear) complementarity problem (NCP) arises among the constraints (e. g. [5, 3, 17, 9, 13, 10]). In some papers the authors have employed the generalized differential calculus of B. Mordukhovich ([6, 7]) which proved to be an efficient tool especially in deriving optimality conditions. This calculus has been used also in case of so-called mathematical programs with equilibrium constraints ([19, 11]), where one has to do with general equilibrium models including NCPs as a particular case.

M.C. Ferris et al. (eds.), Complementarity: Applications, Algorithms and Extensions, 253–271.

This paper deals with the mathematical program

$$\begin{array}{l} \text{minimize} \quad f(x,y) \\ \text{subject to} \\ F(x,y) = \lambda - \mu,\ \langle \lambda, y - a \rangle = 0,\ \langle \mu, b - y \rangle = 0,\ a \le y \le b \\ \qquad \lambda, \mu \in \mathcal{R}^m_+ \\ \qquad (x,y) \in \omega, \end{array} \tag{12.1}$$

where $x \in \mathcal{R}^n$ is the *control* or *design* variable, $y \in \mathcal{R}^m$ is the *state* variable, $a, b \in \mathcal{R}^m$ with $a < b$ are the vectors of lower and upper bounds for y, and λ, μ are the multipliers associated with the constraints $y \ge a$, $y \le b$, respectively. Further, $f[\mathcal{R}^n \times \mathcal{R}^m \to \mathcal{R}]$ is the objective and $\omega \subset \mathcal{R}^n \times \mathcal{R}^m$ is the set of *feasible control-state pairs.* The equilibrium constraint in (12.1) is given by the control-dependent mixed complementarity problem (MCP):

For a given $x \in \mathcal{R}^n$ find $y \in \{z \in \mathcal{R}^m \,|\, a \le z \le b\}$ such that

$$\left.\begin{array}{ll} F^i(x,y) \ge 0 & \text{provided} \quad y^i = a^i \\ F^i(x,y) \le 0 & \text{provided} \quad y^i = b^i \\ F^i(x,y) = 0 & \text{provided} \quad a^i < y^i < b^i, \\ & \qquad i = 1, 2, \ldots, m. \end{array}\right\} \tag{12.2}$$

Such problems can be encountered e.g. in shape optimization, provided we have to do with a bilateral contact problem and perform a suitable discretization. Also some economic models lead to problems of type (12.1); the mixed complementarity problem models then e.g. the Cournot–Nash equilibrium on an oligopolistic market and the respective problem (12.1) can be found in [2].

Our main aims are

(i) to specialize the general 1st-order necessary optimality conditions from [11] to problem (12.1);

(ii) to reformulate the respective constraint qualification to a more workable form enabling its comparison with some standard conditions, and

(iii) to propose a numerical method based on the exact penalization of the equilibrium constraint.

The results can thus be viewed as generalizations of their counterparts in [10], where mathematical programs with standard NCPs among the constraints are investigated. The paper consists essentially of three main sections (2–4) devoted to the above three aims, respectively. The appendix contains, for the reader's convenience, two essential notions from the Mordukhovich calculus which are used throughout the whole paper.

The following notation is employed: x^i is the ith component of a vector $x \in \mathcal{R}^n$, E is the unit matrix, $\mathcal{B}$ is the unit ball in $\mathcal{R}^n$ and $\mathcal{R}^n_+$ denotes the nonnegative orthant of $\mathcal{R}^n$. For an $[m \times n]$ matrix A and index sets $I \subset \{1, 2, \ldots, m\}$, $J \subset \{1, 2, \ldots, n\}$, $A_{I,J}$ denotes the submatrix of A with rows and columns, specified by I and J, respectively. A_I is the submatrix of A with rows specified by I. Similarly, for a vector $d \in \mathcal{R}^n$, d_I is the subvector composed from the components d^i, $i \in I$. Furthermore, *Gph* Φ is the graph of a multifunction Φ and *epi* f is the epigraph of a function f. If D is a cone with vertex at the origin, then D^0 is its negative polar cone. For x, $y \in \mathcal{R}^n$ the inequalities $x \geq y$, $x > y$ mean $x^i \geq y^i$ and $x^i > y^i$ for all i, respectively.

2 OPTIMALITY CONDITIONS

Throughout the whole paper we will assume that the objective is locally Lipschitz, the function $F[\mathcal{R}^n \times \mathcal{R}^m \to \mathcal{R}^m]$ is continuously differentiable and ω is nonempty and closed. Denote

$$\Gamma := \{y \in \mathcal{R}^m \,|\, a \leq y \leq b\}. \tag{12.3}$$

Then the MCP (12.2) can be equivalently written down as the *generalized equation* (cf. [14])

$$0 \in F(x, y) + N_\Gamma(y), \tag{12.4}$$

where

$$N_\Gamma(y) := \begin{cases} \text{the standard normal cone to } \Gamma \text{ at } y \text{ provided } y \in \Gamma \\ \emptyset \text{ otherwise.} \end{cases}$$

In terms of the generalized normal cones to ω and *Gph* Γ and the Mordukhovich's subdifferential of f, given in Definitions A.7 and A.8 of the Appendix, the optimality conditions from [11] attain the following form.

Theorem 12.1 *Let $(\hat{x}, \hat{y})$ be a (local) solution of (12.1). Assume that the constraint qualification*

$$(CQ) \quad \left. \begin{array}{l} \begin{bmatrix} (\nabla_x F(\hat{x}, \hat{y}))^{\mathrm{T}} z \\ -w + (\nabla_y F(\hat{x}, \hat{y}))^{\mathrm{T}} z \end{bmatrix} \in K_\omega(\hat{x}, \hat{y}) \\ \\ (w, z) \in K_{Gph\, N_\Gamma}(\hat{y}, -F(\hat{x}, \hat{y})) \end{array} \right\} \textit{implies } w = 0,\ z = 0$$

holds true.

Then there exists a pair $(\xi, \eta) \in \partial^- f(\hat{x}, \hat{y})$, *a pair* $(\gamma, \delta) \in K_\omega(\hat{x}, \hat{y})$ *and a Karush – Kuhn – Tucker (KKT) pair* $(\hat{w}, \hat{z}) \in K_{Gph\, N_\Gamma}(\hat{y}, -F(\hat{x}, \hat{y}))$ *such that*

$$0 = \xi - (\nabla_x F(\hat{x}, \hat{y}))^{\mathrm{T}} \hat{z} + \gamma \tag{12.5}$$

$$0 = \eta + \hat{w} - (\nabla_y F(\hat{x}, \hat{y}))^{\mathrm{T}} \hat{z} + \delta. \tag{12.6}$$

If F *is affine with respect to both variables and* ω *is convex polyhedral, then equations (12.5), (12.6) are fulfilled even without assuming the constraint qualification.*

As it has been derived in [6], $K_\omega(\hat{x}, \hat{y})$ can easily be computed provided ω is given by (smooth) equalities and inequalities. Thus, to rewrite the above optimality conditions into a workable form, it is necessary to express the (possibly complicated) nonconvex cone $K_{Gph\, N_\Gamma}(\hat{y}, -F(\hat{x}, \hat{y}))$ in terms of the problem data. Fortunately, this is possible due to the specific structure of Γ.

With each pair (x, y), feasible with respect to the equilibrium constraint, we associate the index sets

$$\begin{aligned}
L(y) &:= \left\{i \in \{1, 2, \dots, m\} \mid a^i < y^i < b^i\right\}, \\
I^1(y) &:= \left\{i \in \{1, 2, \dots, m\} \mid y^i = a^i\right\}, \\
I^1_+(x, y) &:= \left\{i \in I^1(y) \mid F^i(x, y) > 0\right\}, \\
I^1_0(x, y) &:= I^1(y) \setminus I^1_+(x, y), \\
I^2(y) &:= \left\{i \in \{1, 2, \dots, n\} \mid y^i = b^i\right\}, \\
I^2_+(x, y) &:= \left\{i \in I^2(y) \mid F^i(x, y) < 0\right\}, \\
I^2_0(x, y) &:= I^2(y) \setminus I^2_+(x, y),
\end{aligned}$$

and observe that

$$\{1, 2, \dots, m\} = L(y) \cup I^1_+(x, y) \cup I^1_0(x, y) \cup I^2_+(x, y) \cup I^2_0(x, y).$$

Since we will associate these index sets exclusively with the pair $(\hat{x}, \hat{y})$, we will drop their arguments whenever they arise as subscripts.

Lemma 12.2

$$\begin{aligned}
K_{Gph\, N_\Gamma}(\hat{y}, -F(\hat{x}, \hat{y})) = \Big\{&(w, z) \in \mathcal{R}^m \times \mathcal{R}^m \mid w_L = 0,\ z_{I^1_+ \cup I^2_+} = 0, \\
&\textit{for } i \in I^1_0(\hat{x}, \hat{y}) \textit{ either } w^i z^i = 0 \textit{ or } w^i < 0,\ z^i > 0, \textit{ and} \\
&\textit{for } i \in I^2_0(\hat{x}, \hat{y}) \textit{ either } w^i z^i = 0 \textit{ or } w^i > 0,\ z^i < 0\Big\}.
\end{aligned}$$

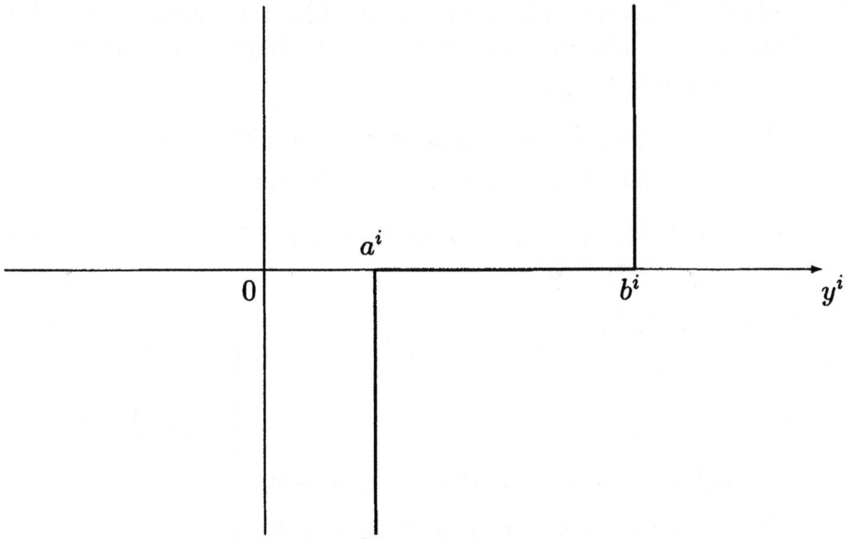

Figure 12.1 $Gph\ N_{\Gamma_i}$

Proof. By virtue of [1], Table 4.5 (3) and the structure of Γ

$$Gph\ N_\Gamma = \text{X}_{i=1}^m\ Gph\ N_{\Gamma_i},$$

where $\Gamma_i := \{\xi \in \mathcal{R} \,|\, a^i \leq \xi \leq b^i\}$, $i = 1, 2, \ldots, m$. Let $(u, v) \in Gph\ N_\Gamma$. By [6], Proposition 1.6 one has then

$$K_{Gph\, N_\Gamma}(u, v) = \text{X}_{i=1}^m K_{Gph\, N_{\Gamma_i}}(u^i, v^i).$$

The cones $K_{Gph\, N_{\Gamma_i}}(u^i, v^i)$ can easily be computed by applying Definition A.7 to the simple polyhedral sets $Gph\ \Gamma_i \subset \mathcal{R}^2$ (depicted in Figure 12.1). One obtains

$$K_{Gph\ N_{\Gamma_i}}(u^i, v^i)$$
$$= \begin{cases} \{(\xi, \eta) \in \mathcal{R}^2 \,|\, \eta = 0\} \text{ if } u^i = a^i,\ v^i < 0 \text{ or } u^i = b^i,\ v^i > 0 \\ \{(\xi, \eta) \in \mathcal{R}^2 \,|\, \xi = 0\} \text{ if } a^i < u^i < b^i \\ \{(\xi, \eta) \in \mathcal{R}^2 \,|\, \text{either } \xi\eta = 0 \text{ or } \xi < 0,\ \eta > 0\} \text{ if } u^i = a^i,\ v^i = 0 \\ \{(\xi, \eta) \in \mathcal{R}^2 \,|\, \text{either } \xi\eta = 0 \text{ or } \xi > 0,\ \eta < 0\} \text{ if } u^i = b^i,\ v^i = 0. \end{cases}$$

Thus it suffices to realize how the position of $(\hat{y}^i, -F^i(\hat{x}, \hat{y}))$ on $Gph\ N_{\Gamma_i}$ is determined by the index set to which i belongs. □

By combining Theorem 12.1 and Lemma 12.2 we obtain a workable form of necessary optimality conditions for problem (12.1) where, to simplify the notation, we put

$$I(\hat{y}) := I^1(\hat{y}) \cup I^2(\hat{y}), \quad I_+(\hat{x},\hat{y}) = I^1_+(\hat{x},\hat{y}) \cup I^2_+(\hat{x},\hat{y})$$
$$\text{and} \quad I_0(\hat{x},\hat{y}) := I^1_0(\hat{x},\hat{y}) \cup I^2_0(\hat{x},\hat{y}).$$

Theorem 12.3 *Let $(\hat{x},\hat{y})$ be (local) solution of (12.1). Assume that the constraint qualification (CQ*)*

$$\left.\begin{array}{l} \begin{bmatrix} (\nabla_x F_{L\cup I_0}(\hat{x},\hat{y}))^{\mathrm{T}} z_{L\cup I_0} \\ -w + (\nabla_y F_{L\cup I_0}(\hat{x},\hat{y}))^{\mathrm{T}} z_{L\cup I_0} \end{bmatrix} \in K_\omega(\hat{x},\hat{y}) \\ w_L = 0 \\ \textit{for } i \in I^1_0(\hat{x},\hat{y}) \textit{ either } w^i z^i = 0 \textit{ or } w^i < 0,\ z^i > 0, \\ \textit{for } i \in I^2_0(\hat{x},\hat{y}) \textit{ either } w^i z^i = 0 \textit{ or } w^i > 0,\ z^i < 0 \end{array}\right\} \Longrightarrow \left\{ \begin{array}{l} w_I = 0, \\ z_{L\cup I_0} = 0 \end{array}\right.$$

holds true.

Then there exists a pair $(\xi,\eta) \in \partial^- f(\hat{x},\hat{y})$, a pair $(\gamma,\delta) \in K_\omega(\hat{x},\hat{y})$ and a KKT pair $(\hat{w},\hat{z}) \in \mathcal{R}^m \times \mathcal{R}^m$ such that $\hat{w}_L = 0$

$$\begin{aligned} 0 &= \xi - (\nabla_x F_{L\cup I_0}(\hat{x},\hat{y}))^{\mathrm{T}} \hat{z}_{L\cup I_0} + \gamma \\ 0 &= \eta_{L\cup I_0} + \hat{w}_{L\cup I_0} - (\nabla_y F_{L\cup I_0, L\cup I_0}(\hat{x},\hat{y}))^{\mathrm{T}} \hat{z}_{L\cup I_0} + \delta_{L\cup I_0}, \\ & \textit{for } i \in I^1_0(\hat{x},\hat{y}) \textit{ either } \hat{w}^i \hat{z}^i = 0 \textit{ or } \hat{w}^i < 0,\ \hat{z}^i > 0, \quad \textit{and} \\ & \textit{for } i \in I^2_0(\hat{x},\hat{y}) \textit{ either } \hat{w}^i \hat{z}^i = 0 \textit{ or } \hat{w}^i > 0,\ \hat{z}^i < 0. \end{aligned} \tag{12.7}$$

If F is affine with respect to both variables and ω is convex polyhedral, then conditions (12.7) are fulfilled even without assuming the constraint qualification.

Proof. By using the formula, given in Lemma 12.2 for $K_{Gph\, N_\Gamma}(\hat{y}, -F(\hat{x},\hat{y}))$, we easily infer that (CQ) attains the form (CQ*). Since $\hat{z}_{I_+} = 0$, we can restrict ourselves only to the reduced vector $\hat{z}_{L\cup I_0}$. Furthermore, since $\hat{w}_{I_+}$ is arbitrary, we can neglect all equations of (12.6), corresponding to indices $i \in I_+(\hat{x},\hat{y})$. These simplifications directly lead to the (reduced) optimality conditions (12.7). □

Unfortunately, the verification of (CQ*) remains a complicated task whenever $I_0(\hat{x},\hat{y}) \neq \emptyset$. Therefore, we are interested in more easily verifiable conditions implying (CQ*) and devote the next section to this particular problem.

3 CONSTRAINT QUALIFICATION

As a consequence of [11, Prop. 3.2] we conclude that (CQ*) is fulfilled whenever $\omega = \tilde{\omega} \times \mathcal{R}^m$, $\tilde{\omega} \subset \mathcal{R}^n$, and the generalized equation (12.4) is strongly regular at $(\hat{x}, \hat{y})$ in the sense of Robinson, cf. [14]. This is a useful statement because the strong regularity is inherent in a large number of (discretized) mechanical equilibria. To get some other criteria, we reformulate first (CQ*) in the same way as it has been done in [9] in case of NCPs.

Suppose that

$$\omega = \left\{(x, y) \in \mathcal{R}^n \times \mathcal{R}^m \,|\, g^i(x, y) \leq 0,\ i = 1, 2, \ldots, s\right\}, \tag{12.8}$$

where the functions $g^i[\mathcal{R}^n \times \mathcal{R}^m \to \mathcal{R}]$, $i = 1, 2, \ldots, s$, are continuously differentiable. By (A) we denote the standard Mangasarian – Fromowitz constraint qualification for ω at $(\hat{x}, \hat{y})$, i. e.

(A) There exist vectors $\tilde{k} \in \mathcal{R}^n$, $\tilde{h} \in \mathcal{R}^m$ such that

$$\langle \nabla_x g^i(\hat{x}, \hat{y}), \tilde{k}\rangle + \langle \nabla_y g^i(\hat{x}, \hat{y}), \tilde{h}\rangle < 0$$
$$\text{for all } i \in M(\hat{x}, \hat{y}) := \{j \in \{1, 2, \ldots, s\} \,|\, g^j(\hat{x}, \hat{y}) = 0\}.$$

It is well-known that under (A)

$$K_\omega(\hat{x}, \hat{y}) = \left\{ \sum_{i=1}^{s} \mu^i \nabla g^i(\hat{x}, \hat{y}) \,\middle|\, \mu^i \geq 0 \text{ for } i \in M(\hat{x}, \hat{y}),\ \mu^i = 0 \text{ for } i \notin M(\hat{x}, \hat{y}) \right\},$$

cf. [6]. By Lemma 12.2 the index set $I_0^1(\hat{x}, \hat{y})$ splits into the three subsets

$$\begin{aligned} P^1 &:= \{i \in I_0^1(\hat{x}, \hat{y}) \,|\, w^i = 0\} \\ Q^1 &:= \{i \in I_0^1(\hat{x}, \hat{y}) \,|\, z^i = 0\} \\ R^1 &:= \{i \in I_0^1(\hat{x}, \hat{y}) \,|\, w^i < 0,\ z^i > 0\}, \end{aligned}$$

and the index set $I_0^2(\hat{x}, \hat{y})$ splits into the three subsets

$$\begin{aligned} P^2 &:= \{i \in I_0^2(\hat{x}, \hat{y}) \,|\, w^i = 0\} \\ Q^2 &:= \{i \in I_0^2(\hat{x}, \hat{y}) \,|\, z^i = 0\} \\ R^2 &:= \{i \in I_0^2(\hat{x}, \hat{y}) \,|\, w^i > 0,\ z^i < 0\}. \end{aligned}$$

For the sake of notational simplicity we put

$$P := P^1 \cup P^2, \quad Q := Q^1 \cup Q^2 \quad \text{and} \quad R := R^1 \cup R^2.$$

The constraints corresponding to P behave essentially like those from $L(\hat{y})$, the constraints corresponding to Q behave like those from $I_+(\hat{x},\hat{y})$ and only the constraints from R are the really difficult ones. Unfortunately, in (CQ*) each such splitting is possible and must be analyzed.

To unburden the notation, we denote by $\nabla G(\hat{x},\hat{y})$ the Jacobian of the function

$$G(x,y) := \left(g^1(x,y),\, g^2(x,y), \dots, g^s(x,y)\right)^{\mathrm{T}}$$

at $(\hat{x},\hat{y})$. By using the structure of ω and the splitting of $I_0(\hat{x},\hat{y})$, condition (CQ*) amounts to the following two conditions:

(i) For every decomposition of $I_0^1(\hat{x},\hat{y})$ into the sets P^1, Q^1, R^1 and for every decomposition of $I_0^2(\hat{x},\hat{y})$ into the sets P^2, Q^2, R^2 with $R^1 \cup R^2 \neq \emptyset$ the linear system of equations and inequalities

$$\begin{aligned}
-(\nabla_x F_{L\cup P\cup R}(\hat{x},\hat{y}))^{\mathrm{T}} z_{L\cup P\cup R} + (\nabla_x G_M(\hat{x},\hat{y}))^{\mathrm{T}} \mu_M &= 0\\
-(\nabla_y F_{L\cup P\cup R, L\cup P}(\hat{x},\hat{y}))^{\mathrm{T}} z_{L\cup P\cup R} + (\nabla_y G_{M,L\cup P}(\hat{x},\hat{y}))^{\mathrm{T}} \mu_M &= 0\\
w_{I_+\cup Q\cup R}\\
-(\nabla_y F_{L\cup P\cup R, I_+\cup Q\cup R}(\hat{x},\hat{y}))^{\mathrm{T}} z_{L\cup P\cup R} + (\nabla_y G_{M,I_+\cup Q\cup R}(\hat{x},\hat{y}))^{\mathrm{T}} \mu_M &= 0\\
w_{R^1} < 0,\ z_{R^1} > 0,,\ w_{R^2} > 0,\ z_{R^2} < 0,\ \mu_M \geq 0
\end{aligned} \tag{12.9}$$

does not have a solution.

(ii) For every decomposition of $I_0^1(\hat{x},\hat{y})$ into the sets P^1, Q^1 and for every decomposition of $I_0^2(\hat{x},\hat{y})$ into the sets P^2, Q^2 one has

$$\left\{ (w_{I_+\cup Q}, z_{L\cup P}) \,\middle|\, (w_{I_+\cup Q}, z_{L\cup P}, \mu_M) \in Ker \begin{bmatrix} 0 & -(\nabla_x F_{L\cup P}(\hat{x},\hat{y}))^{\mathrm{T}} & (\nabla_x G_M(\hat{x},\hat{y}))^{\mathrm{T}} \\ 0 & -(\nabla_y F_{L\cup P, L\cup P}(\hat{x},\hat{y}))^{\mathrm{T}} & (\nabla_y G_{M,L\cup P}(\hat{x},\hat{y}))^{\mathrm{T}} \\ E & -(\nabla_y F_{L\cup P, I_+\cup Q}(\hat{x},\hat{y}))^{\mathrm{T}} & (\nabla_y G_{M,I_+\cup Q}(\hat{x},\hat{y}))^{\mathrm{T}} \end{bmatrix}, \ \mu_M \geq 0 \right\} = \{0\}. \tag{12.10}$$

By using Motzkin's and Tucker's Theorem of the Alternative, we can reformulate (CQ*) in the following way.

Theorem 12.4 *Let $(\hat{x},\hat{y})$ satisfy the constraints of problem (12.1) with ω given by (12.8). Further suppose that assumption (A) is fulfilled. Then (CQ*) holds true if and only if* *(i)' for every decomposition of $I_0^1(\hat{x},\hat{y})$ into the sets P^1, Q^1, R^1 and for every decomposition of $I_0^2(\hat{x},\hat{y})$ into the sets P^2, Q^2, R^2 with $R^1 \cup R^2 \neq \emptyset$ there exist vectors $k \in \mathcal{R}^n$, $h \in \mathcal{R}^m$ such that $h_{I_+} = 0$, $h_Q = 0$, $h_{R^1} \geq 0$, $h_{R^2} \leq 0$,*

$$\begin{aligned} \nabla_x F_{L\cup P}(\hat{x},\hat{y})\,k + \nabla_y F_{L\cup P}(\hat{x},\hat{y})\,h &= 0 \\ \nabla_x F_{R^1}(\hat{x},\hat{y})\,k + \nabla_y F_{R^1}(\hat{x},\hat{y})\,h &\geq 0 \\ \nabla_x F_{R^2}(\hat{x},\hat{y})\,k + \nabla_y F_{R^2}(\hat{x},\hat{y})\,h &\leq 0 \\ \nabla_x G_M(\hat{x},\hat{y})\,k + \nabla_y G_M(\hat{x},\hat{y})\,h &\leq 0 \end{aligned}$$

and there exists an index $i \in R^1 \cup R^2$ such that

- *either $h^i > 0$ or $\langle \nabla_x F^i(\hat{x},\hat{y}),\, k\rangle + \langle \nabla_y F^i(\hat{x},\hat{y}),\, h\rangle > 0$ if $i \in R^1$,*
- *either $h^i < 0$ or $\langle \nabla_x F^i(\hat{x},\hat{y}),\, k\rangle + \langle \nabla_y F^i(\hat{x},\hat{y}),\, h\rangle < 0$ if $i \in R^2$;*

(ii)' for every decomposition of $I_0^1(\hat{x},\hat{y})$ into the sets P^1, Q^1 and for every decomposition of $I_0^2(\hat{x},\hat{y})$ into the sets P^2, Q^2 the matrix

$$\begin{bmatrix} \nabla_x F_{L\cup P}(\hat{x},\hat{y}) & \nabla_y F_{L\cup P, L\cup P}(\hat{x},\hat{y}) & \nabla_y F_{L\cup P, I_+\cup Q}(\hat{x},\hat{y}) \\ 0 & 0 & E \end{bmatrix} \tag{12.11}$$

has full row rank, and there exist vectors $k \in \mathcal{R}^n$, $h \in \mathcal{R}^m$ such that

$$\left.\begin{aligned} & h_{I_+} = 0, \quad h_Q = 0 \\ & \nabla_x F_{L\cup P}(\hat{x},\hat{y})\,k + \nabla_y F_{L\cup P}(\hat{x},\hat{y})\,h = 0, \\ & \nabla_x G_M(\hat{x},\hat{y})\,k + \nabla_y G_M(\hat{x},\hat{y})\,h < 0. \end{aligned}\right\} \tag{12.12}$$

Proof. Since one can use exactly the same method as in [9], Prop. 3.3, we only sketch the main ideas. To verify condition (i), system (12.9) can be replaced by a simpler system of equations and inequalities only in variables z_L, z_P, z_R and μ_M. The application of Motzkin's Theorem of the Alternative to this simpler system leads directly to the consistency requirement of condition (i)'. To verify the equivalence of (ii) and (ii)', observe first that the set on the left-hand side of (12.10) amounts to the union $\mathcal{A}_1 \cup \mathcal{A}_2$, where

$$\mathcal{A}_1 := Ker \begin{bmatrix} 0 & -(\nabla_x F_{L\cup P}(\hat{x},\hat{y}))^T \\ 0 & -(\nabla_y F_{L\cup P, L\cup P}(\hat{x},\hat{y}))^T \\ E & -(\nabla_y F_{L\cup P, I_+\cup Q}(\hat{x},\hat{y}))^T \end{bmatrix}$$

and

$$\mathcal{A}_2 := \left\{ (w_{I_+ \cup Q}, z_{L \cup P}) \middle| (w_{I_+ \cup Q}, z_{L \cup P}, \mu_M) \right.$$

$$\in Ker \begin{bmatrix} 0 & -(\nabla_x F_{L\cup P}(\hat{x},\hat{y}))^{\mathrm{T}} & (\nabla_x G_M(\hat{x},\hat{y}))^{\mathrm{T}} \\ 0 & -(\nabla_y F_{L\cup P, L\cup P}(\hat{x},\hat{y}))^{\mathrm{T}} & (\nabla_y G_{M, L\cup P}(\hat{x},\hat{y}))^{\mathrm{T}} \\ E & -(\nabla_y F_{L\cup P, I_+\cup Q}(\hat{x},\hat{y}))^{\mathrm{T}} & (\nabla_y G_{M, I_+\cup Q}(\hat{x},\hat{y}))^{\mathrm{T}} \end{bmatrix},$$

$$\left. \mu_M \geq 0,\ \mu_M \neq 0 \right\}.$$

Condition (ii) amounts thus to the requirement

$$\mathcal{A}_1 \cup \mathcal{A}_2 = \{0\}. \tag{12.13}$$

If $\mu_M \neq 0$, then assumption (A) implies in virtue of Gordan's Theorem that

$$(\nabla G_M(\hat{x},\hat{y}))^{\mathrm{T}} \mu_M \neq 0.$$

From this follows that (12.13) is equivalent to the next two conditions:

$$\begin{aligned} \mathcal{A}_1 &= \{0\} \\ \mathcal{A}_2 &= \emptyset. \end{aligned}$$

One readily sees that the former one is equivalent to the requirement that matrix (12.11) has full row rank. Further, the inconsistency of the linear system defining $\mathcal{A}_2$ is equivalent to the consistency of system (12.12) in virtue of Tucker's Theorem of the Alternative. This completes the proof. □

On the basis of the preceding result we can now compare (CQ*) among others with the *linear independence constraint qualification* (LICQ) and the *piecewise Mangasarian – Fromowitz constraint qualification* (PMFCQ). The former one is closely connected with the nonsmooth variant of the Mangasarian – Fromowitz constraint qualification developed in [4] for general nonsmooth programs. For problem (12.1) with ω given by (12.8) (LICQ) attains the following form.

(LICQ) The matrix

$$\begin{bmatrix} \nabla_x F_{L\cup I_0}(\hat{x},\hat{y}) & \nabla_y F_{L\cup I_0}(\hat{x},\hat{y}) \\ 0 & E_{I_+\cup I_0} \\ \nabla_x G_M(\hat{x},\hat{y}) & \nabla_y G_M(\hat{x},\hat{y}) \end{bmatrix}$$

has full row rank.

(LICQ) contains implicitly the requirement that n is not smaller than the cardinality of $I_0(\hat{x},\hat{y})$ plus the cardinality of $M(\hat{x},\hat{y})$. It directly implies that the matrix (12.11) has full row rank. Moreover, under (LICQ) the system in condition (i)' of Theorem 12.4 as well as the system (12.12) are consistent for each decomposition of $I_0(\hat{x},\hat{y})$. Therefore (LICQ) implies (CQ*).

(PMFCQ) arises in connection with the so-called piecewise programming approach to mathematical programs with equilibrium constraints which is thoroughly investigated in [5]. This approach leads to a finite set of optimality conditions in the KKT form, corresponding to each "piece" of the given problem. For (12.1) with ω given by (12.8) (PMFCQ) attains the following form.

(PMFCQ) For every partition of $I_0^1(\hat{x},\hat{y})$ into the sets α_1, $\overline{\alpha}_1$ and for every partition of $I_0^2(\hat{x},\hat{y})$ into the sets α_2, $\overline{\alpha}_2$ the $[m \times (m+n)]$ matrix

$$\Gamma(\hat{x},\hat{y}) := \begin{bmatrix} \tilde{A} & \tilde{B} & \tilde{C} \\ 0 & 0 & E \end{bmatrix} \text{has full row rank,}$$

(where $\tilde{A} = \nabla_x F_{L\cup\overline{\alpha}_1\cup\overline{\alpha}_2}(\hat{x},\hat{y})$, $\tilde{B} = \nabla_y F_{L\cup\overline{\alpha}_1\cup\overline{\alpha}_2, L\cup\overline{\alpha}_1\cup\overline{\alpha}_2}(\hat{x},\hat{y})$ and $\tilde{C} = \nabla_y F_{L\cup\overline{\alpha}_1\cup\overline{\alpha}_2, I_+\cup\alpha_1\cup\alpha_2}(\hat{x},\hat{y})$)
and there exist vectors $k \in \mathcal{R}^n$, $h \in \mathcal{R}^m$ such that

$$\Gamma(\hat{x},\hat{y}) \begin{bmatrix} k \\ h \end{bmatrix} = 0$$

$$h_{\overline{\alpha}_1} > 0, \; h_{\overline{\alpha}_2} < 0$$

$$\nabla_x F_{\alpha_1}(\hat{x},\hat{y})\, k + \nabla_y F_{\alpha_1}(\hat{x},\hat{y})\, h > 0$$

$$\nabla_x F_{\alpha_2}(\hat{x},\hat{y})\, k + \nabla_y F_{\alpha_2}(\hat{x},\hat{y})\, h < 0$$

$$\nabla_x G_M(\hat{x},\hat{y})\, k + \nabla_y G_M(\hat{x},\hat{y})\, h < 0.$$

Proposition 12.5 *Let $(\hat{x},\hat{y})$ satisfy the constraints of problem (12.1) with ω given by Theorem 12.4. Further suppose that assumption (A) is fulfilled. Then (PMFCQ) implies* (CQ*).

Proof. To verify condition (i)' of Theorem 12.4, we put $\alpha_1 := Q^1 \cup R^1$, $\overline{\alpha}_1 = P_1$, $\alpha_2 := Q^2 \cup R^2$ and $\overline{\alpha}_2 = P_2$. By (PMFCQ) there exist vectors $k \in \mathcal{R}^n$, $h \in \mathcal{R}^m$ such that $h_{I_+} = 0$, $h_Q = 0$, $h_R = 0$,

$$\begin{aligned} \nabla_x F_{L\cup P}(\hat{x},\hat{y})\, k + \nabla_y F_{L\cup P}(\hat{x},\hat{y})\, h &= 0 \\ \nabla_x F_{R^1}(\hat{x},\hat{y})\, k + \nabla_y F_{R^1}(\hat{x},\hat{y})\, h &> 0 \\ \nabla_x F_{R^2}(\hat{x},\hat{y})\, k + \nabla_y F_{R^2}(\hat{x},\hat{y})\, h &< 0 \\ \nabla_x G_M(\hat{x},\hat{y})\, k + \nabla_y G_M(\hat{x},\hat{y})\, h &< 0, \end{aligned}$$

and we are done. Concerning condition (ii)' of Theorem 12.4, we observe first that with $\alpha_1 := Q^1$, $\overline{\alpha}_1 := P^1$, $\alpha_2 := Q^1$ and $\overline{\alpha}_2 := P^2$, $\Gamma(\hat{x}, \hat{y})$ is exactly the matrix (12.11). Moreover, with this choice of α_1, $\overline{\alpha}_1$, α_2, $\overline{\alpha}_2$, (PMFCQ) implies the existence of vectors $k \in \mathcal{R}^n$, $h \in \mathcal{R}^m$ such that $h_{I_+} = 0$, $h_Q = 0$ and

$$\begin{aligned} \nabla_x F_{LUP}(\hat{x}, \hat{y})\, k + \nabla_y F_{LUP}(\hat{x}, \hat{y})\, h &= 0 \\ \nabla_x G_M(\hat{x}, \hat{y})\, k + \nabla_y G_M(\hat{x}, \hat{y})\, h &< 0. \end{aligned}$$

Thus, condition (ii)' of Theorem 12.4 is also fulfilled and our statement has been established. □

4 EXACT PENALIZATION

This concluding section is devoted to some issues relating the above theory with the numerical solution of problem (12.1).

As explained in [19], under the assumptions of Theorem 12.3 there exist neighborhoods $\mathcal{U}$ of $(\hat{x}, \hat{y})$ and $\mathcal{V}$ of $0 \in \mathcal{R}^m \times \mathcal{R}^m$ such that $(\hat{x}, \hat{y}, 0)$ solves the optimization problem

$$\begin{aligned} &\text{minimize} \quad f(x, y) + \rho\|v\| \\ &\text{subject to} \\ &\qquad v \in \begin{bmatrix} -y \\ F(x, y) \end{bmatrix} + Gph\, N_\Gamma \\ &\qquad (x, y) \in \omega \cap \mathcal{U} \\ &\qquad v \in \mathcal{V}, \end{aligned} \tag{12.14}$$

whenever the penalty parameter $\rho \geq 0$ is sufficiently large. Problem (12.14) can be decoupled, since for each fixed pair $(\overline{x}, \overline{y})$ we can compute a solution of the (inner) minimum-norm problem

$$\begin{aligned} &\text{minimize} \quad \|v\| \\ &\text{subject to} \\ &\qquad v \in \begin{bmatrix} -\overline{y} \\ F(\overline{x}, \overline{y}) \end{bmatrix} + Gph\, N_\Gamma, \end{aligned} \tag{12.15}$$

where we do not take into account the constraint $v \in \mathcal{V}$. For a suitable norm $\|\cdot\|$ in $\mathcal{R}^m \times \mathcal{R}^m$ we can even find a solution of (12.15) in an explicit form.

Assume that for a vector $\kappa = (\kappa_1, \kappa_2) \in \mathcal{R}^m \times \mathcal{R}^m$ one has

$$\|(\kappa_1, \kappa_2)\| = \sqrt{\sum_{i=1}^{m} \|(\kappa_1^i, \kappa_2^i)\|_2^2}, \tag{12.16}$$

where $\|\cdot\|_2$ is the Euclidean norm.

Proposition 12.6 *Let $\|\cdot\|$ be given by (12.16) and $\overline{v}$ be a solution of problem (12.15) with $\overline{y} \in \Gamma$. Then for $i = 1, 2, \ldots, m$ one has*

$$\|(\overline{v}_1^i, \overline{v}_2^i)\|_2 = \begin{cases} \min\left\{F^i(\overline{x}, \overline{y}), \overline{y}^i - a^i\right\} & \text{if } F^i(\overline{x}, \overline{y}) \geq 0 \\ \min\left\{-F^i(\overline{x}, \overline{y}), b^i - \overline{y}^i\right\} & \text{if } F^i(\overline{x}, \overline{y}) \leq 0. \end{cases} \tag{12.17}$$

Proof. Since $Gph\ N_{\mathcal{R}^m_+} = \mathsf{X}_{i=1}^m\ Gph\ N_{\mathcal{R}_+}$, problem (12.15) can be decoupled into m subproblems of the form

$$\begin{array}{ll} \text{minimize} & \|(v_1^i, v_2^i)\|_2 \\ \text{subject to} & \\ & \begin{bmatrix} v_1^i + \alpha \\ v_2^i - \beta \end{bmatrix} \in Gph\ N_{\Gamma_i} \end{array}$$

over $\mathcal{R}^2$. The reals α, β correspond to $\overline{y}^i$ and $F^i(\overline{x}, \overline{y})$, respectively. We have to examine only the following four cases, corresponding to the four regions depicted in Figure 12.2.

1. $0 \leq -\beta \leq b^i - \alpha$: Then one has $\overline{v}_1^i = 0$, $\overline{v}_2^i = \beta$ and $\|(\overline{v}_1^i, \overline{v}_2^i)\|_2 = |\beta| = \min\{-\beta, b^i - \alpha\}$.

2. $b^i - \alpha \leq -\beta$: Then one has $\overline{v}_1^i = b^i - \alpha$, $\overline{v}_2^i = 0$ and $\|(\overline{v}_1^i, \overline{v}_2^i)\|_2 = b^i - \alpha = \min\{-\beta, b^i - \alpha\}$.

3. $a^i - \alpha \leq -\beta \leq 0$: Then one has $\overline{v}_1^i = 0$, $\overline{v}_2^i = \beta$ and $\|\overline{v}_1^i, \overline{v}_2^i\|_2 = \beta = \min\{\beta, \alpha - a^i\}$.

4. $-\beta \leq a^i - \alpha$: Then one has $\overline{v}_1^i = a^i - \alpha$, $\overline{v}_2^i = 0$ and $\|\overline{v}_1^i, \overline{v}_2^i\|_2 = \alpha - a^i = \min\{\beta, \alpha - a^i\}$. □

Consider now the generalized equation in (12.14). Since $v = 0$ implies $y \in \Gamma$, we observe that under the assumptions of Theorem 12.3 there exist neighborhoods $\tilde{\mathcal{U}}$ of $(\hat{x}, \hat{y})$ and $\tilde{\mathcal{V}}$ of 0 such that $(\hat{x}, \hat{y}, 0)$ solves the optimization problem

$$\begin{array}{ll} \text{minimize} & f(x, y) + \rho\|v\| \\ \text{subject to} & \\ & v \in \begin{bmatrix} -y \\ F(x, y) \end{bmatrix} + Gph\ N_\Gamma \\ & (x, y) \in \omega \cap \tilde{\mathcal{U}} \cap (\mathcal{R}^m \times \Gamma) \\ & v \in \tilde{\mathcal{V}}, \end{array} \tag{12.18}$$

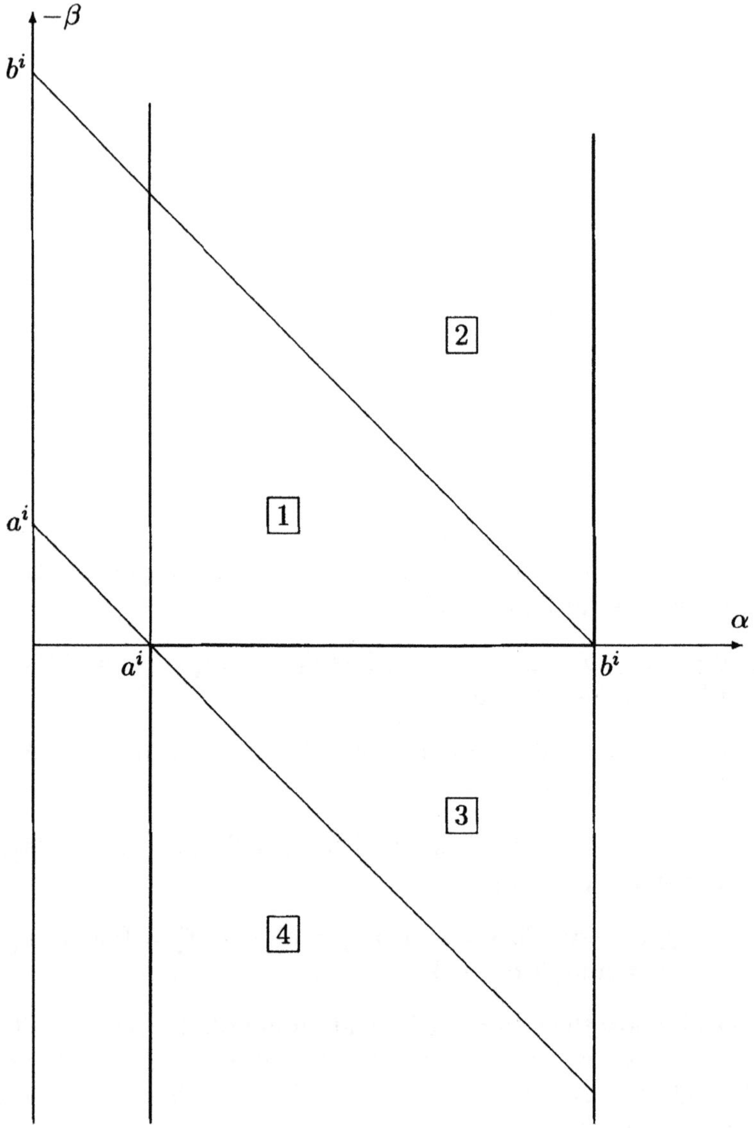

Figure 12.2 Regions given by the metric projection onto *Gph* N_{Γ_i}

whenever ρ is sufficiently large. Therefore we can take (12.16) as the norm in (12.18) and the neighborhood $\tilde{\mathcal{V}}$ as the closed ε-ball in $\mathcal{R}^{2m}$ for some positive real ε. By virtue of Proposition 12.6 problem (12.18) attains then the form

$$\begin{array}{ll} \text{minimize} & f(x,y) + \rho \, \|P(x,y)\|_2 \\ \text{subject to} & \\ & (x,y) \in \omega \cap \tilde{\mathcal{U}} \cap (\mathcal{R}^n \times \Gamma) \\ & \|P(x,y)\|_2 \le \varepsilon, \end{array} \tag{12.19}$$

where

$$P^i(x,y) = \begin{cases} \min\{F^i(x,y),\, y^i - a^i\} & \text{if } \; F^i(x,y) \ge 0 \\ \min\{-F^i(x,y),\, b^i - y^i\} & \text{if } \; F^i(x,y) < 0, \end{cases}$$

$i = 1, 2, \ldots, m$. All function P^i are continuous and nonnegative for $y^i \in [a^i, b^i]$, since the terms $y^i - a^i$, $b^i - y^i$ are nonnegative. They are even locally Lipschitz due to our assumptions on F. Thus, under the assumptions of Theorem 12.3, we can try to solve numerically instead of (12.1) the nonsmooth program

$$\begin{array}{ll} \text{minimize} & f(x,y) + \rho \|P(x,y)\|_2 \\ \text{subject to} & \\ & (x,y) \in \omega \cap (\mathcal{R}^n \times \Gamma), \end{array} \tag{12.20}$$

provided ρ is sufficiently large, the starting point is sufficiently close to $(\hat{x}, \hat{y})$ and the starting value of $\|P(x,y)\|_2$ is sufficiently small.

Remark. As the outer norm in (12.16) one could use also the ℓ_1-norm instead of the Euclidean norm. This leads to the penalty

$$\sum_{i=1}^{n} P^i(x,y)$$

which enables to use different penalty parameters for single terms in the above sum.

As a test example we have used an academic problem, where $n = 1$, $m = 2$,

$$f(x, y^1, y^2) = 20\sqrt{2x + p_1} + p_2 x - \frac{x c_3}{T}$$

and

$$F(x, y^1, y^2) = \begin{bmatrix} \frac{20}{\sqrt{2y^1 + c_1}} + c_2 - c_3 \frac{x + y^2}{T^2} \\ \frac{20}{\sqrt{2y^2 + d_1}} + d_2 - c_3 \frac{x + y^1}{T^2} \end{bmatrix}$$

with $T = x + y^1 + y^2$ and parameters p_1, c_1, d_1, p_2, c_2, d_2, c_3. These parameters and bounds a^1, a^2, b^1, b^2 are given in Table 1.

Table 12.1 Data of the test example

p_1	c_1	d_1	p_2	c_2	d_2	c_3	a^1	a^2	b^1	b^2
64	36	64	4	4	3.2	2000	0	0	100	100

Further, $\omega = \{(x, y) \in \mathcal{R} \times \mathcal{R}^2 \,|\, 1 \le x \le 200\}$.

With the ℓ_1-variant of our penalty and the bundle-trust algorithm described in [16] we reached the approximate solution

$$x_a = 172.7272 \quad y_a^1 = 0.0 \quad y_a^2 = 100.0$$

with the prescribed accuracy $\varepsilon = 0.00001$ in 17 iterations from the starting point (50, 50, 50). The value of the penalty was $6.1 \cdot 10^{-36}$. Since $F^1(x_a, y_a^1, y_a^2) = 8.8 \cdot 10^{-9}$, we can assume that we have found a degenerate solution of the considered MCP.

Then we increased the upper bound b^2 to 1000 and obtained the approximate solution

$$x_a = 163.2218 \quad y_a^1 = 0.0 \quad y_a^2 = 109.5054$$

in 47 iterations. The penalty value was $1.1 \cdot 10^{-9}$. Again $F^1(x_a, y_a^1, y_a^2)$ was very small ($1.1 \cdot 10^{-9}$).

Finally, we changed the parameter c_2 to 3.4 and put $\omega = \{(x, y) \in \mathcal{R} \times \mathcal{R}^2 \,|\, 1 \le x \le 100\}$. After 34 iterations we have computed the approximate solution

$$x_a = 100.0 \quad y_a^1 = 88.5275 \quad y_a^2 = 103.7508.$$

The respective penalty value was $8.7 \cdot 10^{-7}$.

Remark. The above academic example admits an economic interpretation. Namely, x_a, y_a^1 and y_a^2 can be viewed as "optimal" productions of three producers on an oligopolistic market. The first producer applies the Stackelberg strategy whereas the remaining two share the rest of the market according to the Cournot–Nash equilibrium principle. Such problems have been used as test examples already in [2] and [12]. In these works, however, in accordance with [8], the production costs were convex and $F(x, \cdot)$ was strongly monotone. Here we work with concave production costs, which is more realistic from the economic point of view, but makes the problem substantially more difficult. In particular, the considered MCP represents then only a necessary condition for the Cournot–Nash equilibrium, the existence of which is not guaranteed at all. When analyzing the results obtained from this point of view, we easily conclude that only in case of the third data set the pair (y_a^1, y_a^2) is a Cournot–Nash equilibrium. In case of the first two data sets the production $y_a^1 = 0$ is only a stationary point, but not a local maximum in the profit maximization problem of the 2nd producer.

5 CONCLUSION

The optimality conditions stated in Theorem 12.3 can be used for testing of stationarity of approximate solutions to problem (12.1) computed by various available numerical methods. The constraint qualification, needed in the general (nonpolyhedral) case, is implied either by the strong regularity of the considered MCP (provided ω concerns only the control variable), or by (LICQ) or by (PMFCQ). Constraint qualifications corresponding to (CQ*) and to the conditions (i)', (ii)' of Theorem 12.4 have been derived in a forthcoming paper [18] for a slightly different class of equilibria. To derive a counterpart of Theorem 12.4 in [18], however, also the method from [9] has been applied. The application of a bundle method to the numerical solution of the penalized program (12.20) is, according to our opinion, practical only in "small" problems. A better alternative is probably a special method, developed in [15] for nonsmooth programs of the same type, where the penalty parameters are updated during the iteration process.

Appendix

Consider a set $\Pi \subset \mathcal{R}^p$.

Definition A.7 *Let $a \in cl\, \Pi$. The nonempty cone*

$$T_\Pi(a) := \limsup_{t \downarrow 0} \frac{\Pi - a}{t}$$

is called the contingent cone to Π at a. The generalized normal cone to Π at a, denoted $K_\Pi(a)$, is defined by

$$K_\Pi(a) = \limsup_{a' \xrightarrow{cl\, \Pi} a} T_\Pi^0(a'),$$

where $T_\Pi^0(a')$ denotes the negative polar cone to $T_\Pi(a')$.

The "lim sup" in the definitions of $T_\Pi(a)$ and $K_\Pi(a)$ is the upper limit of multifunctions in the sense of Kuratowski–Painlevé, cf. [1].

If Π is convex, one has $K_\Pi(a) = T_\Pi^0(a)$. The cone $K_\Pi(a)$ is generally nonconvex, but the multifunction $K_\Pi(\cdot)$ is upper semicontinuous at each point of *cl* Π (with respect to *cl* Π), which is essential in the calculus of Mordukhovich's subdifferentials introduced below.

Definition A.8 *Let $\varphi[\mathcal{R}^p \to \overline{\mathcal{R}}]$ be an arbitrary extended real-valued function and $a \in dom\ \varphi$. The set*

$$\partial^- \varphi(a) := \{a^* \in \mathcal{R}^p | (a^*, -1) \in K_{epi\ \varphi}(a, \varphi(a))\}$$

is called the Mordukhovich's subdifferential of φ at a.

For convex functions this subdifferential coincides with the classical subdifferential in the sense of convex analysis. If φ is Lipschitz near a, then *conv* $\partial^- \varphi(a)$ coincides with the Clarke's generalized gradient of φ at a.

Acknowledgments

The author is very grateful to the anonymous referees for their valuable suggestions. This research was supported by the grant No. A 1075707 from the Academy of Sciences of the Czech Republic.

References

[1] Aubin, J.-P. and Frankowska H. (1990). *Set–Valued Analysis.* Birkhäuser, Boston.

[2] Harker, P. T. and Choi, S. C. (1987). A penalty function approach to mathematical programs with variational inequality constraints. Technical report, Decision Sciences Department, University of Pennsylvania.

[3] Kočvara M. and Outrata J. V. (1997). A nonsmooth approach to optimization problems with equilibrium constraints. In Ferris, M. and Pang, J.-S., editors. *Complementarity and Variational Problems*, pages 148–164, SIAM, Philadelphia.

[4] Kuntz L. and Scholtes S. (1995). A nonsmooth variant of the Mangasarian – Fromovitz constraint qualification. *J. Optimization Theory and Applications*, 82: 59–75.

[5] Luo, Z.-Q., Pang, J.-S., and Ralph, D. (1996). *Mathematical Programs with Equilibrium Constraints.* Cambridge University Press, Cambridge.

[6] Mordukhovich, B. S. (1988). *Approximation Methods in Problems of Optimization and Control.* Nauka, Moscow. (In Russian; English edition to appear in Wiley – Interscience.)

[7] Mordukhovich, B. S. (1994). Generalized differential calculus for nonsmooth and set-valued mappings. *J. Math. Analysis and Applications*, 183: 250–288.

[8] Murphy, F. H., Sherali, H. D., and Soyster, A. L. (1982). A mathematical programming approach for determining oligopolistic market equilibrium. *Mathematical Programming*, 24: 92–106.

[9] Outrata, J. V. (1999). Optimality conditions for a class of mathematical programs with equilibrium constraints. *Mathematics of Operations Research*, 24, 627–644.

[10] Outrata, J. V. (1999). On mathematical programs with complementarity constraints. Accepted in *J. of Optimization Methods and Software.*

[11] Outrata, J. V. (2000). A generalized mathematical program with equilibrium constraints. *SIAM J. on Control and Optimization*, 38: 1623-1638.

[12] Outrata, J. V. and Zowe, J. (1995). A numerical approach to optimization problems with variational inequality constraints. *Mathematical Programming*, 68: 105–130.

[13] Pang, J.-S. and Fukushima, M. (1999). Complementarity constraint qualifications and simplified B-stationarity conditions for mathematical programs with equilibrium constraints. *Computational Optimization and Applications*, 13: 111–136.

[14] Robinson, S. M. (1980). Strongly regular generalized equations. *Mathematics of Operations Research*, 5: 43–62.

[15] Scholtes, S. and Stöhr, M. (1999). Exact penalization of mathematical programs with equilibrium constraints. *SIAM J. on Control and Optimization*, 37: 617–652.

[16] Schramm, H. and Zowe, J. (1992). A version of the bundle idea for minimizing a nonsmooth function: conceptual idea, convergence analysis, numerical results. *SIAM J. on Optimization*, 2: 121–152.

[17] Ye, J. J. (1999). Optimality conditions for optimization problems with complementarity constraints. *SIAM J. on Optimization*, 9: 374–387.

[18] Ye, J. J. (2000). Constraint qualifications and necessary optimality conditions for optimization problems with variational inequality constraints. *SIAM J. on Optimization*, 10: 943–962.

[19] Ye, J. J. and Ye, X. Y. (1997). Necessary optimality conditions for optimization problems with variational inequality constraints. *Mathematics of Operations Research*, 22: 977–997.

Chapter 13

A GENERATION OPERATION PLANNING MODEL IN DEREGULATED ELECTRICITY MARKETS BASED ON THE COMPLEMENTARITY PROBLEM

Michel Rivier
Instituto de Investigación Tecnológica, Universidad Pontificia Comillas, Alberto Aguilera 23, 28015 Madrid, Spain
michel.rivier@iit.upco.es

Mariano Ventosa
Instituto de Investigación Tecnológica, Universidad Pontificia Comillas, Alberto Aguilera 23, 28015 Madrid, Spain

Andrés Ramos
Instituto de Investigación Tecnológica, Universidad Pontificia Comillas, Alberto Aguilera 23, 28015 Madrid, Spain

Francisco Martínez-Córcoles
IBERDROLA, Hermosilla 3, 28001 Madrid, Spain

Ángel Chiarri Toscano
IBERDROLA, Hermosilla 3, 28001 Madrid, Spain

Abstract Throughout the world, the electricity industry is currently undergoing a significant restructuring process toward deregulation and competition. In the new deregulated power markets, electric firms assume much more risk and become highly responsible for their own decisions, and therefore they need decision-support models that fulfill their new requirements.

M.C. Ferris et al. (eds.), Complementarity: Applications, Algorithms and Extensions, 273–295.

This paper proposes a new approach for long-term operation-planning models, fully adapted to represent an annual or hyperannual power generation scheduling in a competitive framework. The method explicitly states the Cournot market equilibrium by analytically formulating the equations defining the optimal behavior of any strategic generation company. The technical constraints relevant to the time scope studied are also taken into account. The subsequent system of non-linear equations has the structure of a Complementarity Problem and can be solved directly. The model includes a technologically detailed representation of the generation firms, the treatment of hydro operations and of non-linear cost functions. A detailed case study illustrates how this model works, and how different generation technologies are optimally operated. An application to the Spanish electric market is also briefly presented in order to show that the model is also able to deal with large-scale electric systems.

Keywords: Competitive power market, Mixed Complementarity Problem (MCP), market equilibrium.

1 INTRODUCTION

Throughout the world, the electricity industry is currently undergoing a significant restructuring process toward deregulation and competition. In the new context, generation of electricity becomes an unbundled activity subject to strong liberalization in which both expansion and operation decisions no longer depend upon administrative and centralized procedures -usually based on cost-minimization schemes-, but rather on the managerial decisions of the generation companies looking forward to maximizing their own profits. Therefore, in the new deregulated power markets electric firms assume much more risk and become highly responsible for their own decisions. Companies need decision-support models that fulfill these new requirements.

An important developmental effort is needed either to adapt the traditional models based on cost minimization schemes to the new context or to design new models and tools that explicitly represent the electricity market behavior. The market behavior will depend on companies' production cost profiles, on companies' technical operation constraints, and on the economic market equilibrium resulting from the interaction of all market participants.

The interest of the research community regarding the development of models adapted to the new circumstances has grown and has been demonstrated as such in numerous publications. Two great tendencies exist in modeling the market equilibrium [7]. On one hand, the models based on Supply Function Equilibria [5, 11] and, on the other hand, models based on obtaining the Cournot market equilibrium [2, 3, 8, 9, 10, 13]. Modeling advances have been notable, nevertheless most of them still present some limitations in regard to the correct representation of the system (no consideration of time-coupling technical constraints, simplified cost functions, or a limited number of hydro reservoirs).

Most recently, Hobbs [6] utilizes the Linear Complementarity Problem (LCP) to model imperfect competition among electricity producers. His model includes a congestion pricing scheme for transmission. Jing-Yuan and Smeers [15] use a Variational Inequality (VI) approach for computing the market equilibria addressing the problem of transmission pricing.

This paper presents an approach based also on the Mixed Complementarity Problem (MCP) to model Cournot market equilibrium. However, our work focuses on long-term generation-operation planning. It includes the relation among periods, which is important in systems with a significant hydro component such as the Spanish one.

Section 2 discusses the new role and structure of the operation-planning decision-support models in the new competitive environment of the electric industry. Section 3 presents an overview of the proposed methodology to design a long-term operation-planning model. Sections 4 and 5 state the mathematical formulation. Section 6 discusses the implementation and finally section 7 presents and analyses the numerical results for a case study.

2 THE NEW ROLE OF OPERATION PLANNING MODELS

The goal of operation planning, in traditional regulation schemes, is to meet the electricity demand with satisfactory supply security and at minimum cost.

Furthermore, all the energy constraints that for diverse reasons may exist must also be satisfied (environmental, mandatory domestic fuel purchases, development of special generation, etc.). Usually the operation functions are centralized either at the level of an entire integrated system with various companies, which was for example the situation in the Spanish case, or within the territorial operational zone of each company. Therefore the operation models in such a traditional scheme are based on cost minimization problems, which schematically correspond to the following structure:

$$\min_{y} z(y) \quad \text{subject to} \quad h(y) = 0, \quad g(y) \leq 0 \tag{13.1}$$

where y stands for the decision variables vector, z for the cost function to be minimized and h and g for the set of operation and technical constraints of the system.

An important number and variety of such types of models, fully adapted to the specificity and requirements of each particular electrical generation system and commonly used in most companies, can be found in technical literature. These models have reached an important degree of sophistication and detail in their representation of the system behavior. They are usually run in a hierarchic manner, beginning with long-term-scope hydrothermal coordination models and ending with online generation economic dispatch models.

The energy scheduling procedures within a free market regulation are much more decentralized than in a traditional environment. The generation companies must assume not only all the normal functions related to the operation planning of their power plants (start-ups, shut-downs, hydro-thermal coordination, maintenance, etc.) but also must face new functions associated with free market competition. This requires a high level of coordination between them: risk management, long term contracts, fuel purchases, offers to buy and sell on the different markets...

The results of an operation model adapted to the new environment should not only permit the foreseeing of the generation operation, but they should also supply relevant information for the two new main functions: the preparation of daily offers -price and quantity- for the different markets which includes both long term (i.e. hydro scheduling) and short term (i.e. offering curve) decisions, and the risk management (economical value of long term electricity sale contracts, position in the market, influence of hydro resources stochasticity...). In order to achieve these purposes, new models should correctly characterize the behavior of all the implied agents -market operator, demand, and companies- while keeping the same technical constraints included in traditional models.

The simultaneous consideration of the profit maximization objective of each company, that is to say, the calculation of the market equilibrium, constitutes the newest and most complicated issue of these models. The market equilibrium defines the point -price and quantity produced by each company- at which no company improves its profits by modifying its decisions in a unilateral way.

The modeling of the market behavior in these new conditions hardly fits with the traditional structure of cost minimization given as (13.1). Conceptually, the new structure better corresponds to various simultaneous optimizations (for each one of the participating companies the maximization of its profits subject to its particular technical constraints) linked among themselves by the market price (the price resulting from

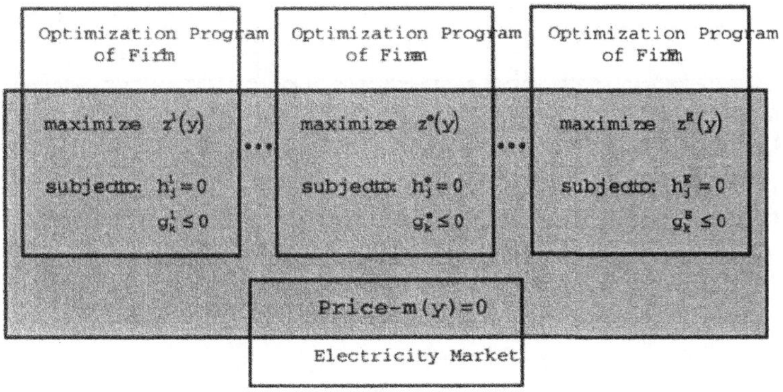

Figure 13.1 Market Equilibrium

the interaction of all of them), as is represented in Figure 13.1. Now, z represents the company's operation profit (market revenues minus production costs) and the set of constraints h and g are particularized for each company. The electricity market is represented by the demand function (m), which models the consumers' behavior. This function relates the supplied demand to the electricity price.

3 MODEL STATEMENT

In this article a new approach is proposed for the design of long-term operation-planning models of a power system, perfectly adapted to the annual or hyperannual operation of generation in a competitive environment. The method consists of an explicit statement of the market equilibrium by analytically formulating the non-linear equations that define the optimal behavior of each firm in a deregulated electricity market, considering the technical constraints that affect the energy scheduling between periods.

The mentioned equations, schematically represented in figure 13.2, are the optimality conditions associated to the set of maximization programs represented in figure 13.1. In this figure, L represents the Lagrangian function of the corresponding optimization problem and λ and μ represent the Lagrange multipliers associated to the set of h and g constraints respectively. The optimality conditions can be written down as three sets of equations. The first one groups together the gradient of the Lagrangian function with respect to the decision variables equalized

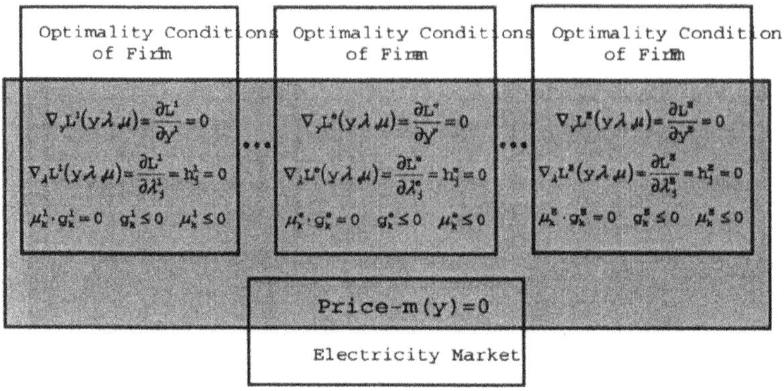

Figure 13.2 Market equilibrium as a mixed complementarity problem

to zero. The second set (gradient of the Lagrangian function with respect to the Lagrangian multipliers λ equalized to zero) coincides with the h constraints themselves. The third one states the complementary slackness conditions associated to the inequality constraints g.

The resulting system of equations can be solved by directly taking advantage of its structure, as a mixed complementarity problem [4], whose particularities allow for the use of special resolution methodologies incorporated nowadays in commercial software packages. Section 4 states in detail the optimization problem that defines the operation of each agent (figure 13.1).

Section 5 analytically derives the associated Karush-Kuhn-Tucker optimality conditions, which form the mentioned system of non-linear equations to be solved (figure 13.2).

The proposed methodology allows the sufficient complexity and flexibility as to correctly model the physical behavior of an electrical system while considering the profit-maximization goal of each company participating in the market.

The described model considers in detail the particular characteristics of each type of generation unit. For each thermal generator, the rated power output and the quadratic fuel consumption are modeled. Each type of fuel is characterized by its price. For thermal plants, fuel purchases and the maximum size of the stockpiles are also considered. For hydro units, it is necessary to consider both their rated power output and the limited capacity of their reservoirs. Regarding pumping units, it is also necessary to characterize the performance in the pumping and generation cycle.

Traditionally, in operation models, demand has been considered to be constant for each load level. Nevertheless, in the new operation models, the modeling of the demand should consider the reaction of the quantity demanded to changes in the price. In the proposed model the total demand at each load level is a linear function of the price.

This long-term model considers a hyperannual scope divided into periods and load levels. Normally, periods coincide with months, while the grouping of peak, plateau, and off-peak hours make up the load levels.

4 PROBLEM FORMULATION FOR EACH GENERATION COMPANY

This section states in detail the optimization problem that defines the behavior of each generation company (schematically represented in figure 13.1) under the new electric regulatory framework. The goal is to maximize their own profits (defined as the market revenues minus the production costs) subject to the constraints that limit their operation decisions. Only the most relevant constraints to the long-term scope considered have been included in this article. However, the methodology can be directly applied and extended to other types of constraints.

In the problem formulation, special attention has been placed on explicitly including the Lagrangian multiplier of each constraint and the number of constraints of each type.

4.1 OBJECTIVE FUNCTION

The objective of each generation company is to maximize its profits for the entire scope of the model. These are calculated as the total gross revenues minus the fuel and O&M variable costs of thermal units. The revenues are computed as the company's power generation times the marginal price. Thermal units' fuel consumption is modeled by a quadratic curve, for each e.

$$\max \sum_{p,n} D_{n,p} \left(\pi_{n,p} g_{n,p,e} - \sum_{t \in e} \left(u_t p_{n,p,t} + \nu_t \left(o'_t \frac{p_{n,p,t}}{k_t} + o''_t \left(\frac{p_{n,p,t}}{k_t} \right)^2 \right) \right) \right)$$

4.1.1 Operation technical constraints.

The constraints considered in a long term model are those related to hydro and fuel management. In the first case, they correspond to reservoirs' operation along time to accommodate the natural hydro inflows to the production of hydro units. In the second case, the operation of thermal plants' fuel stocks are modeled with a take-or-pay fuel purchase contract or with an enforced consumption of determined domestic fuels for strategic energy policy or socioeconomic reasons. The structure of both constraints is similar. Additional constraints are imposed for the pumped-storage units management.

Hydro scheduling. This equation states that the energy produced during each period, expressed as a function of the hydro and pumped-hydro production for each load level, is limited by the hydro inflows and the initial and final reservoir levels of that period. The initial reservoir level of the first period and the final level of the last period are known for the optimization problem.

$$\sum_{n} D_{n,p}\left(h_{n,p,h} - \eta_h b_{n,p,h}\right) \leq A_{p,h} + R_{p,h} - R_{p+1,h} \perp \mu^R_{p,h} \quad \forall p, h \in e, e$$

Fuel scheduling. The fuel consumption in a thermal plant during any period, as a function of the production of their thermal units in each load level, is lower bounded by the purchase in that period and the fuel stocks at the beginning and the end of that period. Note that this inequality constraint has a $\geq$ sign instead of the $\leq$ of the previous one because it represents a mandatory fuel consumption. The initial stock level of the first period and the final level for the last period are known for the optimization problem.

$$\sum_{n}\sum_{t \in C} D_{n,p} p_{n,p,t} \geq C_{p,c} + S_{p,c} - S_{p+1,c} \perp \mu^S_{p,c} \quad \forall p, c \in e, e$$

Pumped-storage scheduling. Pumped-storage units are used by each company to buy energy at lower prices and sell it within the same period at higher prices, making then a price arbitrage. The first constraint establishes for each pumped-storage unit that generated energy is lower than pumped energy, while the second one limits the total pumped energy in each period.

$$\sum_{n} D_{n,p}\left(h_{n,p,b} - \eta_b b_{n,p,b}\right) \leq 0 \perp \mu^R_{p,b} \quad \forall p, b \in e, e$$

$$\sum_{n} D_{n,p} \eta_b b_{n,p,b}) \leq \bar{R}_b \perp \mu^{\bar{R}}_{p,b} \quad \forall p, b \in e, e$$

Variable bounds. These are simple bounds on all the variables involved in the optimization problem.

$$\begin{array}{llll}
\underline{R}_{p,h} \leq R_{p,h} \leq \bar{R}_{p,h} & \perp & \mu^{\underline{R}}_{p,h}; \mu^{\bar{R}}_{p,h} & \forall p, h \in e, e \\
\underline{S}_{p,c} \leq S_{p,c} \leq \bar{S}_{p,c} & \perp & \mu^{\underline{S}}_{p,c}; \mu^{\bar{S}}_{p,c} & \forall p, c \in e, e \\
\underline{p}_{t} \leq p_{n,p,t} \leq \bar{p}_{t} & \perp & \mu^{\underline{p}}_{n,p,t}; \mu^{\bar{p}}_{n,p,t} & \forall n, p, t \in e, e \\
\underline{h}_{b} \leq h_{n,p,b} \leq \bar{h}_{b} & \perp & \mu^{\underline{h}}_{n,p,b}; \mu^{\bar{h}}_{n,p,b} & \forall n, p, b \in e, e \\
\underline{h}_{p,h} \leq h_{n,p,h} \leq \bar{h}_{p,h} & \perp & \mu^{\underline{h}}_{n,p,h}; \mu^{\bar{h}}_{n,p,h} & \forall n, p, h \in e, e \\
\underline{b}_{b} \leq b_{n,p,b} \leq \bar{b}_{b} & \perp & \mu^{\underline{b}}_{n,p,b}; \mu^{\bar{b}}_{n,p,b} & \forall n, p, b \in e, e \\
\underline{b}_{h} \leq b_{n,p,h} \leq \bar{b}_{h} & \perp & \mu^{\underline{b}}_{n,p,h}; \mu^{\bar{b}}_{n,p,h} & \forall n, p, h \in e, e
\end{array}$$

4.1.2 Auxiliary constraints. For the sake of clarity we explicitly state the following two constraints. However, in the model implementation, price and total power generation of each company have been substituted by these expressions:

Total power generation of each firm. The total power generation of each firm represents the effective output that is sold in the market, therefore excluding long-term contract production already sold at a fixed price.

$$g_{n,p,e} = \sum_{t \in e} p_{n,p,t} + \sum_{h \in e} (h_{n,p,h} - b_{n,p,h}) - L_{p,e} \quad \forall n, p, e$$

Price equation. The marginal price is obtained as a linear function of the total purchased power in the market.

$$\pi_{n,p} = d'_{n,p}(h_{n,p} - \sum_{e} g_{n,p,e}) \quad \forall n, p$$

5 FORMULATION OF THE ELECTRICITY MARKET EQUILIBRIUM PROBLEM

This section states in detail the optimality conditions of each generation company (schematically represented in figure 13.2), which define the Cournot market equilibrium. The first set of the Karush-Kuhn-Tucker optimality conditions of the generation company e is the gradient of the Lagrangian function with respect to the decision variables equalized to zero. Some of the resulting equations provide useful information about the optimal operation of each kind of technology for each firm. The most important ones will be explained.

$$\frac{\partial L_e}{\partial p_{n,p,t}} = D_{n,p}\left(\pi_{n,p} - g_{n,p,e}d'_{n,p}\right) - D_{n,p}\left(u_t + \nu_t\left(\frac{o'_t}{k_t} + 2\frac{o''_t}{k_t^2}p_{n,p,t}\right)\right)$$
$$-D_{n,p}\mu^S_{p,c} + \left(\mu^{\bar{p}}_{n,p,t} - \mu^{\underline{p}}_{n,p,t}\right) = 0$$

for all n, p, $t \in e$ and e; this equation indicates that a firm must operate each owned thermal unit looking forward to equalizing its own marginal revenue (first term of the equation) with thermal unit's marginal cost (second term) whenever units rated capacity allows it (third term). Then, the thermal unit will operate at its maximum power capacity whenever its marginal cost is lower than the firm's marginal revenue, and it will operate at its minimum power whenever its marginal cost is greater than the firm's marginal revenue.

$$\frac{\partial L_e}{\partial h_{n,p,h}} = D_{n,p}\left(\pi_{n,p} - g_{n,p,e}d'_{n,p} + \mu^R_{p,h}\right) + \left(\mu^{\bar{h}}_{n,p,h} - \mu^{\underline{h}}_{n,p,h}\right)$$
$$= 0 \qquad (13.2)$$

for all n, p, $h \in e$ and e; this equation expresses that hydro units also operate looking forward to equalizing the firm's marginal revenue. The water value is now the economic parameter to be compared with the firm's marginal revenue in order to decide the hydro unit operation scheduling.

$$\frac{\partial L_e}{\partial b_{n,p,h}} = -D_{n,p}\left(\pi_{n,p} - g_{n,p,e}d'_{n,p} + \eta_h\mu^R_{p,h}\right) + \left(\mu^{\bar{b}}_{n,p,h} - \mu^{\underline{b}}_{n,p,h}\right) = 0$$

for all n, p, $h \in e$ and e. Pumped units operate trying to arbitrage firm's marginal revenue. They pump water (buying energy in the market) during off-peak hours, when the firm's marginal revenue is lower than the water value times the performance of the pumping cycle. The water stored is then produced within the same period as any other hydro unit (13.2).

$$\frac{\partial L_e}{\partial R_{p,h}} = \mu^R_{p-1,h} - \mu^R_{p,h} + \left(\mu^{\bar{R}}_{p,h} - \mu^{\underline{R}}_{p,h}\right) = 0$$

for all $p > 1$, $h \in e$ and e. This equation indicates that the hydro must be scheduled to equalize water values between periods (therefore trying to equalize the firm's marginal revenue between periods).

$$\frac{\partial L_e}{\partial h_{n,p,b}} = D_{n,p}\left(\pi_{n,p} - g_{n,p,e}d'_{n,p} + \mu^R_{p,b}\right) + \left(\mu^{\bar{h}}_{n,p,b} - \mu^{\underline{h}}_{n,p,b}\right) = 0$$

for all n, p, $b \in e$ and e.

$$\begin{aligned}\frac{\partial L_e}{\partial b_{n,p,b}} &= -D_{n,p}\left(\pi_{n,p} - g_{n,p,e}d'_{n,p} + \eta_b(\mu^{R}_{p,b} - \mu^{\bar{R}}_{p,b})\right) + \left(\mu^{\bar{b}}_{n,p,b} - \mu^{\underline{b}}_{n,p,b}\right) \\ &= 0\end{aligned}$$

for all n, p, $b \in e$ and e.

$$\frac{\partial L_e}{\partial S_{p,c}} = -\mu^{S}_{p-1,c} + \mu^{S}_{p,c} + \left(\mu^{\bar{S}}_{p,c} - \mu^{\underline{S}}_{p,c}\right) = 0$$

for all $p > 1$, $c \in e$ and e.

There are no equality constraints. Therefore, in order to complete the set of non-linear equations that define the optimization problem of each company, the following three sets of equations must be added. First of all, the inequality constraints above multiplied by their corresponding Lagrange multiplier (complementary slackness conditions); next, the Lagrange multiplier are less than zero; and finally, the inequality constraints themselves.

As was previously established in section 3, the previous set of equations that define the market equilibrium make a system of non-linear equations with the structure of a complementarity problem.

6 MODEL IMPLEMENTATION

This model was implemented in GAMS version 2.50, a specific mathematical language designed to efficiently create and solve optimization problems. There exist two specific commercial solvers available for large scale complementarity problems, MILES and PATH [12]. MILES is the solver used, which is based on a generalization of the classic Newton's method, in which each subproblem is solved as a linear complementarity problem using an extension of the Lemke algorithm [1].

With the aim of facilitating the convergence, all the variables and equations of the problem have been scaled in such a way so that all of them take values around one. The base units are GW for power, k$ for currency, kh for time, TWh for energy and kTcal for fuel consumption. With the same purpose, special attention has been paid on obtaining a good initial point by previously solving the market equilibrium in a simplified and fast manner [14].

The main results obtained with the model for each load level are market prices, energy produced and marginal economic information about revenues and costs. Additionally, the water value for hydro units can be determined.

Table 13.1 Demand data

d (MW)	P_1	P_2	P_3	P_4	P_5
N_1	14400	16000	19200	17800	18000
N_2	10800	13200	16800	14000	14000
N_3	8000	8400	13500	9600	10000

7 NUMERICAL RESULTS

The model described in this article has been successfully applied to a large-scale electric system to represent the annual operation within a competitive market context. In particular it has been applied to the Spanish electric system with very promising results. A very brief description of the main results is presented in section 7.2. It is interesting to point out the size of the system solved and its associated Mixed Complementarity Problem, showing the powerful capabilities of the approach and of the commercial MCP solvers available.

However, in order to obtain accurate results for real systems it is necessary to include in the model several other specific features of the regulatory framework that can strongly condition the market behavior (e.g., in Spain it is necessary to model Competition Transition Charges, domestic coal quotas, etc.). As this kind of features has not been presented in the more general mathematical approach described in this paper, the case study shown in detail in section 7.1 corresponds to a smaller example prepared to explicitly illustrate the way the model operates different type of plants within the adopted Cournot equilibrium scheme and how the theoretical results stated in the paper are corroborated by the results. The chosen case study is complex enough to show the capabilities of the model.

7.1 CASE STUDY

7.1.1 System description. The system has 11700 MW of installed generation capacity belonging to 3 generation firms (firm A, B and C). The model scope is 5 months.

Load data. Demand is approximated by load duration curves which are divided into 5 periods (one for each month) and three load levels within each period (peak, plateau and off-peak). A homogeneous duration of 248 hours is assumed for all 15 load levels considered in this example.

Table 13.2 Thermal plants parameters

Firm	Name	Fuel	Power (MW)	ν ($/Gcal)	o' (Gcal/ MWh)	o'' (Gcal/ MW^2h)
A	TA1	Nuclear	1000	5	1	0.0001
	TA2	Coal	500	6.25	1.6	0.0005
	TA3	Gas	500	10	1.8	0.0007
	TA4	Oil	350	13.75	2	0.0012
B	TB1	Nuclear	1000	5	1	0.0001
	TB2	Nuclear	1000	5.3125	1	0.0001
	TB3	Coal	500	6.25	1.6	0.0005
	TB4	Coal	500	8.125	1.6	0.0005
	TB5	Gas	500	10	1.8	0.0005
	TB6	Oil	420	11.875	2	0.0010
	TB7	Oil	330	13.75	2.2	0.0012
C	TC1	Nuclear	1000	5	1	0.0001
	TC2	Coal	500	6.25	1.6	0.0005
	TC3	Oil	400	13.75	2	0.0012

A linear demand function represents the consumers price reaction. For the sake of simplicity a homogeneous slope (5 $/MWh/GW) for this function is assumed for all load levels in all periods. Table 13.1 presents the maximum load (associated to a zero market price) for each load level.

Generation data. There are 19 units (14 thermal, 4 hydro and 1 pumped-storage). Tables 13.2 and 13.3 summarize the technical and economic data for each thermal unit (firm owner, name, maximum power output and quadratic heat rate function) and each hydro unit (firm owner, name, maximum power output, reservoir capacity, initial and final reservoir levels for the scope, estimated hydro inflows for each one of the 5 periods).

Firm A groups together a wide variety of power station technologies. It operates base load units (low fuel costs) and peaking thermal units (high fuel costs). Firms A also manages different hydro units: one with a large reservoir capacity (HA1), able to store water from period to period; another with a small reservoir capacity (HA2), able only to move water among load levels within the same period; and a pumped-storage unit (BA1). Because of this unit diversity, the firm A operation scheduling will be the most interesting to be analyzed.

Table 13.3 Hydroelectric plants parameters (measured in GWh unless otherwise stated)

Firm	Name	Power (MW)	Reser-voir	$R_1 = R_6$	A_1	A_2	A_3	A_4	A_5
A	HA1	2000	250	100	150	150	50	50	150
	HA2	300	0	0	20	10	1	10	20
	BA1	300	15	0	0	0	0	0	0
B	HB1	300	20	10	10	10	10	10	10
C	HC1	300	20	10	10	10	10	10	10

Table 13.4 Market results

Period	Level	π (\$/MWh)	P_A (MW)	P_B (MW)	P_C (MW)
	N_1	36.625	2291	3242	1543
P_1	N_2	26.108	1532	2546	1500
	N_3	18.088	1108	2000	1275
	N_1	39.948	2855	3490	1665
P_2	N_2	33.643	1929	3043	1500
	N_3	19.154	1200	2000	1369
	N_1	50.578	3129	3940	2015
P_3	N_2	45.396	2303	3644	1774
	N_3	34.582	1978	3105	1500
	N_1	47.564	2737	3720	1831
P_4	N_2	36.251	2000	3217	1533
	N_3	23.115	1200	2277	1500
	N_1	47.850	2794	3763	1873
P_5	N_2	36.251	2000	3217	1533
	N_3	24.423	1222	2393	1500

7.1.2 Results. Table 13.4 presents the results obtained for the whole system. Marginal market prices and total power output for each firm for each load level and period. It is interesting to note that firm A, which owns a lot of flexible hydro generation, produces proportionally more during peak hours, while firm C, which is smaller and with more base load plants, produces proportionally more during off-peak hours.

Figure 13.3 represents the evolution of both the market prices and the load. A clear correlation can be observed between both variables. How-

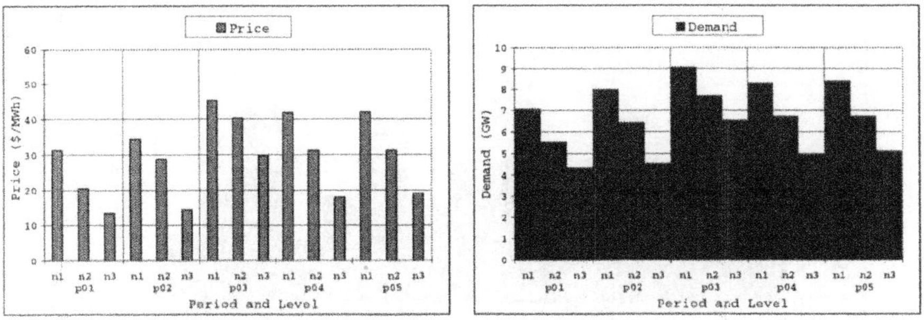

(a) Price (b) Demand

Figure 13.3 Prices and productions of the case study

Table 13.5 Firm A economic results

	P_1	P_2	P_3	P_4	P_5
Energy (GWh)	1222.8	1484.0	1837.8	1472.3	1492.0
Costs (M$)	10.634	13.038	19.349	14.806	14.931
Profits (M$)	25.061	37.040	62.796	42.338	43.608

ever this conclusion can not been generalized. Different hydro inflows patterns or demand function slopes can lead to other correlation figures.

Results for firm A, that are shown below, can illustrate the optimal operation for a company owning a wide variety of power stations. Figure 13.4 shows the production profile for each plant, while Table 13.5 summarizes the economic results for firm A. As it is clearly identifiable, the cheapest thermal units (TA1 and TA2) produce at base load at their maximum power capacity for the 5 periods. On the other hand, the most expensive thermal units (TA3 and TA4) produce at higher load levels. Hydro units (HA1 and HA2) produce during peak hours, therefore substituting the most expensive thermal units. Finally, pumped-storage unit (BA1) pumps during off-peak hours at lower energy market price and uses this pumped water to produce during peak hours.

This kind of behavior does not seem to differ from the behavior in more traditional contexts based on total system-cost-minimization schemes. However, a more detailed analysis of the results will reveal how firm A uses its resources to maximize its own benefits (revenues minus costs). As has been clearly stated in previous sections of this pa-

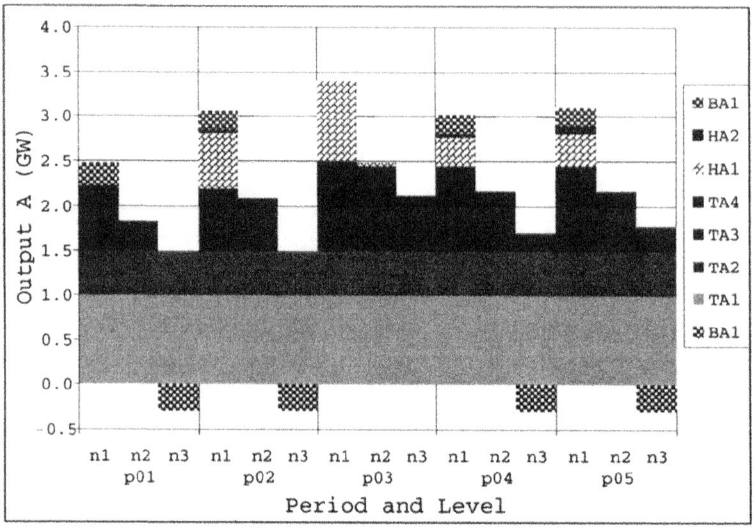

Figure 13.4 Detailed firm A production

per, each company will compare its marginal revenues (instead of price), MR_A, and the marginal cost of each thermal plant it owns, λ_{TAi}, to decide whether this plant will produce or not. Those units (infra-marginal) with $\lambda_{TAi} < MR_A$ will produce at their maximum output, those units (supra-marginal) with $\lambda_{TAi} > MR_A$ will produce at their minimum output (zero in this example), and those units (marginal) with $\lambda_{TAi} = MR_A$ will produce at some value in between their minimum and maximum outputs. A similar rule is applicable to the hydro unit, where the water value plays the role of the marginal cost in thermal units. Tables 13.6 and 13.7 detail those figures for period 4 (firm A).

For load level N_1 (N_1 corresponds to a peak load level) the marginal revenue for firm A amounts to 33.88 \$/MWh. Units TA1, TA2 and TA3 are infra-marginal and therefore produce at their maximum output, while unit TA4 is marginal and produces 193 MW. For lower load levels N_2 and N_3, the marginal revenue for firm A descends and first unit TA4 and then unit TA3 stop producing, as their marginal cost becomes higher than MR_A.

For load level N_1 all hydro units are marginal and therefore their water value is equal to the firm's marginal revenue. The performance of the pumping-production cycle of the pumped-storage unit is 0.7. Therefore, whenever the ratio marginal revenue in N_3 versus marginal revenue in N_1 is smaller than 0.7 it will be economically suitable for the firm to

Table 13.6 Firm A thermal units operation results for period 4 (MR_A in \$/MWh, P_i in MW, λ_i in \$/MWh)

Level	MR_A	P_{TA1}	λ_{TA1}	P_{TA2}	λ_{TA2}	P_{TA3}	λ_{TA3}	P_{TA4}	λ_{TA4}
N_1	33.88	1000	6.0	500	13.13	500	25.0	193	33.88
N_2	26.25	1000	6.0	500	13.13	500	25.0	0	27.50
N_3	17.11	1000	6.0	500	13.13	0	18.0	0	27.50

Table 13.7 Firm A hydro units operation results for period 4 (MR_A in \$/MWh, P_i in MW, λ_i in \$/MWh)

Level	MR_A	P_{HA1}	λ_{HA1}	P_{HA2}	λ_{HA2}	P_{BA1}	λ_{BA1}	$-P_{BA1}$	$0.7\lambda_{BA1}$
N_1	33.88	293	33.88	40	33.88	210	33.88	0	23.72
N_2	26.25	0	33.88	0	33.88	0	33.88	0	23.72
N_3	17.11	0	33.88	0	33.88	0	33.88	-300	23.72

Table 13.8 Water value of HA1

	P_1	P_2	P_3	P_4	P_5
Reserve (GWh)	100	250	250	49.58	26.87
λ_{HA1} (k\$/MWh)	25.39	25.68	33.88	33.88	33.88

pump during off-peak hours and to produce during peak hours. In this case $17.11/23.72 = 0.505 < 0.7$.

Table 13.8 illustrates how firm A operates its hydro resources to balance the marginal revenues across all periods. Firm A tries to keep water in its reservoir in order to use it in the last periods, when its marginal revenue (and its water value) is higher. It succeeds in equalizing the marginal revenue only during the last periods because the reservoir maximum capacity (250 GWh) prevents from keeping more water for later on.

7.2 SPANISH SYSTEM

The model presented in this article has been applied to the Spanish electricity market, in which four main companies operate: Endesa, IBERDROLA, Unión Fenosa and Hidrocantábrico. The thermal subsystem is composed of 84 units, grouped into 42 power plants. Hydro units have been grouped into 10 equivalent generation units plus 10

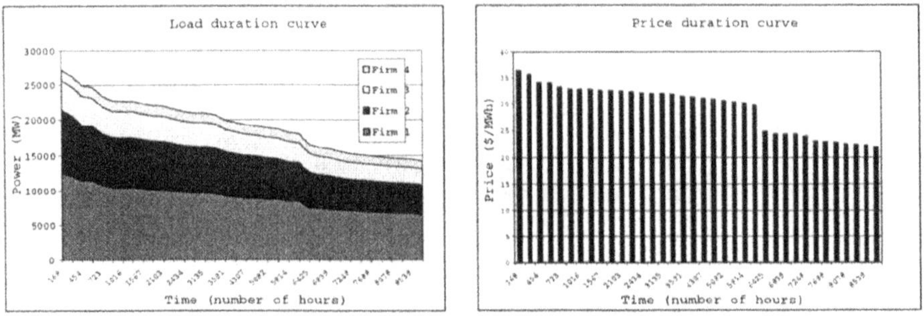

(a) Load duration curve

(b) Price duration curve

Figure 13.5 Outputs and prices of the case study

other pumping units, with a total production capacity of 28337 GWh. Demand is modeled as a linear function of the price with a slope of 7 ($/MWh)/GW. This is the average value resulting from a simple econometric study. A more detailed estimation could not be achieved because of a lack of available data. The annual scope of the model has been divided into twelve periods (months) with 3 load levels (peak, plateau and off-peak) for each period.

A proper market representation is achieved not only by modeling a Cournot firms behavior, but also by considering additional relevant market features (i.e. Competition Transition Charges, long-term contracts and mandatory fuel purchases).

Figure 13.5 shows the output of each one of the companies and the system marginal price obtained from the model for the year 1998. These results in terms of average values (due to non-stochastic modeling) are very promising as they are very close to the actual market results.

The size of the presented problem is 5079 equations and 5099 variables, and it is solved in approximately 2 minutes on a PC Pentium-II 233 MHz with 64 MB.

8 CONCLUSION

A new approach to design generation operation-planning models within the context of deregulated electricity markets has been presented in this article. This method, based on the complementarity problem, allows for the correct modeling of the electrical system explicitly considering all the technical system constraints and the profit maximization behavior of the companies, i.e., the market equilibrium.

This article describes mathematically in detail an annual or hyper-annual operation model based on this methodology which considers the profit-maximizing behavior of the generation companies in an imperfect competitive environment and that explicitly includes the special characteristics of each type of generation.

The mathematical model has been implemented in GAMS, taking advantage of the specific solvers designed to deal with complementarity problems. A detailed case study shows how different generation technologies are optimally operated in a competitive environment. This model has shown to be robust and able to manage large-scale electric systems.

References

[1] S.C. Billups, S. P. Dirkse, and M. C. Ferris. "A comparison of large scale mixed complementarity problem solvers". Computational Optimization and Applications, 7:3-25, 1997.

[2] S. Borenstein, and J. Bushnell. "An Empirical Analysis of the Potential for Market Power in California's Electricity Industry". Journal of Industrial Economics, Vol 47, No. 3, September, 1999.

[3] J. Bushnell, Water and Power: "Hydroelectric Resources in the Era of Competition in the Western US". POWER Conference on Electricity Restructuring, University of California, Energy Institute. 1998.

[4] R.W. Cottle, J.S. Pang, and R.E. Stone. The Linear Complementarity Problem. Academic Press, Boston, 1992.

[5] R.J. Green, and D. M. Newbery, "Competition in the British Electricity Spot Market". Journal of Political Economy, Vol. 100 No. 5 pp 929-953, 1992.

[6] B. F. Hobbs. "LCP Models of Nash-Cournot Competition in Bilateral and POOLCO-Based Power Markets". Proceedings, IEEE Winter Power Meeting, NY City, Feb. 1999.

[7] E. Kahn, "Numerical techniques for analyzing market power in electricity". The Electricity Journal. pp. 34-43. July 1998.

[8] S. Oren, "Economic Inefficiency of Passive Transmission Rights in Congested Electrical System with Competitive Generation". The Energy Journal 18 (1), pp 63-83, 1997.

[9] I. Otero-Novas, C. Meseguer, C. Batlle and J. J. Alba. "A simulation model for a competitive generation market". IEEE Power Engineering Society, Paper PE-380-PWRS-1-09-1998.

[10] A. Ramos, M. Ventosa, M. Rivier. "Modeling Competition in Electric Energy Markets by Equilibrium Constraints". Utilities Policy, Vol. 7 Issue 4 pp 233-242, Dec. 98.

[11] A. Rudkevich, M. Duckworth and R. Rosen. "Modeling electricity pricing in a deregulated generation industry: The potential for oligopoly pricing in a Poolco", The Energy Journal, Vol. 19 No. 3. pp 19-48. 1998.

[12] T. F. Rutherford, "Extension of GAMS for complementarity problems arising in applied economic analysis". Journal of Economic Dynamics and Control, 19:1299-1324, 1995.

[13] T. J. Scott, and E.G. Read, "Modelling Hydro Reservoir Operation in a Deregulated Electricity Market". International Transactions in Operational Research, Vol. 3 pp 243-253, 1996.

[14] M. Ventosa, A. Ramos, M. Rivier. "Modeling Profit Maximisation in Deregulated Power Markets by Equilibrium Constraints". 13th PSCC Conference, Norway. Vol. 1 pp. 231-237. 1999.

[15] W. Jing-Yuan, Y. Smeers. "Spatial Oligopolistic Electricity Models with Cournot Generators and Regulated Transmission Prices". Operations Research, Vol. 47, No. 1, January - February 1999, pp. 102 - 112.

Appendix

In this appendix all symbols used in the model formulation are identified and classified according to their use in indices, parameters, variables and Lagrange multipliers.

Table 13.A.1 Indices and sets

Index	Description
b	Pumped-storage units (pumped units able to pump and produce energy with daily or weekly cycle)
c	Thermal plants (several thermal units located in the same site)
e	Firms or companies
h	Hydro and/or pumped-hydro units with a seasonal cycle
n	Load levels (different levels of demand into a period)
p	Periods (time interval considered for the hydro and fuel scheduling)
t	Thermal units (nuclear, coal, oil and gas thermal generators)

Table 13.A.2 Decision and auxiliary variables

Variable	Description
$b_{n,p,b}$, $h_{n,p,b}$	Power consumption and generation by pumped-storage unit b in load level n of period p [GW]
$b_{n,p,h}$, $h_{n,p,h}$	Power consumption and generation by pumped-hydro unit h in load level n of period p [GW]
$p_{n,p,t}$	Power generation by thermal unit t in load level n of period p [GW]
$R_{p,h}$	Hydro energy reserve of hydro unit h at the beginning of period p [TWh]
$S_{p,c}$	Fuel stock of thermal plant c at the beginning of period p [TWh]
$g_{n,p,e}$	Total power generation of firm e in load level n of period p [GW]
$\pi_{n,p}$	System marginal price in load level n of period p [k$/TWh]

Table 13.A.3 Parameters

Parameter	Description
$A_{p,h}$	Natural hydro inflows for hydro unit h in period p [TWh]
$\bar{b}_b, \underline{b}_b$	Maximum and minimum capacity of pumped-storage unit b when pumping [GW]
$\bar{b}_h, \underline{b}_h$	Maximum and minimum capacity of pumped-hydro unit h when pumping [GW]
$C_{p,c}$	Contracted fuel purchase for thermal plant c at the beginning of period p [TWh]
$d_{n,p}, d'_{n,p}$	Power demand purchased in the market at zero price [GW] and constant slope of the demand function in load level n of period p [(k$/TWh)/GW] (they are the zero intercept and slope of the linear demand function)
$D_{n,p}$	Duration of load level n of period p [kh]
$\bar{h}_{p,h}, \underline{h}_{p,h}$	Maximum and minimum capacity of pumped-hydro unit h in period p [GW]
$\bar{h}_b, \underline{h}_b$	Maximum and minimum capacity of pumped-storage unit b when producing [GW]
k_t	Self-consumption coefficient of thermal unit t [p.u.]
$L_{p,e}$	Long term contract for power of firm e in period p [GW]
o'_t, o''_t	Net heat rate (linear [kTcal/TWh] and quadratic [kTcal/(GW2kh)] terms) of thermal unit t.
$\bar{p}_t, \underline{p}_t$	Maximum and minimum rated capacity of thermal unit t [GW]
$\bar{R}_b$	Maximum hydro energy reserve of pumped-storage unit b [TWh]
$\bar{R}_{p,h}, \underline{R}_{p,h}$	Maximum and minimum hydro energy reserve of hydro unit h in period p [TWh]
$\bar{S}_{p,c}, \underline{S}_{p,c}$	Maximum and minimum fuel stock of thermal plant c in period p [TWh]
u_t	O&M variable cost of thermal unit t [k$/GW]
ν_t	Fuel cost of thermal unit t [k$/kTcal]
η_b, η_h	Performance of pumped-storage unit b or of pumped-hydro unit h [p.u.]

Table 13.A.4 Lagrange Multipliers

Multiplier	Associated constraint
$\mu^{R}_{p,h}$	Hydro scheduling of hydro unit h in period p [k$/TWh]
$\mu^{S}_{p,c}$	Fuel scheduling of thermal plant c in period p [k$/TWh]
$\mu^{R}_{p,b}$	Scheduling of pumped-storage unit b in period p [k$/TWh]
$\mu^{\bar{R}}_{p,b}$	Maximum hydro energy reserve of pumped-storage unit b in period p [k$/TWh]
$\mu^{\bar{R}}_{p,h}, \mu^{\underline{R}}_{p,h}$	Maximum and minimum hydro energy reserve of hydro unit h in period p [k$/TWh]
$\mu^{\bar{S}}_{p,c}, \mu^{\underline{S}}_{p,c}$	Maximum and minimum fuel stock of thermal plant c in period p [k$/ TWh]
$\mu^{\bar{p}}_{n,p,t}, \mu^{\underline{p}}_{n,p,t}$	Maximum and minimum rated capacity of thermal unit t in load level n of period p [k$/GW]
$\mu^{\bar{h}}_{n,p,h}, \mu^{\underline{h}}_{n,p,h}$	Maximum and minimum capacity of pumped-hydro unit h in load level n of period p [k$/GW]
$\mu^{\bar{b}}_{n,p,h}, \mu^{\underline{b}}_{n,p,h}$	Maximum and minimum capacity of pumped-hydro unit h in load level n of period p [k$/GW]
$\mu^{\bar{h}}_{n,p,b}, \mu^{\underline{h}}_{n,p,b}$	Maximum and minimum capacity of pumped-storage unit b in load level n of period p [k$/GW]
$\mu^{\bar{b}}_{n,p,b}, \mu^{\underline{b}}_{n,p,b}$	Maximum and minimum capacity of pumped-storage unit b in load level n of period p [k$/GW]

Chapter 14

A CLASS OF GLOBALLY CONVERGENT ALGORITHMS FOR PSEUDOMONOTONE VARIATIONAL INEQUALITIES

M. V. Solodov

Instituto de Matemática Pura e Aplicada, Estrada Dona Castorina 110, Jardim Botânico, Rio de Janeiro, RJ 22460-320, Brazil *

solodov@impa.br

Abstract We describe a fairly broad class of algorithms for solving variational inequalities, global convergence of which is based on the strategy of generating a hyperplane separating the current iterate from the solution set. The methods are shown to converge under very mild assumptions. Specifically, the problem mapping is only assumed to be continuous and pseudomonotone with respect to at least one solution. The strategy to obtain (super)linear rate of convergence is also discussed. The algorithms in this class differ in the tools which are used to construct the separating hyperplane. Our general scheme subsumes an extragradient-type projection method, a globally and locally superlinearly convergent Josephy-Newton-type method, a certain minimization-based method, and a splitting technique.

1 INTRODUCTION

Given a function $F : \boldsymbol{R}^n \to \boldsymbol{R}^n$ and a set $C \subset \boldsymbol{R}^n$, the classical variational inequality problem [1, 3, 8, 9, 5], abbreviated VIP(F, C), is to find a point x such that

$$x \in C, \quad \langle F(x), y - x \rangle \geq 0 \quad \text{for all } y \in C,$$

where $\langle \cdot, \cdot \rangle$ denotes the usual inner product in $\boldsymbol{R}^n$. Throughout this paper we assume that C is closed convex, F is continuous, and the solution

*This research is supported in part by CNPq Grant 300734/95-6 and by PRONEX–Optimization.

M.C. Ferris et al. (eds.), Complementarity: Applications, Algorithms and Extensions, 297–315.

set of VIP(F, C), denoted by S, is nonempty. When C is the nonnegative orthant $\boldsymbol{R}^n_+$, VIP(F, C) reduces to the nonlinear complementarity problem NCP(F), which is to find a point $x \in \boldsymbol{R}^n$ such that

$$x \geq 0, \quad F(x) \geq 0, \quad \langle F(x), x \rangle = 0 .$$

More generally, when C is a box (i.e., a Cartesian product of intervals), VIP(F, C) is the mixed complementarity problem [6].

For the convergence of our general algorithm, we further assume that there exists $\bar{x}$, some element of the solution set S, such that

$$\langle F(y), y - \bar{x} \rangle \geq 0 \quad \text{for all } y \in C . \tag{14.1}$$

Under this condition, we prove that the sequence of iterates generated by our method is bounded and each of its accumulation points is a solution of VIP(F, C). Furthermore, if condition (14.1) holds for all $\bar{x} \in S$ then the accumulation point is unique, i.e., the whole sequence of iterates converges to a solution.

Condition (14.1) is very mild. It can be thought of as pseudomonotonicity of F on the feasible set C with respect to *one* solution of VIP(F, C). Note that pseudomonotonicity of F with respect to the whole solution set S is a stronger, but still fairly mild assumption. It is clear that the latter is satisfied if F is monotone, i.e.,

$$\langle F(x) - F(y), x - y \rangle \geq 0 \quad \text{for all } x, y \in \boldsymbol{R}^n .$$

More generally, it also holds if F is pseudomonotone (as defined in [12]), i.e., for all $x, y \in \boldsymbol{R}^n$

$$\langle F(y), x - y \rangle \geq 0 \quad \Longrightarrow \quad \langle F(x), x - y \rangle \geq 0 .$$

Moreover, it is possible to construct examples where (14.1) is satisfied but F is not monotone or pseudomonotone everywhere. Of course, since this condition involves (unknown!) solution of the problem, in general it is not possible to verify it a priori. Still, one might expect (14.1) to hold under some kind of generalized monotonicity assumptions on F, some of which are not difficult to check (see [20, 4]). And at any rate, (14.1) obviously holds on the class of (pseudo)monotone problems.

The general framework to be presented is essentially based on simple geometric considerations. If given a point $x \in \boldsymbol{R}^n$ we could find (by whichever means) some point $y \in C$ such that

$$\langle F(y), x - y \rangle > 0 , \tag{14.2}$$

then condition (14.1) implies that the hyperplane

$$H_y := \{ w \in \boldsymbol{R}^n \mid \langle F(y), w - y \rangle = 0 \}$$

separates the point x from any solution $\bar{x}$ satisfying (14.1). It then follows that moving from x in the direction of this hyperplane decreases the distance to the solution point $\bar{x}$. Formally, this can be stated as follows.

Lemma 14.1 *Let $x \in \boldsymbol{R}^n$ and $y \in C$ be any two points satisfying (14.2), and let*

$$\hat{x}(\beta) := x - \beta \frac{\langle F(y), x - y \rangle}{\|F(y)\|^2} F(y),$$

where $\beta \geq 0$. Then for any $\bar{x} \in S$ satisfying (14.1) it holds that

$$\|\hat{x}(\beta) - \bar{x}\|^2 \leq \|x - \bar{x}\|^2 - \beta(2 - \beta) \left(\frac{\langle F(y), x - y \rangle}{\|F(y)\|} \right)^2 .$$

Proof Denote

$$t := \frac{\langle F(y), x - y \rangle}{\|F(y)\|^2} > 0,$$

so that $\hat{x}(\beta) = x - \beta t F(y)$. With this notation, we obtain

$$\begin{aligned} \|\hat{x}(\beta) - \bar{x}\|^2 &= \|x - \bar{x}\|^2 + t^2\beta^2\|F(y)\|^2 - 2t\beta \langle F(y), x - \bar{x} \rangle \\ &\leq \|x - \bar{x}\|^2 + t^2\beta^2\|F(y)\|^2 - 2t\beta \langle F(y), x - y \rangle \\ &= \|x - \bar{x}\|^2 - \beta(2 - \beta)t \langle F(y), x - y \rangle , \end{aligned}$$

where the inequality follows from (14.1), and the last equality follows from the relation $t\|F(y)\|^2 = \langle F(y), x - y \rangle$. Using now the definition of t concludes the proof. □

Lemma 14.1 implies that if we choose $\beta \in (0, 2)$ then the point $\hat{x}(\beta)$ is closer to $\bar{x} \in S$ than the point x. Note that taking $\beta = 1$ corresponds to $\hat{x}(\beta)$ being the projection of x onto the separating hyperplane H_y, while choosing $\beta \in (1, 2)$ or $\beta \in (0, 1)$ corresponds to a step with over-relaxation or under-relaxation, respectively. Alternatively, $\hat{x}(\beta)$ can also be viewed as a projection of x onto a shifted hyperplane $H_y(\beta)$, with $H_y(1) = H_y$. If furthermore the point y is obtained in such a way that the quantity $\langle F(y), x - y \rangle$ has a meaningful connection to "optimality" of x, then Lemma 14.1 can be used as the basis for devising algorithms for solving VIP(F, C) under the assumption (14.1).

Of course, the ways to compute (given some x) an acceptable point y, is the central question for this algorithmic development. In this paper, after presenting a general framework (Section 2), we shall give a number of examples of how this can be done (Section 3). These include a projection algorithm, a (Josephy-)Newton method, a splitting method, and a certain minimization-based algorithm. The separation-projection methodology appears to be useful for devising some new computational techniques, forcing convergence of iterative schemes that otherwise might diverge, and for relaxing error tolerance requirements in solving subproblems within some convergent algorithms. Each of those issues will be illustrated by the given specific instances of the general method.

2 GENERAL ALGORITHMIC FRAMEWORK

A common approach to solving VIP(F, C) consists of solving a sequence of subproblems VIP(F_k, C) approximating the original problem (see, for example, [9]). This is a natural view, which we shall adopt also for our development. For the resulting algorithm to be potentially practical, it is clear that F_k must be such that subproblems are structurally easier to solve than the original problem. It is also reasonable to say that F_k should provide some kind of approximation for F around the current iterate x^k. It is therefore natural to write

$$F_k(x) := F(x^k) + A_k(x^k, x), \tag{14.3}$$

where $A_k : \mathbf{R}^n \times \mathbf{R}^n \to \mathbf{R}^n$ is some mapping whose choice will define each specific algorithm.

For $x \in \mathbf{R}^n$ define

$$r(x) := \|x - P_C[x - F(x)]\|,$$

where $P_C[\cdot]$ stands for the orthogonal projection operator onto the set C. It is well known that some x solves VIP(F, C) if, and only if, $r(x) = 0$. We are now ready to formally state our general method.

Algorithm 14.2 Choose parameters $t > 0$, $0 \le \rho_u < 1$, $0 < \sigma_l \le \sigma_u < 1$, $0 < \beta_l \le \beta_u < 2$, $0 < \gamma < 1$ and $0 < \mu_l \le \mu_u$. Choose any $x^0 \in \mathbf{R}^n$. Set $k := 0$.

Variational inequality subproblem. If $r(x^k) = 0$, stop. Otherwise, choose a mapping A_k such that $A_k(x^k, \cdot)$ is strongly monotone with modulus $\mu_k > 0$, where

$$\min\{\mu_l\,,\ r(x^k)^t\} \le \mu_k \le \mu_u\,. \tag{14.4}$$

Find $z^k \in C$ as an approximate solution of VIP(F_k, C) in the sense that

$$\langle F_k(z^k) + e^k, x - z^k \rangle \ge 0 \quad \forall x \in C\,, \tag{14.5}$$

where

$$\|e^k\| \leq \rho_k \mu_k \|z^k - x^k\|, \quad \rho_k \in [0, \rho_u]. \tag{14.6}$$

Linesearch step. Choose $\sigma_k \in [\sigma_l, \sigma_u]$. Find $y^k = x^k + \alpha_k(z^k - x^k)$, where $\alpha_k = \gamma^{m_k}$ with m_k being the smallest nonnegative integer m such that

$$\langle F(x^k + \gamma^m(z^k - x^k)), x^k - z^k \rangle \geq \sigma_k \mu_k (1 - \rho_k) \|z^k - x^k\|^2. \tag{14.7}$$

Projection step. Choose $\beta_k \in [\beta_l, \beta_u]$. Compute

$$x^{k+1} = P_{C_k}[\hat{x}^k], \tag{14.8}$$

where

$$\hat{x}^k = x^k - \beta_k \frac{\langle F(y^k), x^k - y^k \rangle}{\|F(y^k)\|^2} F(y^k), \tag{14.9}$$

and

$$C_k = C \quad \text{or} \quad C_k = C \cap \{w \in \boldsymbol{R}^n \mid \langle F(y^k), w - y^k \rangle \leq 0\}. \tag{14.10}$$

Set $k := k + 1$; and repeat.

Before proceeding with the convergence analysis, some comments are in order. The requirement that $A_k(x^k, \cdot)$ be strongly monotone is natural and not restrictive. We can always satisfy it by adding a regularization term with a suitable parameter (for monotone problems, this parameter can be exactly μ_k). Note that the regularization parameter is allowed to go to zero (but no faster than some power of the problem residual $r(x^k)$). This feature is important for ensuring fast local convergence of Newton-type methods.

Our condition (14.5) for inexact solution of subproblems is in the spirit of conditions used in [16, 15, 14]. For implementation purposes, it might be preferable to to re-write it in terms of some computable subproblem residual, as in [21]. Here, we shall keep it in the form of (14.5) because it simplifies somewhat the subsequent analysis.

The role of the linesearch step is to ensure the separation property (i.e., inequality (14.2)), so that Lemma 14.1 comes into play, and global convergence of the method can be obtained. Observe that our linesearch step is computationally inexpensive, as it only requires evaluation of F at the trial points. In particular, in contrast to many projection algorithms for solving VIP(F, C), it does not require a projection operation for each trial point. This is an advantage when C does not have a special structure and making projections is costly.

Note that if one is to use the second choice for the set C_k in (14.10), then $P_{C_k}[\hat{x}^k] = P_{C_k}[x^k]$ (see [22, Lemma 2.2]). That is, in that case x^{k+1} is the projection of x^k onto the intersection of the feasible set C with the halfspace containing the solution $\bar{x} \in S$. In particular, one does not have to compute $\hat{x}^k$. Whether it is better to use the first or second choice for C_k in (14.10) depends mostly on the structure of the set C. Theoretically, it is clear that projecting onto the intersection yields faster convergence (because we get closer to the solution). This also seems to have been confirmed by some numerical experiments in [22]. However, if projection onto C is explicit then it might be better to choose $C_k = C$, because in that case x^{k+1} would also be computed explicitly. Introducing an additional constraint, even though it is linear, may ruin this structure.

Finally, we formally specify requirements regarding the choice of the algorithmic mapping A_k.

1. For each k, $A_k(x^k, \cdot)$ is strongly monotone with modulus $\mu_k > 0$:

$$\langle A_k(x^k, z_1) - A_k(x^k, z_2), z_1 - z_2 \rangle \geq \mu_k \|z_1 - z_2\|^2 .$$

2. For each k, $A_k(x^k, x^k) = 0$.

3. For each k, there exists $M_k > 0$ such that

$$\|A_k(x, z)\| \leq M_k \|x - z\| \quad \forall\, x, z \in \mathbf{R}^n .$$

   Furthermore, for any bounded set $D \subset \mathbf{R}^n$ there exists some $M > 0$ (independent of k) such that

$$\|A_k(x, z)\| \leq M \|x - z\| \quad \forall\, x, z \in D .$$

As we shall see in Section 3, these conditions are natural and are satisfied in many cases of interest.

We next establish convergence properties of Algorithm 14.2.

Theorem 14.3 *Suppose F is continuous, and let $\{x^k\}$ be any sequence generated by Algorithm 14.2.*

If condition (14.1) holds for some $\bar{x} \in S$ then the sequence $\{x^k\}$ is bounded and each of its accumulation points is a solution of VIP(F, C).

If (14.1) holds for all $\bar{x} \in S$ then the sequence $\{x^k\}$ converges to a solution of VIP(F, C).

Proof We start with showing that the method is well-defined. First note that by strong monotonicity of $A_k(x^k, \cdot)$, VIP(F_k, C) always has (unique) solution. Hence, it also has approximate solutions satisfying (14.5), (14.6) (this is because the exact solution always satisfies these conditions, for any value of ρ_k).

We next show that the linesearch procedure (14.7) always terminates with a positive stepsize α_k. Suppose that for some iteration index k this is not the case. That is, for all integers m we have

$$\langle F(x^k + \gamma^m(z^k - x^k)), x^k - z^k\rangle < \sigma_k\mu_k(1-\rho_k)\|z^k - x^k\|^2 .$$

Since F is continuous, passing onto the limit as $m \to \infty$, we obtain

$$\langle F(x^k), x^k - z^k\rangle \le \sigma_k\mu_k(1-\rho_k)\|z^k - x^k\|^2 . \tag{14.11}$$

On the other hand, by (14.5), we obtain that

$$\begin{aligned}\langle F(x^k), x^k - z^k\rangle &\ge \langle A_k(x^k, z^k) + e^k, z^k - x^k\rangle \\ &\ge \langle A_k(x^k, z^k) - A_k(x^k, x^k), z^k - x^k\rangle - \|e^k\|\|z^k - x^k\| \\ &\ge \mu_k(1-\rho_k)\|z^k - x^k\|^2 ,\end{aligned} \tag{14.12}$$

where the assumptions on A_k and (14.6) were used in the last two relations. From the assumptions on the parameters of the method, it follows that (14.11) and (14.12) can be consistent only if $z^k = x^k$.

Note that by (14.5),

$$z^k - P_C[z^k - F_k(z^k) - e^k] = 0 .$$

By the triangle and Cauchy-Schwarz inequalities, the nonexpansiveness of the projection operator, the assumption on A_k, and (14.6), we have

$$\begin{aligned}\|z^k - x^k\| &= \|P_C[z^k - F_k(z^k) - e^k] - x^k\| \\ &\ge r(x^k) - \|P_C[x^k - F(x^k)] - P_C[z^k - F_k(z^k) - e^k]\| \\ &\ge r(x^k) - \|x^k - z^k + A_k(x^k, z^k) + e^k\| \\ &\ge r(x^k) - (1 + M_k + \rho_k\mu_k)\|x^k - z^k\| .\end{aligned}$$

Hence,

$$(2 + M_k + \rho_k\mu_k)\|z^k - x^k\| \ge r(x^k) . \tag{14.13}$$

Now, if the linesearch is activated then $r(x^k) > 0$ (otherwise the method would have stopped), and so necessarily $z^k \ne x^k$, by (14.13). The latter contradicts the conclusion above, obtained assuming that the linesearch procedure does not terminate in a finite number of steps. Hence, that assumption was invalid, and so the linesearch always terminates with some positive stepsize α_k.

It also follows immediately from $z^k \neq x^k$ and (14.7) that $F(y^k) \neq 0$, and so the point $\hat{x}^k$ is well-defined. Furthermore, C_k is a closed convex set which is nonempty (for either choice in (14.10), $\bar{x} \in C_k$), which implies that the projection step is also well-defined. This concludes the proof that the whole algorithm is well-defined.

We assume, from now on, that $r(x^k) > 0$ for each k, so that an infinite sequence $\{x^k\}$ is generated. Take any $\bar{x} \in S$ satisfying (14.1). Since $\bar{x} \in C_k$, (14.8) clearly implies that

$$\|x^{k+1} - \bar{x}\| \leq \|\hat{x}^k - \bar{x}\| .$$

By (14.7) we have that

$$\langle F(y^k), x^k - y^k\rangle = \alpha_k \langle F(y^k), x^k - z^k\rangle \geq \sigma_k \mu_k (1-\rho_k)\alpha_k \|z^k - x^k\|^2 ,$$

and so Lemma 14.1 is readily applicable. In particular,

$$\begin{aligned} \|x^{k+1} - \bar{x}\|^2 &\leq \|\hat{x}^k - \bar{x}\|^2 \\ &\leq \|x^k - \bar{x}\|^2 - \beta_k(2-\beta_k)\left(\frac{\langle F(y^k), x^k - y^k\rangle}{\|F(y^k)\|}\right)^2 \\ &\leq \|x^k - \bar{x}\|^2 - \beta_k(2-\beta_k)\left(\sigma_k\mu_k(1-\rho_k)\alpha_k \frac{\|z^k - x^k\|^2}{\|F(y^k)\|}\right)^2 \end{aligned}$$

and it follows that the sequence $\{\|x^k - \bar{x}\|\}$ converges, and so the sequence $\{x^k\}$ is bounded. Taking further into account assumptions on the parameters, the last relation implies that

$$0 = \lim_{k\to\infty} \mu_k \alpha_k \frac{\|z^k - x^k\|^2}{\|F(y^k)\|} . \qquad (14.14)$$

Suppose there exists $\tilde{x}$, an accumulation point of the sequence $\{x^k\}$, which is not a solution of VIP(F, C). In that case, $r(\tilde{x}) > 0$. Let $\{x^{k_i}\}$ be the subsequence of $\{x^k\}$ converging to $\tilde{x}$. By (14.4) we have that in this case $\liminf_{i\to\infty} \mu_{k_i} > 0$. Note that, by (14.12),

$$\|F(x^{k_i})\| \geq \mu_{k_i}(1-\rho_{k_i})\|z^{k_i} - x^{k_i}\| ,$$

and, from boundedness of $\{x^k\}$, it follows that $\{z^{k_i}\}$ is bounded. From (14.13) and the assumption on A_k, we now have that

$$(2 + M + \rho_u \mu_u)\|z^{k_i} - x^{k_i}\| \geq r(x^{k_i}) ,$$

where M is a positive constant (depending on $\{x^{k_i}\}, \{z^{k_i}\}$). Hence, $\liminf_{i\to\infty} \|z^{k_i} - x^{k_i}\| > 0$. Furthermore, since $\{y^{k_i}\}$ is obviously bounded, so is $\{F(y^{k_i})\}$. Using all these facts, (14.14) implies that

$$0 = \lim_{i\to\infty} \alpha_{k_i} .$$

The latter is equivalent to saying that $m_{k_i} \to \infty$. It follows that for every sufficiently large i, the stepsize is decreased at least twice, i.e., $m_{k_i} \geq 2$. This means that the stepsize rule (14.7) is not satisfied for the value of $\gamma^{m_{k_i}-1}$, i.e.

$$\langle F(x^{k_i} + \gamma^{m_{k_i}-1}(z^{k_i} - x^{k_i})), x^{k_i} - z^{k_i}\rangle < \sigma_{k_i}(1 - \rho_{k_i})\mu_{k_i}\|z^{k_i} - x^{k_i}\|^2 .$$

Taking into account boundedness of all the sequences involved, and passing onto a subsequence if necessary, as $i \to \infty$ we obtain

$$\langle F(\tilde{x}), \tilde{x} - \tilde{z}\rangle \leq \tilde{\sigma}(1 - \tilde{\rho})\tilde{\mu}\|\tilde{z} - \tilde{x}\|^2 ,$$

where $\tilde{x}$, $\tilde{\mu}$, $\tilde{\rho}$ and $\tilde{z}$ are limits of corresponding subsequences. On the other hand, passing onto the limit along the same subsequences in (14.12), we have that

$$\langle F(\tilde{x}), \tilde{x} - \tilde{z}\rangle \geq (1 - \tilde{\rho})\tilde{\mu}\|\tilde{z} - \tilde{x}\|^2 .$$

But the last two relations are a contradiction because $\tilde{\sigma} \in (0,1)$, $\tilde{\rho} < 1$, while $\tilde{z} \neq \tilde{x}$, $\tilde{\mu} > 0$ by the assumption that $r(\tilde{x}) > 0$. Hence, this case is not possible. It follows that every accumulation point of $\{x^k\}$ solves VIP(F, C).

Now, if condition (14.1) holds for all $\bar{x} \in S$, then we can choose above as $\bar{x}$ any accumulation point of $\{x^k\}$. As before, we then have that the sequence $\{\|x^k - \bar{x}\|\}$ converges. Since $\bar{x}$ is actually an accumulation point of $\{x^k\}$, it must then be the case that $\{x^k\}$ converges to $\bar{x} \in S$. This completes the proof. □

Note that strong monotonicity of $A_k(x^k, \cdot)$ was used mainly to ensure that subproblems VIP(F_k, C) are solvable. In the proof, we essentially need the following property:

$$\langle A_k(x^k, z^k), z^k - x^k\rangle \geq \mu_k\|z^k - x^k\|^2 , \qquad (14.15)$$

which may hold without strong monotonicity of $A_k(x, \cdot)$ (more on this in Section 3). So for the convergence result to hold, we could alternatively *assume* that subproblems have (approximate) solutions, and that (14.15) holds.

3 SOME SPECIFIC ALGORITHMS

We next turn our attention to some specific choices of the algorithmic mapping A_k. The list of possibilities discussed in this section is by no means exhaustive. Our objective here is to highlight only some of the situations where our general strategy can be useful.

3.1 A PROJECTION ALGORITHM

Projection algorithm corresponds to the most simple choice of

$$A_k(x,z) = z - x \quad \text{for all } k.$$

Obviously, this A_k satisfies all the assumptions imposed in Section 2. It is easy to see that the (exact) solution of VIP(F_k, C) in this case is given by

$$z^k = P_C[x^k - F(x^k)],$$

and the point defining the separating hyperplane has the form

$$y^k = \alpha_k P_C[x^k - F(x^k)] + (1-\alpha_k)x^k, \quad \alpha_k \in (0,1].$$

The next iterate x^{k+1} is obtained by projecting x^k either on this hyperplane and then onto C, or directly onto the intersection of C with the halfspace containing a solution point. We thus recover some methods studied in [10, 22]. Note that in [10] F was assumed to be monotone, and in [22] pseudomonotonicity with respect to the whole solution set was used. In Section 2, our assumption is weaker. Furthermore, inexact computation of z^k is allowed, which might be a useful feature in practice. [In principle, we could also analyze inexact computing of x^{k+1}, so that to make things consistent in this projection method. This is certainly possible. We shall not pursue this issue, mainly for the sake of clarity of exposition.]

One advantage of this algorithm compared to most other projection techniques (for example, see [23] and references therein), is that no projections are necessary within the linesearch procedure. Thus only two projections per iteration are required.

Note that the strong monotonicity constant in the simplest implementation is $\mu_k = 1$ for all k. In principle, no adaptive regularization is needed in the projection method. However, as a practical matter, it is important to use an adaptive stepsize (or regularization, in the terminology of the present paper) to compute the first projection point, i.e., $z^k = P_C[x^k - \mu_k^{-1}F(x^k)]$, which corresponds to taking $A_k(x,z) = \mu_k(z-x)$. In [22] some ways to coordinate the two stepsizes μ_k and α_k were proposed. In the numerical experiments of [22] this issue appeared to be of importance.

3.2 A JOSEPHY-NEWTON METHOD

The Josephy-Newton (or linearization) method consists of solving successive approximations VIP(F_k, C), where $F_k(x) = F(x^k) + F'(x^k)(x -$

x^k), assuming F is differentiable. While this scheme possesses fast local convergence [11, 17, 2] in a neighborhood of a regular [18] solution, there are serious problems with ensuring global convergence. In particular, far from a regular solution of the problem, there is no guarantee that the linearized subproblems are solvable. And even if the subproblem solution exists, there is no guarantee that it actually constitutes some progress towards solving the original problem, VIP(F, C).

A modified Josephy-Newton-type method was developed in [21] to address these two problems, (this was done in the context of monotone NCP(F), but global convergence analysis readily extends to the more general setting). In the context of the present paper, this algorithm corresponds to taking

$$A_k(x, z) = (F'(x) + \mu_k I)(z - x),$$

where $\mu_k > 0$ is a suitable regularization parameter. It is easy to see that this choice satisfies all the assumptions of Section 2 if F is monotone (so that $F'(x)$ is positive semidefinite). If $F'(x)$ is not positive semidefinite, one could add an extra regularization term, and then the global convergence analysis still goes through.

By applying the framework of Section 2 to a monotone VIP(F, C), we obtain a Newton-type method which is globally well-defined and convergent to a solution without making any regularity assumptions. Note that the solution set S may even be unbounded. To see why this is a very strong property (in the context of Newtonian methods), it is enough to think of the more typical globalization strategies based on merit functions. For methods based on those strategies, when S is unbounded, even boundedness of iterates cannot be established in general (this is because the level sets of merit functions are necessarily unbounded in this case). Furthermore, global convergence of the whole sequence to a solution typically requires assumptions which imply (global) uniqueness of the solution point. We refer the reader to [21] for a more detailed discussion.

The issue of the local rate of convergence of this method will be addressed in Section 4.

3.3 A MINIMIZATION-BASED METHOD

Consider, for the moment, a monotone affine VIP(F, C), i.e., C is a polyhedral set, and $F(x) = Mx + q$ where M is a positive semidefinite matrix. An interesting iterative procedure was suggested in [7] for solving VIP(F, C) in this special case. Given $x^k \in C$, compute

$$x^{k+1} = \arg\min_{z \in C} \langle Mz + q, z - x^k \rangle + \frac{\mu_k}{2} \|z - x^k\|^2, \tag{14.16}$$

where $\mu_k > 0$. One advantage of this approach is the induced symmetry of the quadratic term which permits the use of reliable and efficient quadratic programming techniques for solving the subproblems, such as the successive overrelaxation methods [13]. This is indeed an appealing feature. Unfortunately, a sequence defined by (14.16) need not converge (we note that no convergence proof was given in [7]). To see this, consider the rotation operator in $\boldsymbol{R}^2$ given by

$$M = \begin{pmatrix} 0 & 1 \\ -1 & 0 \end{pmatrix},$$

and let $C = \boldsymbol{R}^2$, $q = 0$. In this case, VIP(F, C) reduces to the system of linear equations $Mx = 0$ which has the unique solution $\bar{x} = (0, 0)^\top$. Take any $x^k \neq \bar{x}$. It is easy to check that (14.16) yields

$$x^{k+1} = \left(I + \frac{1}{\mu_k} M^\top \right) x^k,$$

and

$$\|x^{k+1}\|^2 = \left(1 + \frac{1}{\mu_k^2} \right) \|x^k\|^2.$$

Therefore for any choice of $\mu_k > 0$, $\|x^{k+1}\| > \|x^k\|$, and the sequence $\{x^k\}$ generated this way does not converge to $\bar{x} = 0 \in \boldsymbol{R}^2$. In fact, it diverges (at least if $\{\mu_k\}$ is bounded above).

We next show that if, instead of accepting the solution of (14.16) as our next iterate, we use it within the framework of Algorithm 14.2, then a convergent algorithm is obtained. In fact, a solution of (14.16) immediately gives a separating hyperplane, so no linesearch is necessary. Furthermore, we extend the idea of (14.16) to the general VIP(F, C) setting. Let F be monotone and differentiable, and consider the mapping

$$A_k(x, z) = F(z) - F(x) + (F'(z)^\top + \mu_k I)(z - x).$$

It is easy to see that in this case VIP(F_k, C) is equivalent to the first-order necessary optimality conditions of the optimization problem

$$\min_{z \in C} \langle F(z), z - x^k \rangle + \frac{\mu_k}{2} \|z - x^k\|^2, \tag{14.17}$$

which is a formal extension of (14.16). If F is affine, then A_k satisfies the assumptions of Section 2 for any $\mu_k > 0$. If F is not affine, then making $A_k(x, \cdot)$ strongly monotone may not be so easy (when C is compact, this can still be done by taking the regularization parameter large enough). Note however that we could use the weaker condition discussed in the end of Section 2 (see (14.15)). Indeed, we have that

$$\begin{aligned} &\langle A_k(x^k, z^k), z^k - x^k \rangle \\ &= \langle F(z^k) - F(x^k) + (F'(z^k)^\top + \mu_k I)(z^k - x^k), z^k - x^k \rangle \\ &\geq \mu_k \|z^k - x^k\|^2, \end{aligned}$$

by monotonicity of F, and so (14.15) holds. In that case, we have to make a separate assumption that the optimization problem (14.17) has stationary points.

By the first-order necessary optimality condition, z^k, an approximate solution (or stationary point) of (14.17), satisfies

$$\begin{aligned} \langle F(z^k), x^k - z^k \rangle &\geq \langle (F'(z^k)^\top + \mu_k I)(z^k - x^k) - e^k, z^k - x^k \rangle \\ &\geq \mu_k \|z^k - x^k\|^2 - \langle e^k, z^k - x^k \rangle \\ &\geq \mu_k (1 - \rho_k) \|z^k - x^k\|^2, \end{aligned}$$

where the second inequality follows from positive semidefiniteness of $F'(z^k)$, and the third from (14.6). In particular, we conclude that the linesearch rule (14.7) is always satisfied with $m = 0$ for any $\sigma_k \leq 1$. Hence, $\alpha_k = 1$ and $y^k = z^k$ for all k (no linesearch is necessary).

Theorem 14.3 implies that any sequence based on solving (14.17) and modified using the strategy of Section 2 converges to a solution of VIP(F, C). Let us go back to the example with the rotation operator considered above. For simplicity, in Algorithm 14.2 set $\mu_k = \beta_k = 1$ and $\rho_k = 0$ for all k. It is easy to see that with this choice of parameters $y^k = z^k = (I + M^\top)x^k$, $F(y^k) = (I + M)x^k$, and $x^k - y^k = -M^\top x^k$. Furthermore,

$$\langle F(y^k), x^k - y^k \rangle = \|x^k\|^2, \quad \|F(y^k)\|^2 = 2\|x^k\|^2,$$

where we have used the facts that $M = -M^\top$, and $\|(I + M)x^k\| = \|x^k\| = \|Mx^k\|$. Hence,

$$x^{k+1} = x^k - \frac{\langle F(y^k), x^k - y^k \rangle}{\|F(y^k)\|^2} F(y^k) = \frac{1}{2}(I - M)x^k.$$

Therefore,

$$\|x^{k+1}\|^2 = \frac{1}{2}\|x^k\|^2,$$

which means that now $\{x^k\}$ converges to $\bar{x} = 0$. In fact, the rate of convergence is linear. The question of convergence rate will be further discussed in Section 4.

3.4 A SPLITTING ALGORITHM

Suppose now that the function F defining VIP(F, C) is monotone and has the structure

$$F(x) = F^1(x) + F^2(x),$$

where the part $-F^2$ is strongly monotone. Note that this implies that F^1 is also strongly monotone (and we can take the same modulus). Suppose further that VIP(F_k, C) with

$$F_k(x) = F^1(x) + F^2(x^k),$$

where x^k is fixed, and $-F^2$ is strongly monotone with modulus $\mu_k > 0$, is in some sense easier to solve than the original problem. In this setting the splitting method often proves to be useful. This method consists of solving a sequence of subproblems VIP(F_k, C) with the structure defined above. In our framework, this corresponds to taking

$$A_k(x, z) = F^1(z) - F^1(x^k).$$

Obviously, this choice satisfies the assumptions of Section 2. Hence, we have a convergent splitting-type algorithm, with the error tolerance criterion given by

$$\|e^k\| \le \mu_k \rho_k \|z^k - x^k\|,$$

where the relaxation quantity $\mu_k \rho_k$ can be kept bounded away from zero. This criterion compares favorably to conditions often used in this setting, which are of the type

$$\sum_{k=0}^{\infty} \|e^k\|^2 < \infty.$$

Issues related to constructive approximation rules are discussed in more detail in [26, 24] (the proximal point techniques studied there are in this sense quite close to the splitting methods).

Finally, we note that no linesearch is necessary when the splitting method is used within our Algorithm 14.2. Indeed,

$$\begin{aligned}
\langle F(z^k), x^k - z^k \rangle &= \langle F^1(z^k) + F^2(z^k), x^k - z^k \rangle \\
&= \langle F^1(z^k) + F^2(x^k) + F^2(z^k) - F^2(x^k), x^k - z^k \rangle \\
&\ge \langle F^2(z^k) - F^2(x^k), x^k - z^k \rangle - \langle e^k, x^k - z^k \rangle \\
&\ge \mu_k (1 - \rho_k) \|x^k - z^k\|^2,
\end{aligned}$$

where strong monotonicity of $-F^2$ was used. This shows that the unit stepsize always satisfies the linesearch rule (14.7).

Finally, note that we could easily consider dynamic splitting, where the choice of F^1 and F^2 depends on the iteration index k.

4 SPEEDING UP CONVERGENCE

In this section, we briefly outline a technique that can be applied to speed up convergence of Algorithm 14.2 (here, F has to be monotone). It is based on a hybrid inexact proximal point method of [26]. In essence, this approach consists of checking whether z^k, solution of subproblem VIP(F_k, C), provides an acceptable (in the sense of [26]) approximate solution to the proximal point subproblem defined at x^k:

$$0 \in (F + N_C)(z) + \mu_k(z - x^k), \tag{14.18}$$

where N_C stands for the normal cone to the set C. If this is the case, a hybrid proximal step is performed. This proximal step certainly preserves global convergence, but it is also (locally) a linear or superlinear step under standard assumptions. More specifically, Algorithm 14.2 can be modified the following way.

For simplicity, we shall assume here that z^k is the exact solution of VIP(F_k, C), i.e., $e^k = 0$. We then have that $0 \in (F_k + N_C)(z^k)$, and hence

$$v^k = F(z^k) - F_k(z^k) \in (F + N_C)(z^k).$$

According to [26], a pair $z^k, v^k \in (F + N_C)(z^k)$ is an acceptable approximate solution of the proximal point subproblem (14.18), if

$$\|v^k + \mu_k(z^k - x^k)\| \leq \theta \max\{\|v^k\|, \mu_k\|z^k - x^k\|\}, \quad \theta \in [0, 1). \tag{14.19}$$

Note that all information needed for checking this condition is readily available after solving VIP(F_k, C), and so no additional work is required. If this condition is satisfied, then the next iterate is obtained by the same projection step as in Algorithm 14.2, but with $F(y^k)$ replaced by v^k (and also setting $y^k = z^k$). Now, if one can show that (14.19) is always satisfied in a neighborhood of a solution with certain properties, then the resulting iterates can be viewed as iterates generated by the method of [26]. This method has the same rate of convergence properties as the classical exact/inexact proximal point method [19]. In particular, if $(F + N_C)^{-1}$ is Lipschitz continuous at zero then convergence is linear. If, in addition, $\mu_k \to 0$ then it is superlinear. We next show that this indeed happens for some methods considered in Section 3.

For the Josephy-Newton method,

$$\begin{aligned}\varepsilon^k &= v^k + \mu_k(z^k - x^k)\\ &= F(z^k) - F(x^k) - A_k(x^k, z^k) + \mu_k(z^k - x^k)\\ &= F(z^k) - F(x^k) - F'(x^k)(z^k - x^k).\end{aligned}$$

Assuming F' is locally Lipschitz continuous with constant $L > 0$, for k large enough we have that

$$\|\varepsilon^k\| \le L\|z^k - x^k\|^2 \le L\|z^k - x^k\|^{1+t'}$$

for any $t' \in (t, 1)$. Since in Algorithm 14.2 we have $\mu_k \ge r(x^k)^t$,

$$\theta \max\{\|v^k\|, \mu_k\|z^k - x^k\|\} \ge \theta r(x^k)^t\|z^k - x^k\|,$$

and to verify (14.19) it remains to show that

$$\|z^k - x^k\| = O(r(x^k)). \tag{14.20}$$

We already know that the sequence converges to some $\bar{x}$, a solution of VIP(F, C). If this solution satisfies conditions for the superlinear convergence of the (pure) Josephy-Newton method, it is easy to see that

$$\|z^k - x^k\| = O(\|x^k - \bar{x}\|).$$

In this setting also

$$\|x^k - \bar{x}\| = O(r(x^k))$$

(the projection residual provides a local error bound on the distance to the solution), and so (14.20) holds. This essentially implies that our Newton method (modified this way) converges superlinearly if we choose $\mu_k \to 0$. We refer the reader to [25, 21], where the case of systems of equations ($C = \boldsymbol{R}^n$) and nonlinear complementarity problems ($C = \boldsymbol{R}^n_+$) are studied in detail, including the effect of inexact subproblems solution.

Finally, consider the sequential minimization algorithm given by (14.17). In this case,

$$\begin{aligned}\varepsilon^k &= v^k + \mu_k(z^k - x^k)\\ &= -F'(z^k)^\top(z^k - x^k).\end{aligned}$$

It is easy to see that

$$\|\varepsilon^k\| \le \|F'(z^k)^\top\|\|z^k - x^k\| \le \theta\mu_k\|z^k - x^k\|,$$

if $\mu_k \ge \|F'(z^k)^\top\|/\theta$. Since $\{z^k\}$ is bounded, μ_k can be kept bounded above, and the rate of convergence is linear (under the standard assumption that $(F + N_C)^{-1}$ is Lipschitz continuous at zero). Note that it does not seem possible to drive the regularization parameter to zero in this case, and so superlinear convergence cannot be guaranteed.

5 CONCLUDING REMARKS

We presented a class of algorithms for solving variational inequalities under the mild assumption that the problem mapping is continuous and pseudomonotone with respect to at least one solution. The algorithms are based on generating a certain hyperplane separating the current iterate from the solution set. Different ways of constructing this hyperplane give rise to different methods within our general framework. A projection method, a Josephy-Newton method, a certain minimization-based method, and a splitting technique, were considered. Obtaining (super)linear convergence rate was also discussed.

Acknowledgments

I wish to thank Patrice Marcotte whose question after my talk at ICCP99 led me to consider the condition of pseudomonotonicity with respect to only one solution point. I also thank the two anonymous referees for their helpful comments.

References

[1] A. A. Auslender. *Optimisation Méthodes Numériques.* Masson, Paris, 1976.

[2] J. F. Bonnans. Local analysis of Newton-type methods for variational inequalities and nonlinear programming. *Applied Mathematics and Optimization*, 29:161–186, 1994.

[3] R.W. Cottle, F. Giannessi, and J.-L. Lions. *Variational Inequalities and Complementarity Problems: Theory and Applications.* Wiley, New York, 1980.

[4] J.-P. Crouzeix. Characterizations of generalized convexity and generalized monotonicity, A survey. In J.-P. Crouzeix et al., editor, *Generalized convexity, generalized monotonicity: Recent results*, pages 237–256. Kluwer Academic Publishers, 1998.

[5] M. C. Ferris and J.-S. Pang (editors). *Complementarity and variational problems: State of the Art.* SIAM Publications, 1997.

[6] M. C. Ferris and J.-S. Pang. Engineering and economic applications of complementarity problems. *SIAM Review*, 39:669–713, 1997.

[7] M.C. Ferris and O.L. Mangasarian. Error bounds and strong upper semicontinuity for monotone affine variational inequalities. *Annals of Operations Research*, 47:293–305, 1993.

[8] R. Glowinski, J.-L. Lions, and R. Trémolières. *Numerical Analysis of Variational Inequalities.* North-Holland, Amsterdam, 1981.

[9] P.T. Harker and J.-S. Pang. Finite-dimensional variational inequality problems : A survey of theory, algorithms and applications. *Mathematical Programming*, 48:161–220, 1990.

[10] A.N. Iusem and B.F. Svaiter. A variant of Korpelevich's method for variational inequalities with a new search strategy. *Optimization*, 42:309–321, 1997.

[11] N.H. Josephy. Newton's method for generalized equations. Technical Summary Report 1965, Mathematics Research Center, University of Wisconsin, Madison, Wisconsin, 1979.

[12] S. Karamardian. Complementarity problems over cones with monotone and pseudomonotone maps. *Journal of Optimization Theory and Applications*, 18:445–455, 1976.

[13] R. De Leone, O.L. Mangasarian, and T.-H. Shiau. Multi-sweep asynchronous parallel successive overrelaxation for the nonsymmetric linear complementarity problem. *Annals of Operations Research*, 22:43–54, 1990.

[14] W. Li. Remarks on matrix splitting algorithms for symmetric linear complementarity problems. *SIAM Journal on Optimization*, 3:155–163, 1993.

[15] Z.-Q. Luo and P. Tseng. Error bound and convergence analysis of matrix splitting algorithms for the affine variational inequality problem. *SIAM Journal on Optimization*, 2:43–54, 1992.

[16] O.L. Mangasarian. Convergence of iterates of an inexact matrix splitting algorithm for the symmetric monotone linear complementarity problem. *SIAM Journal on Optimization*, 1:114–122, 1991.

[17] J.-S. Pang. Inexact Newton methods for the nonlinear complementarity problem. *Mathematical Programming*, 36:54–71, 1986.

[18] S. M. Robinson. Strongly regular generalized equations. *Mathematics of Operations Research*, 5:43–62, 1980.

[19] R.T. Rockafellar. Monotone operators and the proximal point algorithm. *SIAM Journal on Control and Optimization*, 14:877–898, 1976.

[20] S. Schaible, S. Karamardian, and J.-P. Crouzeix. Characterizations of generalized monotone maps. *Journal of Optimization Theory and Applications*, 76:399–413, 1993.

[21] M. V. Solodov and B. F. Svaiter. A truly globally convergent Newton-type method for the monotone nonlinear complementarity problem, 1998. *SIAM Journal on Optimization* 10 (2000), 605-625.

[22] M. V. Solodov and B. F. Svaiter. A new projection method for variational inequality problems. *SIAM Journal on Control and Optimization*, 37:765–776, 1999.

[23] M. V. Solodov and P. Tseng. Modified projection-type methods for monotone variational inequalities. *SIAM Journal on Control and Optimization*, 34:1814–1830, 1996.

[24] M.V. Solodov and B.F. Svaiter. A hybrid approximate extragradient–proximal point algorithm using the enlargement of a maximal monotone operator, 1998. *Set-Valued Analysis* 7 (1999), 323-345.

[25] M.V. Solodov and B.F. Svaiter. A globally convergent inexact Newton method for systems of monotone equations. In M. Fukushima and L. Qi, editors, *Reformulation - Nonsmooth, Piecewise Smooth, Semismooth and Smoothing Methods*, pages 355–369. Kluwer Academic Publishers, 1999.

[26] M.V. Solodov and B.F. Svaiter. A hybrid projection – proximal point algorithm. *Journal of Convex Analysis*, 6:59–70, 1999.

Chapter 15

SUCCESSIVE CONVEX RELAXATION APPROACH TO BILEVEL QUADRATIC OPTIMIZATION PROBLEMS

Akiko Takeda
Department of Mathematical and Computing Science
Tokyo Institute of Technology
takeda@is.titech.ac.jp

Masakazu Kojima
Department of Mathematical and Computing Science
Tokyo Institute of Technology
kojima@is.titech.ac.jp

Abstract The bilevel quadratic optimization problem is an instance of a hierarchical decision process where the lower level constraint set is dependent on decisions taken at the upper level. By replacing the lower level problem by the corresponding KKT optimality condition, the entire problem is transformed into a single level yet non-convex quadratic optimization problem involving the complementarity condition. In this paper, we adopt the successive convex relaxation method given by Kojima and Tunçel for approximating a nonconvex feasible region. By further exploiting the special structure of the bilevel quadratic optimization problem, we present new techniques which enable the efficient implementation of the successive convex relaxation method for the problem. The performance of these techniques is tested in a number of problems, and compared with some other procedures.

Keywords: Bilevel Programming, Nonconvex Quadratic Program, Semi-Infinite LP Relaxation, Reformulation-Linearization Technique, Lift-and-Project Procedure.

M.C. Ferris et al. (eds.), Complementarity: Applications, Algorithms and Extensions, 317–340.

1 INTRODUCTION

Bilevel programming (abbreviated by BLP) belongs to a class of nonconvex global optimization problems. It arises where decisions are made in a two-level hierarchical order and each decision maker has no direct control or influence upon the decision of the other, but actions taken by one decision maker affect returns from the other. Such problems can be formulated as two levels of nested mathematical programs as follows:

$$\begin{array}{ll} \max\limits_{\boldsymbol{x}} & F(\boldsymbol{x},\boldsymbol{y}) \\ \text{subject to} & \boldsymbol{y} \in \operatorname*{argmin}\limits_{\boldsymbol{y}}\{G(\boldsymbol{x},\boldsymbol{y}) : \boldsymbol{g}(\boldsymbol{x},\boldsymbol{y}) \leq 0\}, \\ & \boldsymbol{f}(\boldsymbol{x},\boldsymbol{y}) \leq 0, \\ & \boldsymbol{w} = \begin{pmatrix} \boldsymbol{x} \\ \boldsymbol{y} \end{pmatrix} \in G_0, \end{array} \tag{15.1}$$

where $\boldsymbol{g}(\boldsymbol{x},\boldsymbol{y}) : R^n \to R^{m_1}$, $\boldsymbol{f}(\boldsymbol{x},\boldsymbol{y}) : R^n \to R^{m_2}$, $G(\boldsymbol{x},\boldsymbol{y}) : R^n \to R$, $F(\boldsymbol{x},\boldsymbol{y}) : R^n \to R$, and G_0 denotes a nonempty compact polyhedral subset of R^n. Given an action $\boldsymbol{x}$ at the upper level, the lower level decision maker returns a minimizer $\boldsymbol{y}(\boldsymbol{x})$ of $G(\boldsymbol{x},\boldsymbol{y})$ subject to the constraints $\boldsymbol{g}(\boldsymbol{x},\boldsymbol{y}) \leq \boldsymbol{0}$ to the upper level. As a whole, the upper level decision maker needs to maximize his objective function $F(\boldsymbol{x},\boldsymbol{y}(\boldsymbol{x}))$ subject to the constraints $\boldsymbol{f}(\boldsymbol{x},\boldsymbol{y}(\boldsymbol{x})) \leq \boldsymbol{0}$ and $(\boldsymbol{x},\boldsymbol{y}(\boldsymbol{x})) \in G_0$.

Applications of BLPs are numerous; for example, (i) hierarchical decision making policy problems in mixed economies, where policy makers at the top level influence the decisions of private individuals and companies, (ii) network facility location with delivered price competition, (iii) the Portland Aluminium Smelter in Victoria, Australia [13], aiming to maximize the aluminium production while minimizing the main cost associated with the production. See Vicente and Calamai [20] for a recent comprehensive review of the literature.

We call a BLP lower-convex if the lower level objective function $G(\boldsymbol{x},\boldsymbol{y})$ and constraint functions $g_i(\boldsymbol{x},\boldsymbol{y})$ $(i = 1, 2, \ldots, m_1)$ are all convex in $\boldsymbol{y}$ for each fixed value of $\boldsymbol{x}$. Among the BLPs, lower-convex BLPs have received most of the attention in the literature. The advantage of dealing with lower-convex BLPs is that under an appropriate constraint qualification, the lower level problem can be replaced by its Karush-Kuhn-Tucker (KKT) optimality (or equilibrium) condition to obtain an equivalent (one-level) mathematical program, a special case of the Mathematical Program with Equilibrium Constraints (MPEC) which has been studied intensively in recent years. See Luo, Pang and Ralph [11] for more details about MPEC.

There are three important classes of lower-convex BLPs, namely:

(i) linear BLPs, where all functions involved are affine,

(ii) linear-quadratic BLPs, where $G(\boldsymbol{x}, \boldsymbol{y})$ is convex quadratic and all remaining functions are affine,

(iii) quadratic BLPs, which differ from linear-quadratic BLPs in that $F(\boldsymbol{x}, \boldsymbol{y})$ can also be a quadratic function.

In this paper, we propose new algorithms, based on the successive convex relaxation method [7, 8, 9, 18] for a more general class of BLPs than the class (iii) by allowing that some of $g_i(\boldsymbol{x}, \boldsymbol{y})$ $(i = 1, 2, \ldots, m_1)$ are convex quadratic functions and some of $f_j(\boldsymbol{x}, \boldsymbol{y})$ $(j = 1, 2, \ldots, m_2)$ are (not necessarily convex) quadratic functions. We could further weaken the assumption of quadratic functions to a wide class of nonlinear functions according to the technique proposed by [6], but for simplicity of discussion, we will restrict ourselves to BLQOPs, bilevel quadratic optimization problems where all functions involved in BLPs are linear or quadratic. In these cases, the application of the KKT optimality condition to the lower level problem results in a one-level nonconvex QOP (quadratic optimization problem) including complementarity constraints. We further transform the QOP into a bounded constraint QOP having a linear objective function in order to utilize the successive convex relaxation method.

There are many variants of the successive convex relaxation method for a general QOP having a linear objective function $\boldsymbol{c}^T\boldsymbol{x}$ to be maximized over a compact feasible region F represented by linear or quadratic functions. The original method was proposed by Kojima and Tunçel [8]. It can be seen as an extension of the lift-and-project procedure [10] with the use of the SDP (semidefinite programming) relaxation or the SILP (semi-infinite linear programming) relaxation for 0-1 integer programs to general QOPs. (The latter relaxation is also called the RLT (reformulation-linearization technique) in the literature [15, 16]). The method generates a sequence of compact convex relaxations of F converging to the convex hull of F. This original method is conceptual in the sense that we need to solve "*infinitely many* convex programs (linear programs or semidefinite programs) having *infinitely many* inequality constraints" at each iteration. See also the paper [7] for complexity analysis of the conceptual successive relaxation method.

In their succeeding paper [9], Kojima and Tunçel proposed two types of techniques which discretize and localize (along the objective direction $\boldsymbol{c}$) "*infinitely many* convex programs having *infinitely many* inequality constraints" in order to resolve the difficulty of implementing their original conceptual method of [8]. Given an accuracy $\epsilon > 0$, their discretized-localized variant attains an upper bound for the optimal value within ϵ in a finite number of iterations.

More recently, Takeda, Dai, Fukuda and Kojima [18] presented practical variants of the successive convex relaxation method by slimming down the discretized-localized variant of [9] to overcome a rapid explosion in the number of convex programs to be solved at each iteration. Although the methods of [18] no longer achieve the approximate optimality within a prescribed accuracy, the numerical results reported in the paper [18] look promising. In this paper, we adapt the discretized-localized SILP relaxation variant of [18] to BLQOPs taking full advantage of a special structure of BLQOPs, *i.e.*, complementarity constraints induced from the KKT optimality condition on the lower level problem. The main purpose of this paper is to attain a relatively tight upper bound for the optimal objective value of the BLP (15.1) with reduced computational expense. But the method we will propose has no theoretical guarantee of convergence to the optimal objective value. To compute approximate optimal solutions in practice, our method needs to be combined with some other methods which compute feasible solutions with smaller objective values in the branch and bound framework.

The paper is organized as follows. In the next section, we formally define a BLQOP, and present its reformulation via the KKT optimality condition. Section 3 is devoted to a brief description of the successive convex relaxation method. In Section 4, we explain two types of techniques exploiting the special structure of the BLQOP, and illustrate those techniques using a small example. In Section 5, we report some numerical results and finally, conclude the paper by giving some discussions and possible future work in Section 6.

2 BILEVEL QUADRATIC OPTIMIZATION PROBLEM

Consider a bilevel quadratic optimization problem (BLQOP) having a linear objective function at the upper level:

$$\begin{array}{ll} \max\limits_{\boldsymbol{u}} & \boldsymbol{c}_1^T \mathbf{v} + \boldsymbol{c}_2^T \boldsymbol{u} \\ \text{subject to} & \mathbf{v} \in \operatorname*{argmin}\limits_{\mathbf{v}} \{g_0(\mathbf{v}, \boldsymbol{u}) : g_i(\mathbf{v}, \boldsymbol{u}) \leq 0 \ (i = 1, \dots, m_1)\}, \\ & f_j(\mathbf{v}, \boldsymbol{u}) \leq 0 \ (j = 1, \dots, m_2), \quad \boldsymbol{w} = \begin{pmatrix} \mathbf{v} \\ \boldsymbol{u} \end{pmatrix} \in G_0, \end{array} \tag{15.2}$$

where

$$
\begin{aligned}
\mathbf{v} \in R^{n_1} &: \text{ the lower level variable vector,} \\
\boldsymbol{u} \in R^{n_2} &: \text{ the upper level variable vector,} \\
n &\equiv n_1 + n_2, \\
G_0 &: \text{ a nonempty compact polyhedral subset of } R^n, \\
\mathbf{c}_1 &\in R^{n_1},\ \mathbf{c}_2 \in R^{n_2}, \\
g_i(\cdot,\cdot), f_j(\cdot,\cdot) \in \mathcal{Q} &: \text{ a quadratic function on } R^n \\
&\quad (i = 0, 1, \ldots, m_1,\ j = 1, 2, \ldots, m_2), \\
\boldsymbol{g}(\cdot,\cdot) &= \begin{pmatrix} g_1(\cdot,\cdot) \\ \vdots \\ g_{m_1}(\cdot,\cdot) \end{pmatrix} : R^n \to R^{m_1}, \\
\boldsymbol{f}(\cdot,\cdot) &= \begin{pmatrix} f_1(\cdot,\cdot) \\ \vdots \\ f_{m_2}(\cdot,\cdot) \end{pmatrix} : R^n \to R^{m_2}.
\end{aligned}
$$

If a given BLQOP has a quadratic objective function at the upper level, we could transform the problem into the form (15.2) above by adding a constraint "the objective function$-t = 0$" and replacing the objective function by t.

Condition 15.1

(i) The quadratic function $g_i(\cdot, \boldsymbol{u})$ is convex on $R^{n_1} \times \{\boldsymbol{u}\}$ for fixed $\boldsymbol{u} \in R^{n_2}$ $(i = 0, \ldots, m_1)$.

(ii) If

$$
\begin{aligned}
&\sum_{i=1}^{m_1} \lambda_i \nabla_v g_i(\mathbf{v}, \boldsymbol{u}) = \mathbf{0}, \\
&g_i(\mathbf{v}, \boldsymbol{u}) \leq 0,\ \lambda_i g_i(\mathbf{v}, \boldsymbol{u}) = 0,\ 0 \leq \lambda_i\ (i = 1, \ldots, m_1), \\
&f_j(\mathbf{v}, \boldsymbol{u}) \leq 0\ (j = 1, \ldots, m_2),\ \boldsymbol{w} = \begin{pmatrix} \mathbf{v} \\ \boldsymbol{u} \end{pmatrix} \in G_0,
\end{aligned}
$$

then $\boldsymbol{\lambda} = (\lambda_1, \ldots, \lambda_{m_1})^T = \mathbf{0}$ (A strong version of the Mangasarian Fromovitz constraint qualification).

Applying the Karush-Kuhn-Tucker (KKT) optimality condition to the lower level convex problem of the BLQOP (15.2) under Condition 15.1, we can reduce the constraint

$$
\mathbf{v} \in \operatorname*{argmin}_{\mathbf{v}} \{g_0(\mathbf{v}, \boldsymbol{u}) : g_i(\mathbf{v}, \boldsymbol{u}) \leq 0\ (i = 1, \ldots, m_1)\}
$$

to the constraints

$$\nabla_v g_0(\mathbf{v}, \boldsymbol{u}) + \sum_{i=1}^{m_1} \lambda_i \nabla_v g_i(\mathbf{v}, \boldsymbol{u}) = \mathbf{0},$$
$$g_i(\mathbf{v}, \boldsymbol{u}) \leq 0, \;\; \lambda_i g_i(\mathbf{v}, \boldsymbol{u}) = 0, \;\; 0 \leq \lambda_i \;\; (i = 1, \ldots, m_1),$$

where λ_i denotes the Lagrange multiplier of the ith constraint $g_i(\mathbf{v}, \boldsymbol{u}) \leq 0$ $(i = 1, \ldots, m_1)$. Thus, introducing slack variables $s_i \geq 0$ $(i = 1, \ldots, m_1)$ for the quadratic inequality constraints $g_i(\mathbf{v}, \boldsymbol{u}) \leq 0$ $(i = 1, \ldots, m_1)$, we can rewrite the BLQOP (15.2) as a QOP (quadratic optimization problem)

$$\begin{array}{ll} \max\limits_{\boldsymbol{\lambda}, \boldsymbol{s}, \boldsymbol{w}} & \boldsymbol{c}_1^T \mathbf{v} + \boldsymbol{c}_2^T \boldsymbol{u} \\ \text{subject to} & \nabla_v g_0(\mathbf{v}, \boldsymbol{u}) + \sum_{i=1}^{m_1} \lambda_i \nabla_v g_i(\mathbf{v}, \boldsymbol{u}) = \mathbf{0}, \\ & g_i(\mathbf{v}, \boldsymbol{u}) + s_i = 0, \;\; 0 \leq s_i \leq \bar{s}_i \;\; (i = 1, \ldots, m_1), \\ & \lambda_i s_i = 0, \;\; 0 \leq \lambda_i \;\; (i = 1, \ldots, m_1), \\ & f_j(\mathbf{v}, \boldsymbol{u}) \leq 0 \;\; (j = 1, \ldots, m_2), \quad \boldsymbol{w} = \begin{pmatrix} \mathbf{v} \\ \boldsymbol{u} \end{pmatrix} \in G_0. \end{array} \tag{15.3}$$

Here $\bar{s}_i$ denotes a positive number such that

$$\bar{s}_i \geq \max \{ s_i : g_i(\mathbf{v}, \boldsymbol{u}) + s_i = 0, \;\; s_i \geq 0, \;\; \boldsymbol{w} \in G_0 \} \tag{15.4}$$

$(i = 1, 2, \ldots, m_1)$. Since G_0 is a compact polyhedral set, such a finite positive number $\bar{s}_i$ always exists. If some inequality constraint $g_i(\mathbf{v}, \boldsymbol{u}) \leq 0$ $(i \in \{1, \ldots, m_1\})$ is linear, it is not necessary to introduce the slack variable s_i since the complementarity constraint $\lambda_i g_i(\mathbf{v}, \boldsymbol{u}) = 0$ itself is a quadratic equality constraint. We also know from (ii) of Condition 15.1 that the Lagrange multipliers λ_i $(i = 1, 2, \ldots, m_1)$ involved in the QOP (15.3) are bounded. So we assume to know a sufficiently large number M which bounds the Lagrange multipliers. The bound is necessary because the successive SILP relaxation method which we will present in the next section requires an initial compact polyhedral relaxation C_0 of the feasible region of the problem to be solved. We must say, however, that the tightness of upper bounds for the Lagrange multipliers deeply affects the performance of the successive SILP relaxation method. In Section 4, we present an additional technique to confine the Lagrange multipliers into a bounded convex set without setting *a priori* explicit bound M for them.

To meet the formulation for the successive SILP relaxation, we now rewrite the QOP (15.3) as

$$\text{maximize } \boldsymbol{c}^T \boldsymbol{x} \text{ subject to } \boldsymbol{x} \in F, \tag{15.5}$$

where

$$\begin{aligned}
\boldsymbol{x} &= \begin{pmatrix} \boldsymbol{\lambda} \\ \boldsymbol{s} \\ \boldsymbol{w} \end{pmatrix} \in R^{2m_1+n}, \\
\boldsymbol{c} &= \begin{pmatrix} \mathbf{0} \\ \mathbf{0} \\ \bar{\boldsymbol{c}} \end{pmatrix} \in R^{2m_1+n}, \ \bar{\boldsymbol{c}} = \begin{pmatrix} \boldsymbol{c}_1 \\ \boldsymbol{c}_2 \end{pmatrix} \in R^n, \\
F &= \{\boldsymbol{x} \in C_0 : p(\boldsymbol{x}) \le 0 \ \ (\forall p(\cdot) \in \mathcal{P}_F)\}, \\
\mathcal{P}_F &= \left\{ \begin{array}{l} q_i(\cdot) \ (i = 1, \dots, m_1 + n_1 + m_1 + m_2), \\ -q_j(\cdot) \ (j = 1, \dots, m_1 + n_1 + m_1) \end{array} \right\}, \\
\boldsymbol{q}(\boldsymbol{x}) &= \begin{pmatrix} q_1(\boldsymbol{x}) \\ \vdots \\ q_{m_1}(\boldsymbol{x}) \\ \vdots \\ q_{m_1+n_1}(\boldsymbol{x}) \\ q_{m_1+n_1+1}(\boldsymbol{x}) \\ \vdots \\ q_{2m_1+n_1}(\boldsymbol{x}) \\ q_{2m_1+n_1+1}(\boldsymbol{x}) \\ \vdots \\ q_{2m_1+n_1+m_2}(\boldsymbol{x}) \end{pmatrix} = \begin{pmatrix} \lambda_1 s_1 \\ \vdots \\ \lambda_{m_1} s_{m_1} \\ \nabla_v g_0(\mathbf{v}, \boldsymbol{u}) + \sum_{i=1}^{m_1} \lambda_i \nabla_v g_i(\mathbf{v}, \boldsymbol{u}) \\ g_1(\mathbf{v}, \boldsymbol{u}) + s_1 \\ \vdots \\ g_{m_1}(\mathbf{v}, \boldsymbol{u}) + s_{m_1} \\ f_1(\mathbf{v}, \boldsymbol{u}) \\ \vdots \\ f_{m_2}(\mathbf{v}, \boldsymbol{u}) \end{pmatrix}, \\
C_0 &= \left\{ \boldsymbol{x} = \begin{pmatrix} \boldsymbol{\lambda} \\ \boldsymbol{s} \\ \boldsymbol{w} \end{pmatrix} \in R^{2m_1+n} : \begin{array}{l} 0 \le \lambda_i \le M \ \ (i = 1, \dots, m_1), \\ 0 \le s_i \le \bar{s}_i \ (i = 1, \dots, m_1), \\ \boldsymbol{w} \in G_0 \end{array} \right\}.
\end{aligned}$$

Note that $\mathcal{P}_F$ is a set of quadratic functions on R^{2m_1+n}. In particular, it includes nonconvex quadratic functions induced from the complementarity constraints.

3 SUCCESSIVE SEMI-INFINITE LINEAR PROGRAMMING RELAXATION METHOD

The successive convex relaxation method which we will present in this section is based on a discretized-localized SILP (semi-infinite linear programming) relaxation variant given in [18]. Numerical results on this variant were also reported in the paper [18]. See [7, 8, 9] for more general variants with the use of the SDP and the SILP relaxations. We are concerned with a general QOP (quadratic optimization problem):

$$\text{maximize } \boldsymbol{c}^T\boldsymbol{x} \text{ subject to } \boldsymbol{x} \in F, \tag{15.6}$$

where

$$\begin{array}{rcl} \boldsymbol{c} & \in & R^n, \\ F & = & \{\boldsymbol{x} \in C_0 : p(\boldsymbol{x}) \leq 0 \;\; (\forall p(\cdot) \in \mathcal{P}_F)\}, \\ \mathcal{P}_F & : & \text{a set of finitely many quadratic functions on } R^n, \\ C_0 & : & \text{a nonempty compact polyhedral subset of } R^n. \end{array}$$

To describe the successive SILP relaxation method for the QOP (15.6), we introduce the following notation:

$$\begin{array}{rcl} \mathcal{S}^n & : & \text{the set of } n \times n \text{ symmetric matrices}, \\ qf(\boldsymbol{x}; \gamma, \boldsymbol{q}, \boldsymbol{Q}) & \equiv & \gamma + 2\boldsymbol{q}^T\boldsymbol{x} + \boldsymbol{x}^T\boldsymbol{Q}\boldsymbol{x}, \; \forall \boldsymbol{x} \in R^n, \\ \boldsymbol{Q} \bullet \boldsymbol{X} & \equiv & \text{the (trace) inner product of two symmetric} \\ & & \text{matrices } \boldsymbol{Q} \text{ and } \boldsymbol{X}, \text{ i.e., } \boldsymbol{Q} \bullet \boldsymbol{X} \equiv \sum_i \sum_j Q_{ij} X_{ij}, \\ \overline{D} & = & \{\boldsymbol{d} \in R^n : \|\boldsymbol{d}\| = 1\} \text{ (the set of unit vectors in } R^n), \\ \epsilon_i & : & \text{the } i\text{th unit coordinate vector } (i = 1, \ldots, n). \end{array}$$

Successive SILP relaxation method described below generates a sequence of convex relaxations $C_k \subseteq C_0$ of F ($k = 1, 2, \ldots,$). At each iteration, we first construct a set $\mathcal{P}_k = \{qf(\cdot; \gamma, \boldsymbol{q}, \boldsymbol{Q})\}$ of "rank-2 quadratic functions" each of which induces a valid inequality for the kth iterate C_k:

$$qf(\boldsymbol{x}; \gamma, \boldsymbol{q}, \boldsymbol{Q}) \leq 0 \;\; \text{for every } \boldsymbol{x} \in C_k.$$

Since C_k was chosen to include F in the previous iteration, each relation $qf(\boldsymbol{x}; \gamma, \boldsymbol{q}, \boldsymbol{Q}) \leq 0$ serves as a (redundant) valid inequality for F; hence F is represented as

$$F = \{\boldsymbol{x} \in C_0 : qf(\boldsymbol{x}; \gamma, \boldsymbol{q}, \boldsymbol{Q}) \leq 0 \;\; (\forall qf(\cdot; \gamma, \boldsymbol{q}, \boldsymbol{Q}) \in \mathcal{P}_F \cup \mathcal{P}_k)\}. \tag{15.7}$$

We then apply the SILP relaxation (also called the RLT (reformulation-linearization technique) in the literature [15, 16]) to the set F with the representation in (15.7), which results in the next iterate

$$C_{k+1} = \left\{ \boldsymbol{x} \in C_0 : \begin{array}{c} \exists \boldsymbol{X} \in \mathcal{S}^n \text{such that} \\ \gamma + 2\boldsymbol{q}^T\boldsymbol{x} + \boldsymbol{Q} \bullet \boldsymbol{X} \leq 0 \\ (\forall qf(\cdot; \gamma, \boldsymbol{q}, \boldsymbol{Q}) \in \mathcal{P}_F \cup \mathcal{P}_k) \end{array} \right\}. \tag{15.8}$$

By definition, C_{k+1} is a convex subset of C_0 since it is the intersection of C_0 with the $\boldsymbol{x}$-space projection of a subset of $R^n \times S^n$ represented by the linear inequalities

$$\gamma + 2\boldsymbol{q}^T\boldsymbol{x} + \boldsymbol{Q} \bullet \boldsymbol{X} \leq 0 \quad (\forall qf(\cdot;\gamma,\boldsymbol{q},\boldsymbol{Q}) \in \mathcal{P}_F \cup \mathcal{P}_k).$$

To see $F \subseteq C_{k+1}$ under the assumption of $F \subseteq C_k$, take $\boldsymbol{X} = \boldsymbol{x}\boldsymbol{x}^T$ for some $\boldsymbol{x} \in F$. Then, $(\boldsymbol{x}, \boldsymbol{X})$ satisfies all inequalities of (15.8), so that $F \subseteq C_{k+1}$ is derived.

Algorithm 15.2 (Successive SILP relaxation method)

Step 0: Let $D_0 = \{\epsilon_1, \dots, \epsilon_n, -\epsilon_1, \dots, -\epsilon_n\} \subset \overline{D}$. Compute

$$\alpha_0(\boldsymbol{d}) = \max\{\boldsymbol{d}^T\boldsymbol{x} : \boldsymbol{x} \in C_0\} \ (\forall \boldsymbol{d} \in D_0).$$

Let $C_1 = C_0$ and $k = 1$.

Step 1: If $C_k = \emptyset$ then stop. Compute an upper bound ζ_k of the maximum objective value of QOP (15.6) by $\zeta_k = \max\{\boldsymbol{c}^T\boldsymbol{x} : \boldsymbol{x} \in C_k\}$.

Step 2: Choose a set D_k of vectors in $\overline{D}$. Compute

$$\alpha_k(\boldsymbol{d}) = \max\{\boldsymbol{d}^T\boldsymbol{x} : \boldsymbol{x} \in C_k\} \ (\forall \boldsymbol{d} \in D_k).$$

Step 3: Let

$$\begin{aligned}\mathcal{P}_k &= \{\boldsymbol{d}^T\boldsymbol{x} - \alpha_0(\boldsymbol{d}) : \boldsymbol{d} \in D_0\} \\ &\quad \bigcup \{-(\boldsymbol{d}^T\boldsymbol{x} - \alpha_0(\boldsymbol{d}))((\boldsymbol{d}')^T\boldsymbol{x} - \alpha_k(\boldsymbol{d}')) : \boldsymbol{d} \in D_0,\ \boldsymbol{d}' \in D_k\}\end{aligned}$$

Step 4: Define C_{k+1} by (15.8).

Step 5: Let $k = k + 1$, and go to Step 1.

Algorithm 15.2 generates a sequence of convex relaxations $C_k \subseteq C_0$ of F $(k = 1, 2, \dots)$ and a sequence of real numbers ζ_k $(k = 1, 2, \dots)$ satisfying

$$\begin{aligned}&C_0 \supseteq C_k \supseteq C_{k+1} \supseteq F \ (k = 1, 2, \dots), \\ &\zeta_k \geq \zeta_{k+1} \geq \zeta^* \equiv \sup\{\boldsymbol{c}^T\boldsymbol{x} : \boldsymbol{x} \in F\} \ (k = 1, 2, \dots).\end{aligned}$$

If in addition we take $D_k = \overline{D}$ $(k = 1, 2, \dots)$, then C_k and ζ_k converge to the convex hull of F and ζ^*, respectively. See [6, 8, 9] for more details on the convergence of C_k and ζ_k.

To implement Algorithm 15.2, it is necessary to choose a finite set of directions for D_k at each iteration. As in the paper [18], we take a set of $2n+1$ vectors

$$D(\theta) \equiv \{\boldsymbol{c},\ \boldsymbol{b}_i(\theta),\ \bar{\boldsymbol{b}}_i(\theta)\ \ (i \in \{1,2,\ldots,n\})\} \tag{15.9}$$

for D_k, where

$$\begin{aligned} \boldsymbol{b}_i(\theta) &\equiv (\boldsymbol{c}\cos\theta + \epsilon_i \sin\theta)/\|\boldsymbol{c}\cos\theta + \epsilon_i \sin\theta\|, \\ \bar{\boldsymbol{b}}_i(\theta) &\equiv (\boldsymbol{c}\cos\theta - \epsilon_i \sin\theta)/\|\boldsymbol{c}\cos\theta - \epsilon_i \sin\theta\|, \end{aligned}$$

and $\theta \in (0, \pi/2]$ denotes a parameter. We will explain in Section 5 how we dynamically decrease θ from $\theta_0 = \pi/2$ to a prescribed positive value as the iteration proceeds. When we take $\theta = \pi/2$ during some iterations of Algorithm 15.2, the vectors $\boldsymbol{b}_i(\theta)$ and $\bar{\boldsymbol{b}}_i(\theta)$ in $D(\theta)$ turn out to be the unit vectors ϵ_i and $-\epsilon_i$, and the values $\alpha_k(\epsilon_i)$ and $-\alpha_k(-\epsilon_i)$ give upper and lower bounds for the variable x_i, respectively $(i = 1, 2, \ldots, n)$. In this case, the set $\{-(\boldsymbol{d}^T\boldsymbol{x} - \alpha_0(\boldsymbol{d}))((\boldsymbol{d}')^T\boldsymbol{x} - \alpha_k(\boldsymbol{d}')) : \boldsymbol{d} \in D_0,\ \boldsymbol{d}' \in D_k = D(\pi/2)\}$ constructed in Step 3 of Algorithm 15.2 contains all rank-2 quadratic functions induced from the pairwise products of lower and upper bound constraints for variables x_i $(i = 1, 2, \ldots, n)$. These constraints correspond to underestimators and overestimators for quadratic terms $x_i x_j$, $(i, j \in \{1, 2, \ldots, n\})$ which have been used in many branch-and-bound methods (for instance, [14]). We also see that both $\boldsymbol{b}_i(\theta)$ and $\bar{\boldsymbol{b}}_i(\theta)$ approach to the objective direction $\boldsymbol{c}$ as $\theta \to 0$. See [18] for more details.

4 ADDITIONAL TECHNIQUES FOR BILEVEL QUADRATIC OPTIMIZATION PROBLEMS

In this section, we present two types of techniques to enhance the efficiency of Algorithm 15.2; one technique is a reformulation of the QOP (15.3) into an equivalent scaled problem with explicit bounds for the Lagrange multipliers, and the other is a technique to tighten bounds for complementary pairs of variables.

4.1 SCALING LAGRANGE MULTIPLIERS

In order to apply Algorithm 15.2 to the QOP (15.3), we need an initial compact polyhedral relaxation C_0 of the feasible region F. Among the variable vectors $\boldsymbol{\lambda}$, $\boldsymbol{w}$ and $\boldsymbol{s}$, $\boldsymbol{w}$ is confined into the compact polyhedral set G_0 and $\boldsymbol{s}$ into the compact polyhedral set $\Pi_{i=1}^{m_1}[0, \bar{s}_i]$. Here $\bar{s}_i$ denotes a finite positive number given in (15.4). We have assumed to know a positive number M which bounds the Lagrange multipliers λ_i

$(i = 1, 2, \ldots, m_1)$ in Section 2. Such a positive number certainly exists in view of (ii) of Condition 15.1. It is usually difficult, however, to estimate (tight) upper bounds for the Lagrange multipliers from the QOP formulation (15.3) of the BLQOP (15.2).

To avoid such a difficulty, we artificially restrict the range of values for Lagrange multipliers by introducing a scaling variable α into the QOP (15.3):

$$\begin{array}{ll} \max\limits_{\boldsymbol{\mu},\boldsymbol{s},\boldsymbol{w},\alpha} & \boldsymbol{c}_1^T \mathbf{v} + \boldsymbol{c}_2^T \boldsymbol{u} \\ \text{subject to} & \alpha \nabla_v g_0(\mathbf{v}, \boldsymbol{u}) + \sum_{i=1}^{m_1} \mu_i \nabla_v g_i(\mathbf{v}, \boldsymbol{u}) = \mathbf{0}, \\ & g_i(\mathbf{v}, \boldsymbol{u}) + s_i = 0, \ 0 \le s_i \le \bar{s}_i \ (i = 1, \ldots, m_1), \\ & \mu_i s_i = 0, \ 0 \le \mu_i \le 1 \ (i = 1, \ldots, m_1), \\ & \alpha + \sum_{i=1}^{m_1} \mu_i = 1, \ 0 \le \alpha \le 1, \\ & f_i(\mathbf{v}, \boldsymbol{u}) \le 0 \ (i = 1, \ldots, m_2), \ \boldsymbol{w} = \begin{pmatrix} \mathbf{v} \\ \boldsymbol{u} \end{pmatrix} \in G_0. \end{array} \tag{15.10}$$

Lemma 15.3 *The QOP (15.3) is equivalent to the scaled QOP (15.10) in the sense that $(\boldsymbol{\lambda}, \boldsymbol{s}, \boldsymbol{w}) \in R^{2m_1+n}$ is a feasible solution of the QOP (15.3) if and only if $(\boldsymbol{\mu}, \boldsymbol{s}, \boldsymbol{w}, \alpha) \in R^{2m_1+n+1}$ is a feasible solution of the scaled QOP (15.10) with $\boldsymbol{\mu} = \alpha\boldsymbol{\lambda}$ and some $\alpha > 0$.*

Proof Suppose that $(\boldsymbol{\lambda}, \boldsymbol{s}, \boldsymbol{w}) \in R^{2m_1+n}$ is a feasible solution of the QOP (15.3). Let

$$\alpha = \frac{1}{1 + \sum_{i=1}^{m_1} \lambda_i} > 0 \text{ and } \boldsymbol{\mu} \ = \ \alpha\boldsymbol{\lambda}.$$

The above equalities derive $\alpha\,(1 + \sum_{i=1}^{m_1} \lambda_i) = \alpha + \sum_{i=1}^{m_1} \mu_i = 1$, which corresponds to one of the constraints in (15.10). Clearly, $\boldsymbol{\mu}$ and α defined above satisfy the other constraints of (15.10), and hence, $(\boldsymbol{\mu}, \boldsymbol{s}, \boldsymbol{w}, \alpha) \in R^{2m_1+n+1}$ is a feasible solution of the scaled BLQOP (15.10). Now suppose that $(\boldsymbol{\mu}, \boldsymbol{s}, \boldsymbol{w}, \alpha)$ is a feasible solution of the scaled QOP (15.10). If $\alpha = 0$, then the constraints $\alpha + \sum_{i=1}^{m_1} \mu_i = 1$ and $\alpha \nabla_v g_0(\mathbf{v}, \boldsymbol{u}) + \sum_{i=1}^{m_1} \mu_i \nabla_v g_i(\mathbf{v}, \boldsymbol{u}) = \mathbf{0}$ of (15.10) contradict to (ii) of Condition 15.1. Hence $\alpha > 0$. Let $\boldsymbol{\lambda} = \boldsymbol{\mu}/\alpha$. Then $(\boldsymbol{\lambda}, \boldsymbol{s}, \boldsymbol{w})$ turns out to be a feasible solution of the QOP (15.3). □

We rewrite the scaled QOP (15.10) as

$$\text{maximize } \boldsymbol{c}^T\boldsymbol{x} \text{ subject to } \boldsymbol{x} \in F, \tag{15.11}$$

where

$$\begin{aligned}
\boldsymbol{x} &= \begin{pmatrix} \boldsymbol{\mu} \\ \boldsymbol{s} \\ \boldsymbol{w} \\ \alpha \end{pmatrix} \in R^{2m_1+n+1}, \ \boldsymbol{c} = \begin{pmatrix} \mathbf{0} \\ \mathbf{0} \\ \bar{\boldsymbol{c}} \\ 0 \end{pmatrix} \in R^{2m_1+n+1}, \\
\bar{\boldsymbol{c}} &= \begin{pmatrix} \boldsymbol{c}_1 \\ \boldsymbol{c}_2 \end{pmatrix} \in R^n, \\
F &= \{\boldsymbol{x} \in C_0 : p(\boldsymbol{x}) \leq 0 \ \ (\forall p(\cdot) \in \mathcal{P}_F)\}, \\
\mathcal{P}_F &= \left\{ \begin{array}{l} q_i(\cdot) \ (i = 1, \dots, m_1 + n_1 + m_1 + m_2), \\ -q_j(\cdot) \ (j = 1, \dots, m_1 + n_1 + m_1) \end{array} \right\}, \\
\boldsymbol{q}(\boldsymbol{x}) &= \begin{pmatrix} q_1(\boldsymbol{x}) \\ \vdots \\ q_{m_1}(\boldsymbol{x}) \\ \vdots \\ q_{m_1+n_1}(\boldsymbol{x}) \\ q_{m_1+n_1+1}(\boldsymbol{x}) \\ \vdots \\ q_{2m_1+n_1}(\boldsymbol{x}) \\ q_{2m_1+n_1+1}(\boldsymbol{x}) \\ \vdots \\ q_{2m_1+n_1+m_2}(\boldsymbol{x}) \end{pmatrix} = \begin{pmatrix} \mu_1 s_1 \\ \vdots \\ \mu_{m_1} s_{m_1} \\ \alpha\nabla_v g_0(\mathbf{v}, \boldsymbol{u}) + \sum_{i=1}^{m_1} \mu_i \nabla_v g_i(\mathbf{v}, \boldsymbol{u}) \\ g_1(\mathbf{v}, \boldsymbol{u}) + s_1 \\ \vdots \\ g_{m_1}(\mathbf{v}, \boldsymbol{u}) + s_{m_1} \\ f_1(\mathbf{v}, \boldsymbol{u}) \\ \vdots \\ f_{m_2}(\mathbf{v}, \boldsymbol{u}) \end{pmatrix}, \\
C_0 &= \left\{ \boldsymbol{x} = \begin{pmatrix} \boldsymbol{\mu} \\ \boldsymbol{s} \\ \boldsymbol{w} \\ \alpha \end{pmatrix} \in R^{2m_1+n+1} : \begin{array}{l} 0 \leq \mu_i \leq 1 \ (i = 1, \dots, m_1), \\ \alpha + \sum_{i=1}^{m_1} \mu_i = 1, \ 0 \leq \alpha \leq 1, \\ 0 \leq s_i \leq \bar{s}_i \ (i = 1, \dots, m_1), \\ \boldsymbol{w} \in G_0 \end{array} \right\}.
\end{aligned}$$

Now C_0 is a compact polyhedral set, so that we can apply Algorithm 15.2 to the problem (15.11).

4.2 TIGHTENING UPPER BOUNDS FOR COMPLEMENTARY PAIRS OF VARIABLES

If $\boldsymbol{x}$ is a feasible solution of the QOPs (15.5) or (15.11), then it must satisfy the complementarity condition:

$$x_i \geq 0, \ x_{m_1+i} \geq 0 \ \text{ and } x_i x_{m_1+i} = 0, \ \ i = 1, 2 \dots m_1. \tag{15.12}$$

Exploiting the complementarity condition, we can tighten upper bounds on some of the nonnegative variables x_i $(i = 1, 2, \ldots, 2m_1)$. In fact, if the set $\{\boldsymbol{x} \in C_k : x_{m_1+i} = 0\}$ is empty, we can conclude that $x_{m_1+i} > 0$ and $x_i = 0$ for every $\boldsymbol{x} \in F$. Otherwise

$$x_i \leq \max\{\epsilon_i^T \boldsymbol{x} : \boldsymbol{x} \in C_k,\ x_{m_1+i} = 0\} \ \text{ for every } \boldsymbol{x} \in F.$$

Therefore, in Step 0 and Step 2 with $D_k = D(\pi/2)$ of Algorithm 15.2, we can replace $\alpha_k(\epsilon_i)$ by

$$\alpha'_k(\epsilon_i) = \begin{cases} 0 & \text{if } \{\boldsymbol{x} \in C_k : x_{m_1+i} = 0\} = \emptyset, \\ \max\{\epsilon_i^T \boldsymbol{x} : \boldsymbol{x} \in C_k,\ x_{m_1+i} = 0\} & \text{otherwise,} \end{cases}$$

and similarly $\alpha_k(\epsilon_{m_1+i})$ by

$$\alpha'_k(\epsilon_{m_1+i}) = \begin{cases} 0 & \text{if } \{\boldsymbol{x} \in C_k : x_i = 0\} = \emptyset, \\ \max\{\epsilon_{m_1+i}^T \boldsymbol{x} : \boldsymbol{x} \in C_k,\ x_i = 0\} & \text{otherwise.} \end{cases}$$

It should be noted that $\alpha'_k(\epsilon_i) \leq \alpha_k(\epsilon_i)$ and $\alpha'_k(\epsilon_{m_1+i}) \leq \alpha_k(\epsilon_{m_1+i})$ in general. If the strict inequality holds above, then we can strengthen the SILP relaxation in Step 4 of Algorithms 15.2. We call Algorithm 15.2 combined with this technique as the modified Algorithms 15.2.

4.3 AN ILLUSTRATING EXAMPLE

In this section, we present an example to highlight the main idea of this paper. This example of bilevel quadratic program is taken from Shimizu and Aiyoshi [17].

$$\begin{array}{ll} \max\limits_{x} & F(x, y) = -x^2 - (y - 10)^2 \\ \text{subject to} & y \in \operatorname*{argmin}\limits_{y} \left\{ (x + 2y - 30)^2 : \begin{array}{c} x + y \leq 20, \\ 0 \leq y \leq 20 \end{array} \right\}, \\ & 0 \leq x \leq 15,\ -x + y \leq 0. \end{array} \tag{15.13}$$

Applying the KKT optimality condition to the lower level problem and introducing a scaling variable α, we transform the problem (15.13) into

$$
\begin{array}{ll}
\max & t \\
\text{subject to} & \\
& \left.\begin{array}{l} x^2 + (y-10)^2 + t = 0, \\ x + y \le 20, \ \ -x + y \le 0, \end{array}\right\} \text{(feasibility)} \\
& \alpha(4x + 8y - 120) + \mu_1 + \mu_2 - \mu_3 = 0, \ \text{(stationarity)}, \\
& \left.\begin{array}{l} \mu_1(20 - x - y) = 0, \\ \mu_2(20 - y) = 0, \ \ \mu_3 y = 0, \end{array}\right\} \text{(complementarity)} \\
& \left.\begin{array}{l} \alpha + \sum_{i=1}^{3} \mu_i = 1, \ \ 0 \le \alpha \le 1, \\ 0 \le \mu_i \le 1 \ (i = 1,2,3), \\ 0 \le x \le 15, \ \ 0 \le y \le 20. \end{array}\right\} \text{(bounds)}
\end{array}
\tag{15.14}
$$

In this example (15.13), the upper level problem has a quadratic objective function, so that we have replaced it by a new single variable t to make a linear objective function.

In the scaled QOP formulation (15.14), the scaling technique presented in Section 4.1 has derived the quadratic equality constraint

$$\alpha(4x + 8y - 120) + \mu_1 + \mu_2 - \mu_3 = 0,$$

while if we assume a sufficiently large number M as upper bounds for the Lagrange multipliers λ_i $(i = 1,2,3)$ of the lower level problem of (15.13), we have the linear equality constraint

$$4x + 8y - 120 + \lambda_1 + \lambda_2 - \lambda_3 = 0.$$

Therefore, the scaling technique has created an additional quadratic equality constraint which may worsen the quality of approximation of the maximum objective value, although the technique is important because M chosen may not be guaranteed to bound the Lagrange multipliers λ_i $(i = 1,2,3)$.

Now we apply the technique proposed in Section 4.2 to the complementarity constraints $\mu_1(20 - x - y) = 0, \mu_2(20 - y) = 0$ and $\mu_3 y = 0$. For simplicity of notation, we use a variable itself instead of the corresponding unit coordinate vector ϵ_i below; for example, $\alpha_k(y)$ stands for $\alpha_k(\epsilon_i)$ where ϵ_i denotes the unit coordinate vector corresponding to the y axis. When a set $D_k \subset \overline{D}$ of direction vectors at the kth iterate is equal to $D(\pi/2) = \{\epsilon_1, \dots, \epsilon_n, -\epsilon_1, \dots, -\epsilon_n\}$, we compute

$$\begin{aligned} \alpha'_k(y) &= \max\{y : (x, y, \boldsymbol{\mu}, \alpha, t) \in C_k,\ \mu_3 = 0\}, \\ \alpha'_k(-y) &= \max\{-y : (x, y, \boldsymbol{\mu}, \alpha, t) \in C_k,\ \mu_2 = 0\}, \\ \alpha'_k(\mu_1) &= \max\{\mu_1 : (x, y, \boldsymbol{\mu}, \alpha, t) \in C_k,\ x + y = 20\}, \end{aligned}$$

instead of $\alpha_k(y), \alpha_k(-y)$ and $\alpha_k(\mu_1)$. Moreover, if either $\alpha'_k(y) < 20$ or $\alpha'_k(-y) < 0$ holds in this example, we can set $\mu_2 = 0$ or $\mu_3 = 0$ in the succeeding iterations, respectively. Regarding to upper bounds for μ_2 and μ_3, we also obtain tighter values by

$$\begin{aligned} \alpha'_k(\mu_2) &= \max\{\mu_2 : (x, y, \boldsymbol{\mu}, \alpha, t) \in C_k,\ y = 20,\ \mu_3 = 0\}, \\ \alpha'_k(\mu_3) &= \max\{\mu_3 : (x, y, \boldsymbol{\mu}, \alpha, t) \in C_k,\ y = 0,\ \mu_2 = 0\}. \end{aligned}$$

5 SOME NUMERICAL RESULTS

In this section, we describe our implementation of Algorithm 15.2, and report some encouraging preliminary numerical results. The program was coded in C++ language and run on a DEC Alpha Workstation (600 MHz with 1GB of memory). We used CPLEX Version 6.0 as LP solver to compute $\alpha_k(\boldsymbol{d})$ in Steps 0 and Step 2 of Algorithm 15.2.

5.1 SOME IMPLEMENTATION DETAILS

We start Algorithm 15.2 by constructing a set $D_1 = D(\theta)$ with $\theta = \pi/2$ of direction vectors according to the definition (15.9). If the decrease in the upper bound ζ_k for the maximum objective value becomes little in some iteration, we reduce the value θ to replace $D(\theta)$. Otherwise, we use the same set of direction vectors as that of the previous iteration. Throughout the computational experiments, we use the following replacement rule:

Let $\ell = 0$, $\theta_0 = \pi/2$, $K = 3$ and $\{\sigma_j\}_{j=0}^K$ be a decreasing sequence such that $\{1, \frac{8}{9}, \frac{4}{9}, \frac{2}{9}\}$. If the upper bound ζ_k generated at the kth iteration remains to satisfy

$$\frac{\zeta_{k-1} - \zeta_k}{\max\{|\zeta_k|,\ 1.0\}} \geq 0.001 \times \sigma_\ell,$$

then set $\theta_k = \theta_{k-1}$. Else if $\ell < K$, then set $\ell = \ell + 1$ and $\theta_k = \sigma_\ell \theta_0$, which implies the replacement of $D(\theta_k)$. Otherwise stop the procedure.

For the comparison, we implemented Algorithm 15.2 with another typical algorithm related with the lift-and-project procedure for quadratic programs; the RLT (reformulation-linearization technique) proposed by

[15, 16]. The QOP formulations (15.3) and (15.10) of BLQOPs usually have some linear constraints such as bound constraints in addition to quadratic ones. Following the idea of the RLT, we generate quadratic constraints through the products of pairwise linear constraints. Together with the original quadratic constraints, those new ones are added as input data for the algorithm. We call the expanded input data as "DataRLT" while the original data of the QOP (15.3) or the scaled QOP (15.10) as "DataOrg". We compare the following five cases:

SSILP1 Algorithm 15.2 with input DataOrg;

SSILP1+RLT Algorithm 15.2 with input DataRLT;

SSILP2 the modified Algorithm 15.2 (Section 4.2) with input DataOrg;

SSILP2+RLT the modified Algorithm 15.2 (Section 4.2) with input DataRLT;

RLT the LP relaxation for input DataRLT;

with respect to the following items:

f**up** the solution value found by each algorithm,

R.err the relative error of a solution value, *i.e.*, $\frac{|f\text{up} - f_{\text{opt}}|}{\max\{|f_{\text{opt}}|, 1.0\}}$, where f_{opt} is the global optimum value;

cpu the cpu time in seconds;

itr the number of iterations (Steps 1- 5) the algorithm uses.

5.2 TEST PROBLEMS

Table 15.1 Test problems

Problem	*Source*	n_1	n_2	m_1	m_2	*#conv*	*#non-conv*	f_{opt}
bard1	[4]	1	1	4	2	1	0	-17.00
bard2	–	1	1	4	2	0	1	68.78
bard3	[4]	2	2	4	4	2	1	14.36
shimizu1	[17]	1	1	3	4	1	0	-100.00
shimizu2	[17]	2	2	4	4	1	0	-225.00
aiyoshi1	[2]	2	2	6	5	0	0	-60.00
aiyoshi2	[1]	4	4	12	10	0	1	6600.00

We evaluate the five methods described in Section 5.1 using a set of test problems. Table 15.1 shows some characteristics of the test problems. n_1, n_2, m_1 and m_2 are introduced in the formulation (15.2), and #conv (or #non-conv) indicates the number of convex (or nonconvex, respectively) quadratic constraints in the lower or upper level problem. The transformed one-level mathematical program (15.3) via the KKT optimality condition, includes additional m_1 nonconvex complementarity constraints. Moreover, the technique proposed in Section 4.1 increases the number of nonconvex quadratic constraints by n_1. Therefore, the QOP (15.3) (or the scaled QOP (15.10)) has #non-conv+m_1 (or #non-conv+$m_1 + n_1$, respectively) nonconvex quadratic constraints. The test problems of Table 15.1 are basically chosen from the literature. A nonconvex BLQOP, which we call "bard2", is constructed from Problem "bard1" by multiplying the upper level convex objective function by -1.

Tables 15.2 and 15.3 present numerical results on the test problems when they are reformulated into the QOP (15.3) and the scaled QOP (15.10), respectively. Figures 15.1 and 15.2 show how the upper bound ζ_k for the maximum objective value of Problem "shimizu1" decreases as the iteration proceeds. Problem "shimizu1" of Figure 15.1 takes the QOP formulation (15.3) and that of Figure 15.2 the scaled QOP formulation (15.10). The lines "SSILP1+RLT" and "SSILP2+RLT" in Figure 15.2 designate the similar performance of achieving the global optimum value for Problem "shimizu1", though the initial upper bounds for Lagrange multipliers are different.

Our experiments were conducted in order to see how the following three factors affect the behavior of the algorithms: (i) the scaling technique for Lagrange multipliers (Section 4.1); (ii) the technique for tightening the bound constraints (Section 4.2); (iii) the effect of the RLT.

(i) In the scaled QOP formulation (15.10), we need no deductive upper bounds for Lagrange multipliers, while in the QOP formulation (15.3) we assumed that $\lambda_i \leq 1000,\ i = 1, \ldots, m_2$ by taking $M = 1000$ for all the test problems. Comparing the results in Table 15.2 with those in Table 15.3, we see that the scaling technique works effectively in several instances such as Problems "bard1", "bard2" and "shimizu1" in which better upper bounds for the maximum objective values were attained. As we have discussed in Section 4.3, however, the scaling technique generates new quadratic equality constraints, which influence the performance of Algorithm 15.2 applied to the scaled QOP formulation (15.10). In Problems "shimizu2" and "aiyoshi1" of Tables 15.2 and 15.3, we observe that the scaled QOP formulation (15.10) makes the quality of the upper bound worse.

Table 15.2 Numerical Results of the QOP formulation (15.3)

Problem	*SSILP1*				*SSILP1+RLT*			
	fup	*R.err*	*cpu*	*itr*	*fup*	*R.err*	*cpu*	*itr*
bard1	-3.07	8.2e-1	4.6	46	-14.92	1.2e-1	4.4	30
bard2	100.96	4.7e-1	0.9	16	68.86	1.2e-3	0.8	7
bard3	19.99	3.9e-1	2.8	11	14.54	1.2e-2	3.5	14
shimizu1	-95.37	4.6e-2	0.3	14	-96.45	3.6e-2	0.3	11
shimizu2	-224.13	3.9e-3	0.8	11	-225.00	6.1e-15	0.8	6
aiyoshi1	-55.41	7.7e-2	21.6	53	-59.69	5.2e-3	6.9	11
aiyoshi2	6786.19	2.8e-2	38.1	8	6625.64	3.9e-3	140.0	7

Problem	*SSILP2*				*SSILP2+RLT*			
	fup	*R.err*	*cpu*	*itr*	*fup*	*R.err*	*cpu*	*itr*
bard1	-3.07	8.2e-1	4.2	46	-14.92	1.2e-1	4.2	30
bard2	100.91	4.7e-1	0.8	15	68.86	1.2e-3	0.8	7
bard3	18.90	3.2e-1	5.5	21	14.57	1.5e-2	2.9	12
shimizu1	-95.75	4.3e-2	0.2	11	-96.74	3.3e-2	0.2	10
shimizu2	-225.00	5.3e-15	0.4	6	-225.00	3.8e-16	0.7	6
aiyoshi1	-59.82	3.0e-3	2.3	9	-60.00	1.2e-15	1.4	6
aiyoshi2	6647.18	7.2e-3	20.3	6	6614.68	2.2e-3	100.2	6

Problem	*RLT*		
	fup	*R.err*	*cpu*
bard1	-5.00	7.1e-1	0.0
bard2	75.78	1.0e-1	0.0
bard3	33.64	1.3e+0	0.0
shimizu1	0.00	1.0e+0	0.0
shimizu2	-125.00	4.4e-1	0.0
aiyoshi1	-41.70	3.1e-1	0.0
aiyoshi2	7200.00	9.1e-2	0.3

Table 15.3 Numerical Results of the scaled QOP formulation (15.10)

Problem	*SSILP1*				*SSILP1+RLT*			
	fup	*R.err*	*cpu*	*itr*	*fup*	*R.err*	*cpu*	*itr*
bard1	-14.75	1.3e-1	6.3	40	-16.40	3.6e-2	4.6	26
bard2	72.01	4.7e-2	1.8	24	68.78	3.0e-9	0.8	8
bard3	14.80	3.1e-2	4.2	16	14.57	1.4e-2	3.9	13
shimizu1	-98.39	1.6e-2	0.4	12	-99.36	6.4e-3	0.2	6
shimizu2	-99.54	5.6e-1	1.9	11	-122.71	4.6e-1	6.6	24
aiyoshi1	-59.43	9.5e-3	29.6	26	-55.90	6.8e-2	29.8	31
aiyoshi2	6841.73	3.7e-2	19.5	8	6769.46	2.6e-2	48.1	6

Problem	*SSILP2*				*SSILP2+RLT*			
	fup	*R.err*	*cpu*	*itr*	*fup*	*R.err*	*cpu*	*itr*
bard1	-14.79	1.3e-1	6.1	40	-16.42	3.4e-2	4.2	23
bard2	72.35	5.2e-2	1.7	23	68.78	3.0e-9	0.8	8
bard3	14.80	3.0e-2	4.0	15	14.56	1.4e-2	3.5	13
shimizu1	-100.00	1.4e-15	0.2	6	-100.00	2.7e-15	0.3	6
shimizu2	-168.58	2.5e-1	3.3	30	-210.25	6.6e-2	1.5	10
aiyoshi1	-59.34	1.1e-2	25.3	26	-56.07	6.6e-2	22.0	32
aiyoshi2	6841.72	3.7e-2	18.5	8	6768.79	2.6e-2	46.4	6

Problem	*RLT*		
	fup	*R.err*	*cpu*
bard1	-3.00	8.2e-1	0.0
bard2	94.67	3.8e-1	0.0
bard3	33.64	1.3e+0	0.0
shimizu1	0.00	1.0e+0	0.0
shimizu2	-25.00	8.9e-1	0.0
aiyoshi1	-31.00	4.8e-1	0.0
aiyoshi2	7200.00	9.1e-2	0.1

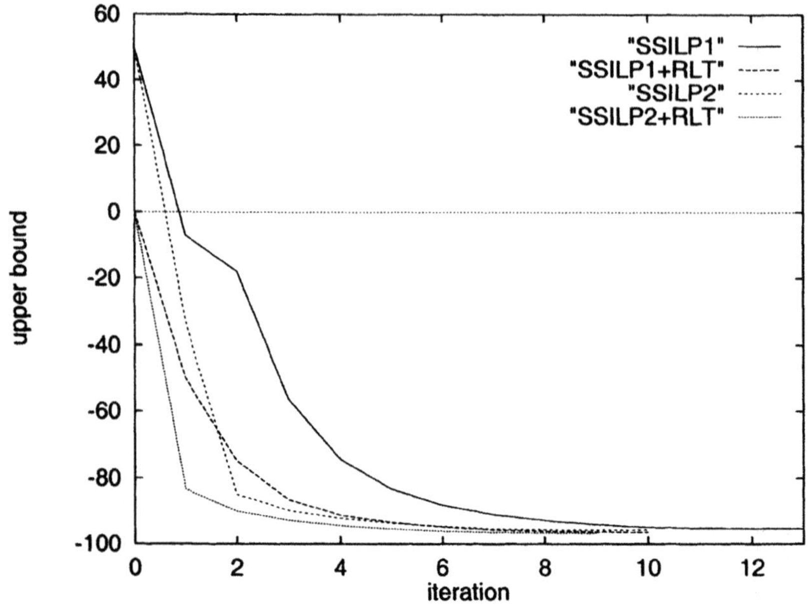

Figure 15.1 Upper bounds for Problem "shimizu1" in the QOP formulation (1.3)

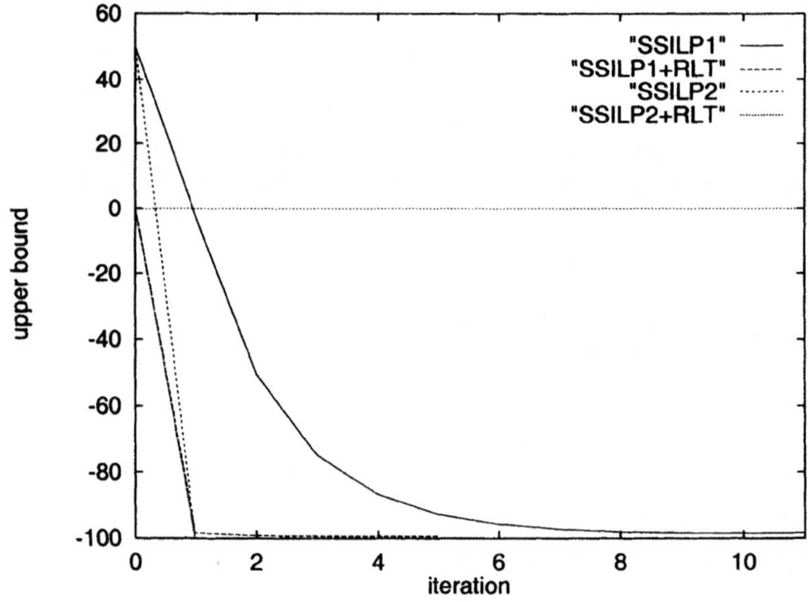

Figure 15.2 Upper bounds for Problem "shimizu1" in the Scaled QOP formulation (1.10)

(ii) Comparing the SSILP1 and SSILP2 cases (or, SSILP1+RLT and SSILP2+RLT) in Tables 15.2 and 15.3, we see the effect of tightening the bounds for some variables. Especially, Problem "aiyoshi1" in Table 15.2 shows the fast convergence to a better upper bound for the maximum objective value due to the tight bound constraints, and also, Problem "shimizu1" in Table 15.3 shows a significant improvement. There are, however, some exceptions, e.g. in Problem "aiyoshi1" of Table 15.3, the SSILP1 case shows a little better performance than the SSILP2 case. This is due to the difference in timing when the parameter θ of Algorithm 15.2 changes in the cases SSILP1 and SSILP2.

(iii) While the sole use of the RLT generates rough upper bounds, the RLT enhances the efficiency of Algorithm 15.2 as Tables 15.2, 15.3 and Figures 15.1, 15.2 show. Although the combined method (SSILP2+RLT) required the greatest computational effort, it also achieved the tightest upper bound with less computing time in several instances, due to its fast convergence.

6 CONCLUSION

In this paper, we have shown two equivalent transformations of a BLQOP (bilevel quadratic optimization problem) into one-level QOPs (quadratic optimization problem), and presented a successive SILP relaxation method to the transformed one-level QOPs. An exploitation of the special structure of the transformed one-level QOPs accelerates the method and generates tighter upper bounds for maximum objective values. The numerical results on a set of test problems have highlighted the effectiveness of the new methods (SSILP2 and SSILP2+RLT), so that the results are satisfactory as a preliminary computational experiments.

Concerning practical implementation of the successive SILP relaxation method, there still remain lots of significant and interesting issues which we need to investigate further. We are above all interested in the following issues:

(a) Our method consumes relatively much computing time in order to achieve tighter upper bounds for maximum objective values. On the one hand, Figures 15.1 and 15.2 show great improvements in upper bounds for the maximum objective values at the first several iterations. Further extension of the research could be to incorporate the first several iterations of the successive convex relaxation method into the branch-and-bound method for solving difficult BLQOPs including several nonconvex quadratic constraints.

(b) If we use multiple computers, we can carry out our method to obtain an upper bound for the maximum objective value of a larger-scaled problem with less computational time. In particular, the $\#D_k$ problems

$$\alpha_k(\boldsymbol{d}) = \max\{\boldsymbol{d}^T\boldsymbol{x} : \boldsymbol{x} \in C_k\} \ (\forall \boldsymbol{d} \in D_k)$$

appeared at the kth iteration of Algorithm 15.2 can be solved independently and simultaneously. Thus, our proposed method is quite suitable for parallel processing.

Acknowledgments

The authors are grateful to Levent Tunçel who suggested us to use the technique presented in Section 4.2 for tightening bounds of a complementary pair of variables, and an anonymous referee for comments which helped to improve the paper's representation.

References

[1] E. Aiyoshi and K. Shimizu, "Hierarchical decentralized systems and its new solution by a barrier method," *IEEE Transactions on Systems, Man and Cybernetics* **SMC-11** (1981) 444–449.

[2] E. Aiyoshi and K. Shimizu, "A solution method for the static constrained Stackelberg problem via penalty method," *IEEE Transactions on Automatic Control* **AC-29** (1984) 1111–1114.

[3] E. Balas, S. Ceria and G. Cornuéjols, "A lift-and-project cutting plane algorithm for mixed 0-1 programs," *Mathematical Programming* **58** (1993) 295–323.

[4] J. F. Bard, "Convex two-level optimization," *Mathematical Programming* **40** (1988) 15–27.

[5] P. H. Calamai and L. N. Vicente, "Generating quadratic bilevel programming problems," *ACM Transactions on Mathematical Software* **20** (1994) 103–122.

[6] M. Kojima, T. Matsumoto and M. Shida, "'Moderate Nonconvexity = Convexity + Quadratic Concavity" Technical Report B-348, Dept. of Mathematical and Computing Sciences, Tokyo Institute of Technology, Meguro, Tokyo, Japan, March 1999, Revised April 1999.

[7] M. Kojima and A. Takeda, "Complexity analysis of successive convex relaxation methods for nonconvex sets," Technical Report B-350, Dept. of Mathematical and Computing Sciences, Tokyo Insti-

tute of Technology, Meguro, Tokyo, Japan, April 1999. Revised July 1999.

[8] M. Kojima and L. Tunçel, "Cones of matrices and successive convex relaxations of nonconvex sets," *SIAM Journal on Optimization* Vol.10, No. 3, 750-778.

[9] M. Kojima and L. Tunçel, "Discretization and localization in successive convex relaxation methods for nonconvex quadratic optimization problems," to appear in Mathematical Programming.

[10] L. Lovász and A. Schrijver, "Cones of matrices and set functions and 0-1 optimization," *SIAM J. on Optimization* **1** (1991) 166–190.

[11] Z.Q. Luo, J.S. Pang and D. Ralph, *Mathematical Programming with Equilibrium Constraints*, (Cambridge University Press, Cambridge, 1996).

[12] R. Horst and H. Tuy, *Global Optimization, Second, Revised Edition*, (Springer-Verlag, Berlin, 1992).

[13] M. G. Nicholls, "The application of non-linear bilevel programming to the aluminium industry," *Journal of Global Optimization* **8** (1996) 245–261.

[14] N. V. Sahinidis, (1998), "BARON : Branch and reduce optimization navigator, User's manual ver. 3.0", Department of Chemical Engineering, University of Illinois at Urbana-Champaign.

[15] H. D. Sherali and A. Alameddine, (1992), "A new reformulation-linearization technique for bilinear programming problems," *Journal of Global Optimization* **2**, 379–410.

[16] H. D. Sherali and C. H. Tuncbilek, "A reformulation-convexification approach for solving nonconvex quadratic programming problems, " *Journal of Global Optimization* **7** (1995) 1–31.

[17] K. Shimizu and E. Aiyoshi, "A new computational method for Stackelberg and min-max problems by use of a penalty method," *IEEE Transactions on Automatic Control* **AC-26** (1981) 460–466.

[18] A. Takeda, Y. Dai, M. Fukuda, and M. Kojima, "Towards implementations of successive convex relaxation methods for nonconvex quadratic optimization problems," in P.M. Pardalos eds., *Approximation and Complexity in Numerical Optimization: Continuous and Discrete Problems,* Kluwer Academic Press, 2000, pp.489-510.

[19] H. Tuy, *Convex analysis and Global Optimization*, (Kluwer, Dordrecht, 1998).

[20] L. N. Vicente and P. H. Calamai, "Bilevel and multilevel programming : A bibliography review," *Journal of Global Optimization* **5** (1994) 291–306.

[21] V. Visweswaran, C. A. Floudas, M. G. Ierapetritou and E. N. Pistikopoulos, (1996), "A decomposition-based global optimization approach for solving bilevel linear and quadratic programs," in C. A. Floudas eds., *State of the Art in Global Optimization*, Kluwer, Dordrecht, 139-162.

Chapter 16

ON A NONSMOOTH NEWTON METHOD FOR NONLINEAR COMPLEMENTARITY PROBLEMS IN FUNCTION SPACE WITH APPLICATIONS TO OPTIMAL CONTROL

Michael Ulbrich
Zentrum Mathematik, Technische Universität München, 80290 München, Germany
mulbrich@ma.tum.de

Abstract Many applications in mathematical modeling and optimal control lead to problems that are posed in function spaces and contain pointwise complementarity conditions. In this paper, a projected Newton method for nonlinear complementarity problems in the infinite dimensional function space L^p is proposed and analyzed. Hereby, an NCP-function is used to reformulate the problem as a nonsmooth operator equation. The method stays feasible with respect to prescribed bound-constraints. The convergence analysis is based on semismoothness results for superposition operators in function spaces. The proposed algorithm is shown to converge locally q-superlinearly to a regular solution. As an important tool for applications, we establish a sufficient condition for regularity. The application of the algorithm to the distributed bound-constrained control of an elliptic partial differential equation is discussed in detail. Numerical results confirm the efficiency of the method.

1 INTRODUCTION

In this paper we investigate a Newton-type method for the infinite dimensional nonlinear complementarity problem (NCP)

$$u \in L^p(\Omega), \quad u \geq 0, \quad F(u) \geq 0, \quad uF(u) \geq 0 \quad \text{on } \Omega.$$

Here, $\Omega \subset \mathbb{R}^n$ is a bounded open domain, and $F : L^p(\Omega) \to L^q(\Omega)$ is an operator from the Lebesgue space $L^p(\Omega)$ into $L^q(\Omega)$, $1 < p, q \leq \infty$. All relations between functions are meant to hold almost everywhere on Ω in the Lebesgue sense. Problems of this form arise frequently in optimal

M.C. Ferris et al. (eds.), Complementarity: Applications, Algorithms and Extensions, 341–360.

control and mathematical modeling [7, 11, 14, 20]. A typical example is the distributed control or the Neumann boundary control of elliptic partial differential equations with pointwise bounds on the control. In particular, we discuss the application of our algorithm to the following distributed optimal control of an elliptic partial differential equation with upper bounds on the control:

$$\begin{aligned} &\underset{w\in L^2(\Omega)}{\text{minimize}} \quad J(w) \stackrel{\text{def}}{=} \frac{1}{2}\|y(w) - y_d\|_{L^2(\Omega)}^2 + \frac{\alpha}{2}\|w - w_d\|_{L^2(\Omega)}^2 \\ &\text{subject to} \quad w \le \psi \quad \text{on } \Omega, \end{aligned}$$

where $y = y(w) \in H_0^1(\Omega)$ is the weak solution of the elliptic state equation

$$-\sum_{i,j=1}^{n} \frac{\partial}{\partial x_i}\left(a_{ij}\frac{\partial y}{\partial x_j}\right) = w \quad \text{on } \Omega.$$

We will apply the general theory for semismooth Newton methods in function spaces recently developed by the author [28] to analyze an infinite dimensional analogue of the successful semismooth Newton methods [19, 21, 22, 24] for finite dimensional NCPs, see, e.g., [5, 9, 17] and the references therein. To this end, an NCP-function is applied pointwise to the pair of functions $(u, F(u))$ to rewrite the NCP as an equivalent nonsmooth operator equation of the form

$$\Phi(u) = 0. \tag{16.1}$$

For the solution of this equation we investigate a Newton-like method that is augmented by a projection to maintain feasibility with respect to upper and lower bounds. In particular, if the lower bounds are set to 0 and the upper bounds to $+\infty$, the generated iterates are all nonnegative. This method is shown to converge locally q-superlinearly to a regular solution of NCP. The feasibility of the iterates with respect to given bound-constraints is desirable since in many applications F is only defined on a (sufficiently large) subset of $L^p(\Omega)$. The bounds can be used to restrict the evaluation of F to those points where it is defined.

In the finite dimensional case, the theory of these and related methods is well developed [5, 9, 17, 19, 21, 22, 24]. On the other hand, very few is known about these methods in the function space context under consideration. Given the fact that many of the NCPs arising in practice are posed in infinite dimensional function spaces, this is especially unsatisfactory.

The numerical solution of infinite dimensional problems requires a discretization. A very desirable property of a solver for the resulting finite dimensional problem is robustness of its performance with respect to the parameters of the discretization, especially the mesh size. Here, infinite dimensional analysis plays a key role. In fact, suppose that an infinite dimensional algorithm with corresponding convergence theory is available. Then the application of a finite dimensional analogue of this method to an appropriate discretization of the problem can be viewed as an inexact version of the infinite dimensional algorithm. Therefore, the numerical method can be expected to inherit asymptotically the good convergence properties of the infinite dimensional algorithm. This shows that the investigation of algorithms in the infinite dimensional context is of significant importance for the development and improvement of numerical algorithms.

The main difficulties and challenges arising in the infinite dimensional context are

(a) The choice of an appropriate vector-valued generalized differential of Φ that replaces the finite dimensional C-subdifferential,

(b) The development of a semismoothness calculus for this generalized differential,

(c) A regularity condition that ensures the invertibility and the uniform boundedness of the elements of the generalized differential in a neighborhood of the solution.

Concerning (a) and (b) we use recent work by the author [28] in which semismoothness-like results for superposition (or Nemyckii) operators in function spaces are developed. Similar as in [28], we address (c) by requiring an abstract regularity condition on which the convergence theory is based.

The paper [28], which deals with the more abstract setting of superposition operators, provides the tools for the development of a Newton-like method for (16.1). We describe this approach in our framework and extend the Newton method by a projection to stay feasible with respect to bound-constraints. The main contribution of this paper is the proof of a sufficient condition that simplifies the verification of the regularity assumption for concrete applications. This condition is used in section 5 to establish the applicability of our results to an elliptic optimal control problem. This problem is also used for a numerical illustration.

Results that are in some sense related to our investigations include the infinite dimensional analysis of primal-dual interior point methods for linear programs [16], and for nonlinear problems the projected gradient method [6, 26], Bertsekas' projected Newton method [18], and affine-scaling algorithms [29, 30].

Throughout, we will assume that the following holds:

Assumption (A):

(a) $F : L^p(\Omega) \to L^q(\Omega)$ *is locally Lipschitz continuous,*

(b) *There exists* $r \in [1, p) \cap [1, q)$ *such that* $u \in L^p(\Omega) \mapsto F(u) \in L^r(\Omega)$ *is continuously Fréchet differentiable with derivative* DF.

The paper is organized as follows: In section 2 the NCP is reformulated as a nonsmooth operator equation by means of the Fischer–Burmeister NCP function. Further, we introduce a generalized differential for the resulting operator and discuss its approximation properties. The Newton-like iteration is formulated and analyzed in section 3. Section 4 is devoted to the derivation of a sufficient condition for the regularity of the generalized differential. In section 5 we apply the developed results to the distributed control of an elliptic partial differential equation. Finally, preliminary numerical results for this problem are reported in section 6.

Notations and Terminology Given the locally Lipschitz continuous function $f : \mathbb{R}^l \to \mathbb{R}$, we denote by

$$\partial f(x) = \text{co}\,\{v;\ \exists (x_k) \subset D_f : x_k \to x,\ \nabla f(x_k) \to v\} \subset \mathbb{R}^l$$

Clarke's [4] generalized gradient at x. Here, 'co' means the convex hull and D_f is the (by Rademacher's theorem) dense set where f is differentiable with gradient ∇f. The function f is called semismooth or strongly semismooth at x if it is directionally differentiable at x and, as $h \to 0$, the residual

$$\max_{v \in \partial f(x+h)} |f(x+h) - f(x) - v^T h|$$

is of the order $o(\|h\|_2)$ or $O(\|h\|_2^2)$, respectively. The C-subdifferential [23] of a locally Lipschitz function $f : \mathbb{R}^l \to \mathbb{R}^m$ at x is given by

$$\partial_C f(x) = \partial f_1(x)^T \times \cdots \times \partial f_m(x)^T \subset \mathbb{R}^{m \times l}.$$

f is (strongly) semismooth at x if all components f_i enjoy this property.

The norm and the inner product on a space X are denoted by $\|\cdot\|_X$ and $(\cdot, \cdot)_X$, respectively. The space of all bounded linear operators from the Banach space X to the Banach space Y is denoted by $\mathcal{L}(X, Y)$.

The Fréchet derivative of a mapping F is denoted by DF. We also use this notation for Jacobians.

By $\mathbf{1}_{\Omega'} : \Omega \to \mathbb{R}$ we denote the characteristic function of the set $\Omega' \subset \Omega$, i.e., $\mathbf{1}_{\Omega'}(x) = 1$ if $x \in \Omega'$, and $\mathbf{1}_{\Omega'}(x) = 0$, otherwise.

The space $H^s(\Omega)$, $s > 0$, is the usual Sobolev space of order s [1], $H_0^s(\Omega)$ is the closure of $C_0^\infty(\Omega)$ in $H^s(\Omega)$, and $H^{-s}(\Omega)$ is the dual space of $H_0^s(\Omega)$ with respect to the dual pairing $\langle \cdot, \cdot \rangle_{H_0^s(\Omega), H^{-s}(\Omega)}$ induced by the inner product $(\cdot, \cdot)_{L^2(\Omega)}$.

Finally, we emphasize that due to the boundedness of Ω we have the continuous embedding $L^{p_1}(\Omega) \subset L^{p_2}(\Omega)$ for $p_1 \geq p_2$.

2 REFORMULATION AS EQUATION

In analogy to the finite dimensional approach suggested by Fischer [12], we apply the Fischer–Burmeister function

$$\phi(a,b) = a + b - \sqrt{a^2 + b^2} \tag{16.2}$$

pointwise to the pair of functions $(u, F(u))$ to define the superposition operator

$$\Phi : u \in L^p(\Omega) \mapsto L^r(\Omega), \quad \Phi(u)(x) = \phi(u(x), F(u)(x)), \quad x \in \Omega. \tag{16.3}$$

The function ϕ is an NCP-function, i.e., satisfies

$$\phi(a,b) = 0 \iff a \geq 0, \quad b \geq 0, \quad ab = 0.$$

Hence, it is obvious that NCP and (16.3) are equivalent. For the investigations to follow we need some facts about ϕ, see [10, Prop. 3.1], [13, Lem. 20].

Proposition 16.1 *The function ϕ defined in* (16.2) *is Lipschitz continuous and strongly semismooth. Furthermore,*

$$\partial\phi(a,b) = \nabla\phi(a,b) = \left(1 - \frac{a}{\|(a,b)\|_2}, 1 - \frac{b}{\|(a,b)\|_2}\right)^T \quad \forall (a,b) \neq 0,$$
$$\partial\phi(0,0) = \left\{(1-\mu_1, 1-\mu_2)^T ;\ \|(\mu_1,\mu_2)\|_2 \leq 1\right\}.$$

Our analysis also works for other suitable NCP-functions. By the Lipschitz continuity of ϕ and the properties of F, the operator Φ is easily seen to be Lipschitz, too. It is well known that the differentiability of superposition operators is a nontrivial matter [2, 31] even for smooth ϕ and F. In our context, ϕ is not even differentiable, but merely strongly semismooth, which makes Φ a nonsmooth operator. The construction of a Newton-like method for the solution of (16.3) requires an appropriate linear approximation of Φ around the actual iteration point. We follow [28] and introduce at $u \in L^p(\Omega)$ the generalized differential $\partial_S\Phi(u) \subset \mathcal{L}(L^p(\Omega), L^r(\Omega))$,

$$\partial_S\Phi(u) \stackrel{\text{def}}{=} \{M(u)\,;\ M(u) = D_1(u) + D_2(u)DF(u)\}\,, \tag{16.4}$$

where the multiplication operators $D_i(u)$ assign to v the value $d_i(u)v$, and the pair $(d_1(u), d_2(u)) \in L^\infty(\Omega)^2$ ranges over all measurable selections [4] of the multifunction

$$x \in \Omega \mapsto \partial\phi(u(x), F(u)(x));$$

in other words,

$$(d_1(u)(x), d_2(u)(x))^T \in \partial\phi(u(x), F(u)(x)) \quad \text{on } \Omega. \tag{16.5}$$

Since $d_i(u) \in L^\infty(\Omega)$, the operators $D_i(u)$ are L^t-automorphisms for all $t \in [1, \infty]$.

The generalized differential $\partial_S\Phi(u)$ can be viewed as a generalization of the finite dimensional C-subdifferential to the function space setting, see [28]. On the basis of this generalized differential it is now straightforward to state (in a preliminary, formal manner) a Newton-like iteration for the solution of (16.3).

Given the current iterate $u^c \in L^p(\Omega)$, choose $M(u^c) \in \partial_S\Phi(u^c)$ and compute the new iterate u^+ by solving

$$M(u^c)(u^+ - u^c) = -\Phi(u^c).$$

If $\bar{u}$ is a solution to the NCP, then $\Phi(\bar{u}) = 0$ and thus

$$M(u^c)(u^+ - \bar{u}) = -\left(\Phi(u^c) - \Phi(\bar{u}) - M(u^c)(u^c - \bar{u})\right). \tag{16.6}$$

An upper bound for the right hand side is provided by the following semismoothness-like estimate, which is a special case of the results in [28]. The rather involved proof can be found in [28].

Theorem 16.2 *Let Assumption* (A) *hold and let ϕ, Φ, and $\partial_S\Phi$, respectively, be defined by* (16.2), (16.3), (16.4), *respectively. Then, as $h \stackrel{L^p}{\to} 0$,*

$$\sup \|\Phi(u+h) - \Phi(u) - M(u+h)h\|_{L^r(\Omega)} = o(\|h\|_{L^p(\Omega)}), \tag{16.7}$$

where the sup *is taken over all $M(u+h) \in \partial_S\Phi(u+h)$.*

Remark 16.3 Under additional assumptions, we can achieve in (16.7) an improved order $O(\|h\|^{1+\beta}_{L^p(\Omega)})$, $\beta > 0$, instead of $o(\|h\|_{L^p(\Omega)})$, see [28]. □

3 NEWTON METHOD WITH PROJECTION

In (16.7) the L^p-norm on the right is stronger than the L^r-norm on the left. In fact, it is not hard to see that in general (16.7) does not hold for $r \geq p$. Hence, for the above Newton iteration only estimates of the form

$$\|u^+ - \bar{u}\|_{L^r(\Omega)} = o(\|u^c - \bar{u}\|_{L^p(\Omega)})$$

with $r < p$ can be expected to hold. This difficulty also arises for other algorithms, e.g., affine-scaling methods [29] and projected Newton methods [18]. As suggested in [18], we will close the gap between the two norms by incorporating a smoothing step into the algorithm.

As a further feature we include a projection step in the algorithm. To this end, we introduce the box

$$\mathcal{B} = \{u \in L^p(\Omega)\,;\ \psi_1 \leq u \leq \psi_2 \text{ on } \Omega\}$$

with measurable functions $\psi_1 : \Omega \to [-\infty, \infty)$ and $\psi_2 : \Omega \to (-\infty, \infty]$. The projection $P : L^p(\Omega) \to \mathcal{B}$ onto $\mathcal{B}$ is given by

$$P(u) = \min\{\max\{\psi_1, u\}, \psi_2\} \quad \text{(pointwise)},$$

and is therefore easy to compute.

Important examples for $\mathcal{B}$ are the trivial choice $\mathcal{B} = L^p(\Omega)$ (i.e., no projection), which corresponds to $\psi_1 \equiv -\infty$, $\psi_2 \equiv +\infty$, and

$$\mathcal{B} = \mathcal{B}_+ \stackrel{\text{def}}{=} \text{feasible set of NCP} = \{u \in L^p(\Omega)\,;\ u \geq 0 \text{ on } \Omega\},$$

i.e., $\psi_1 \equiv 0$, $\psi_2 \equiv +\infty$. The algorithm uses the projection P to generate iterates that are feasible with respect to $\mathcal{B}$. We have now everything at hand to give a complete formulation of the method:

Algorithm 16.4
Choose an initial point $u_0 \in \mathcal{B}$ sufficiently close to a solution $\bar{u} \in \mathcal{B}$ of NCP.

For $k = 0, 1, 2, \ldots$:

1. If $\Phi(u_k) = 0$, STOP with result u_k.

2. Choose $M(u_k) \in \partial_S \Phi(u_k)$ and compute $u^n_{k+1} = u_k + s_k$, where $s_k \in L^r(\Omega)$ solves the Newton equation

$$M(u_k)s_k = -\Phi(u_k).$$

3. Perform a smoothing step:

$$L^r(\Omega) \ni u^n_{k+1} \mapsto u^s_{k+1} \in L^p(\Omega)$$

such that with a constant $C_S > 0$ independent of k holds

$$\left\|u_{k+1}^s - \bar{u}\right\|_{L^p(\Omega)} \leq C_S \left\|u_{k+1}^n - \bar{u}\right\|_{L^r(\Omega)}. \tag{16.8}$$

4. Projection onto $\mathcal{B}$: $\quad u_{k+1} = P(u_{k+1}^s)$.

The projection step serves as a tool to take advantage of available *a priori* knowledge about the solution $\bar{u}$. In fact, if it is known that $\bar{u} \in \mathcal{B}$ holds, it certainly may be attractive to generate iterates that stay feasible with respect to $\mathcal{B}$.

The smoothing step is needed to compensate the discrepancy between the L^r- and the stronger L^p-norm. If (16.8) holds for $u_{k+1}^s = u_{k+1}^n$, then no smoothing step is needed and one can choose $u_{k+1}^s = u_{k+1}^n$. Numerical experiments show that very often this is always satisfied with a moderate value C_S, see the numerical tests in section 6. Of course, condition (16.8) is not implementable since $\bar{u}$ is not known. However, the ratio of $\|s_k\|_{L^p(\Omega)}$ and $\|s_k\|_{L^r(\Omega)}$ is a reasonably good indicator to decide whether a smoothing step is necessary or not [29].

For problems with $F(u) = G(u) + \alpha u$, where $\alpha > 0$ and G maps $L^r(\Omega)$ Lipschitz continuously to $L^p(\Omega)$, it is easy to show that $u_k^n \mapsto u_k^s \stackrel{\text{def}}{=} P_+(u_k^n - \alpha^{-1} F(u_k^n))$ is a smoothing step [18, 29]. Hereby, P_+ denotes the projection onto $\mathcal{B}_+$.

For the local convergence analysis of Algorithm 16.4 we need a regularity property of $\partial_S \Phi(u)$ in a neighborhood of the solution $\bar{u}$.

Assumption (R): Regularity Condition.

There exist constants $\rho > 0$ and $C_{M^{-1}} > 0$ such that, for all $u \in \mathcal{B}$ with $\|u - \bar{u}\|_{L^p(\Omega)} \leq \rho$, all operators $M(u) \in \partial_S \Phi(u)$ are L^r-automorphisms with $\left\|M(u)^{-1}\right\|_{L^r, L^r} \leq C_{M^{-1}}$.

In section 4 we prove sufficient conditions for Assumption (R). Note that in our infinite dimensional setting one can not expect that the mapping $M : u \in L^p(\Omega) \mapsto M(u) \in \mathcal{L}(L^r(\Omega), L^r(\Omega))$ is continuous. One reason for this is that in general the multiplication operators $D_i(u)$ do not depend continuously on $u \in L^p(\Omega)$. Therefore, we have to assume regularity not only at $\bar{u}$ but on a neighborhood of $\bar{u}$.

We now establish the local q-superlinear convergence of Algorithm 16.4. The proof is quite easy and, except for the additional projection, is a special case of the results in [28]. We include it for completeness.

Theorem 16.5 *Let Assumption* (A) *hold and let $\bar{u} \in \mathcal{B}$ be a solution to NCP for which Assumption* (R) *holds. Then there exists $\delta > 0$ such that for all initial points $u_0 \in \mathcal{B}$ with $\|u_0 - \bar{u}\|_{L^p(\Omega)} \leq \delta$ Algorithm* 16.4

either terminates finitely with a solution $u_k \in \mathcal{B}$ of NCP or generates a sequence of iterates $(u_k) \subset \mathcal{B}$ converging q-superlinearly to $\bar{u}$.

Proof For $u_k \in \mathcal{B}$, $\|u_k - \bar{u}\|_{L^p(\Omega)} \leq \rho$, the kth iteration is well-defined according to Assumption (R). Furthermore, by (R) and Theorem 16.2

$$\begin{aligned}\left\|u_{k+1}^n - \bar{u}\right\|_{L^r(\Omega)} &= \left\|M(u_k)^{-1}(\Phi(u_k) - \Phi(\bar{u}) - M(u_k)(u_k - \bar{u}))\right\|_{L^r(\Omega)} \\ &= o(\|u_k - \bar{u}\|_{L^p(\Omega)}) \quad \text{as } \|u_k - \bar{u}\|_{L^p(\Omega)} \to 0.\end{aligned}$$

Therefore,

$$\begin{aligned}\|u_{k+1} - \bar{u}\|_{L^p(\Omega)} &= \left\|P(u_{k+1}^s) - P(\bar{u})\right\|_{L^p(\Omega)} \\ &\leq \left\|u_{k+1}^s - \bar{u}\right\|_{L^p(\Omega)} \leq C_S\left\|u_{k+1}^n - \bar{u}\right\|_{L^r(\Omega)} \qquad (16.9) \\ &= o(\|u_k - \bar{u}\|_{L^p(\Omega)}) \quad \text{as } \|u_k - \bar{u}\|_{L^p(\Omega)} \to 0,\end{aligned}$$

where we have used the nonexpansiveness of P and the fact that $\bar{u} \in \mathcal{B}$. We conclude that for sufficiently small $\delta > 0$ holds

$$\begin{aligned}& u_k \in \mathcal{B}, \quad \|u_k - \bar{u}\|_{L^p(\Omega)} \leq \delta, \quad \Phi(u_k) \neq 0 \\ \Longrightarrow \quad & u_{k+1} \in \mathcal{B}, \quad \|u_{k+1} - \bar{u}\|_{L^p(\Omega)} \leq \frac{1}{2}\|u_k - \bar{u}\|_{L^p(\Omega)} \leq \delta.\end{aligned}$$

Hence, if $\delta > 0$ is sufficiently small and $u_0 \in \mathcal{B}$ satisfies $\|u_0 - \bar{u}\|_{L^p(\Omega)} \leq \delta$, then the algorithm is well defined and either terminates with a solution $u_k \in \mathcal{B}$, or generates a sequence $(u_k) \subset \mathcal{B}$ that converges to $\bar{u}$ in $L^p(\Omega)$. The q-superlinear rate follows from (16.9). □

4 SUFFICIENT REGULARITY CONDITION

The abstract regularity condition (R) is difficult to verify in practice. Hence, our next aim is to establish a sufficient condition for Assumption (R) that can be verified more conveniently.

Assumption (R′):

(a) *Assumption* (A) *is satisfied with $p, q > 2$ and $r = 2$.*

(b) *$DF(u) \in \mathcal{L}(L^2(\Omega), L^2(\Omega))$ holds for all $u \in \mathcal{B}$ in an L^p-neighborhood of $\bar{u}$.*

(c) *$u \in L^p(\Omega) \to DF(u) \in \mathcal{L}(L^2(\Omega), L^2(\Omega))$ is continuous at $\bar{u}$.*

(d) *There exists $\nu > 0$ such that*

$$(v, DF(\bar{u})v)_{L^2(\Omega)} \geq \nu\|v\|_{L^2(\Omega)}^2$$

for all $v \in L^2(\Omega)$.

Theorem 16.6 *If Assumption* (R′) *holds then there exists* $\rho > 0$ *such that Assumption* (R) *holds for the values of* p, q, *and* r *given in* (R′) (a).

Proof For convenience, we set $(\cdot,\cdot) = (\cdot,\cdot)_{L^2(\Omega)}$ and $\|\cdot\| = \|\cdot\|_{L^2(\Omega)}$. Let $\kappa > 0$ and $\rho > 0$ be arbitrary. We will adjust these quantities in the course of the proof. Throughout, we denote by u any

$$u \in L^p(\Omega), \quad \|u - \bar{u}\|_{L^p(\Omega)} \leq \rho,$$

and by $M(u) = D_1(u) + D_2(u)DF(u)$ any element of $\partial_S\Phi(u)$. By Assumption (R′) (b)–(d), we can achieve by reducing $\rho > 0$ that uniformly in u holds with appropriate constant $C_2 > 0$

$$\|DF(u)\|_{L^2,L^2} \leq C_2, \tag{16.10}$$

$$(v, DF(u)v) \geq \frac{\nu}{2}\|v\|^2 \quad \forall v \in L^2(\Omega). \tag{16.11}$$

It is clear from (R′) (b) and the properties of $D_i(u)$, $i = 1, 2$, that $M(u) \in \mathcal{L}(L^2(\Omega), L^2(\Omega))$. Denote by $M(u)^*$ the adjoint of $M(u)$ in $\mathcal{L}(L^2(\Omega), L^2(\Omega))$. In a first step we show that possibly after reducing $\rho > 0$ there is $\gamma > 0$ such that

$$\|M(u)^*v\| \geq \gamma\|v\| \quad \text{for all } v \in L^2(\Omega). \tag{16.12}$$

Let $v \in L^2(\Omega)$, $v \neq 0$ be arbitrary and set $v_1 = D_1(u)v$, $v_2 = D_2(u)v$. We consider two cases:

Case 1: $\|v_2\| \leq \kappa\|v\|$:

By Proposition 16.1 holds $d_i(u) \geq 0$ and $d_1(u) + d_2(u) \geq 1/2$ on Ω. Therefore,

$$\begin{aligned}\|v\| = \left\|\frac{v_1 + v_2}{d_1(u) + d_2(u)}\right\| &\leq \left\|\frac{v_1}{d_1(u) + d_2(u)}\right\| + \left\|\frac{v_2}{d_1(u) + d_2(u)}\right\| \\ &\leq 2\|v_1\| + 2\|v_2\| \leq 2\|v_1\| + 2\kappa\|v\|.\end{aligned}$$

For $\kappa < 1/2$ we obtain

$$\|v\| \leq \frac{2}{1 - 2\kappa}\|v_1\| \quad , \quad \|v_2\| \leq \kappa\|v\| \leq \frac{2\kappa}{1 - 2\kappa}\|v_1\|. \tag{16.13}$$

Hence, using (16.10) we get

$$\begin{aligned}\|v_1\|\|M(u)^*v\| &\geq (M(u)^*v, v_1) = (v, M(u)v_1) \\ &= (v, d_1(u)v_1) + (v, d_2(u)DF(u)v_1) \\ &= \|v_1\|^2 + (v_2, DF(u)v_1) \\ &\geq \|v_1\|^2 - C_2\|v_1\|\|v_2\|.\end{aligned}$$

Therefore, since (16.13) implies $v_1 \neq 0$,

$$\|M(u)^* v\| \geq \|v_1\| - C_2\|v_2\| \geq \left(\frac{1-2\kappa}{2} - \kappa C_2\right)\|v\|.$$

Hence, by choosing κ sufficiently small, we obtain $\|M(u)^* v\| \geq \|v\|/3$ for all v belonging to case 1. From now on, κ is fixed at this value.

Case 2: $\|v_2\| > \kappa\|v\|$:

Then we get by using $(v_1, v_2) \geq 0$ and (16.11)

$$\begin{aligned}\|v_2\|\|M(u)^* v\| &\geq (M(u)^* v, v_2) = (v, M(u)v_2) \\ &= (v, d_1(u)v_2) + (v, d_2(u)DF(u)v_2) \\ &= (v_1, v_2) + (v_2, DF(u)v_2) \\ &\geq \frac{\nu}{2}\|v_2\|^2.\end{aligned}$$

Since by assumption $\|v_2\| > \kappa\|v\| > 0$ we deduce

$$\|M(u)^* v\| \geq \frac{\nu}{2}\|v_2\| \geq \frac{\nu\kappa}{2}\|v\|.$$

Combining both cases we obtain (16.12) with $\gamma = \min\{1/3, \nu\kappa/2\}$.

Hence, $M(u)^* \in \mathcal{L}(L^2(\Omega), L^2(\Omega))$ is injective by (R′) (b) and (16.12). Moreover, (16.12) implies that $M(u)^*$ has closed range $\mathcal{R}(M(u)^*)$ in $L^2(\Omega)$. In fact, let $w \in \mathcal{R}(M(u)^*)^{cl}$ be arbitrary. Then there exists a sequence $(v_k) \subset L^2(\Omega)$ with $M(u)^* v_k \to w$ in $L^2(\Omega)$. Now we have by (16.12)

$$\|v_k - v_l\| \leq \frac{1}{\gamma}\|M(u)^* v_k - M(u)^* v_l\| \to 0 \quad \text{as } k, l \to \infty.$$

Hence, there is $v \in L^2(\Omega)$ with $v_k \xrightarrow{L^2} v$ and we obtain by continuity that $M(u)^* v = w$. Thus, $M(u)^*$ has closed range. Since $M(u)^*$ is injective, the closed range theorem implies that $M(u)$ is surjective, i.e. $\mathcal{R}(M(u)) = L^2(\Omega)$. To show that $M(u)$ is also injective consider $v \in L^2(\Omega)$ with $M(u)v = 0$, i.e.

$$D_1(u)v + D_2(u)DF(u)v = 0. \tag{16.14}$$

Define the measurable sets

$$\Omega_2 = \{x \in \Omega;\ d_2(u)(x) \neq 0\}, \quad \Omega_1 = \Omega \setminus \Omega_2.$$

For $w \in L^1(\Omega)$ we introduce the notation $w_{\Omega_i} = w\mathbf{1}_{\Omega_i}$, $i = 1, 2$. Since $d_i(u) \geq 0$, $i = 1, 2$, and $d_1(u) + d_2(u) \geq 1/2$, we get $d_1(u)_{\Omega_1} \neq 0$,

$d_2(u)_{\Omega_1} = 0$. Hence, (16.14) implies $v_{\Omega_1} = 0$. Since $y \stackrel{\text{def}}{=} DF(u)v \in L^2(\Omega)$, the measurable function defined by

$$w(x) = \begin{cases} d_2(u)(x)^{-1}v(x) & \text{if } x \in \Omega_2, \\ 0 & \text{else,} \end{cases}$$

satisfies $w \in L^2(\Omega)$. To see this, note that if $x \in \Omega_2$ and $d_2(u)(x) \leq 1/4$, then $d_1(u)(x) \geq 1/4$ and, by (16.14), $w(x) = -d_1(u)(x)^{-1}y(x)$.

Now $v_{\Omega_1} = 0$ implies $v = v_{\Omega_2}$ and yields with (16.11), (16.14) and the definition of w

$$\begin{aligned} 0 &= (w, M(u)v) = (w, d_1(u)v) + (wd_2(u), DF(u)v) \\ &= (w, d_1(u)v) + (v_{\Omega_2}, DF(u)v_{\Omega_2}) \geq \frac{\nu}{2}\|v_{\Omega_2}\|^2 = \frac{\nu}{2}\|v\|^2. \end{aligned}$$

Hence, $v = 0$, which proves the injectivity of $M(u)$. Thus, we have shown that $M(u) \in \mathcal{L}(L^2(\Omega), L^2(\Omega))$ is bijective and thus has a bounded inverse $M(u)^{-1} \in \mathcal{L}(L^2(\Omega), L^2(\Omega))$ by the open mapping theorem. We conclude the proof by deriving a bound for $\|M(u)^{-1}\|_{L^2,L^2}$. It is well known that also $(M(u)^*)^{-1} = (M(u)^{-1})^*$ exists and thus

$$\|M(u)^{-1}\|_{L^2,L^2} = \|(M(u)^*)^{-1}\|_{L^2,L^2} \leq \frac{1}{\gamma},$$

where the last inequality follows immediately from (16.12). The proof is complete. □

5 AN ELLIPTIC CONTROL PROBLEM

In this section we discuss the applicability of the algorithm to the following distributed control problem of an elliptic partial differential equation with upper bounds on the control:

$$\begin{aligned} &\text{minimize} \quad J(w) \stackrel{\text{def}}{=} \frac{1}{2}\|y(w) - y_d\|^2_{L^2(\Omega)} + \frac{\alpha}{2}\|w - w_d\|^2_{L^2(\Omega)} \\ &\text{subject to} \quad w \in \mathcal{F} \stackrel{\text{def}}{=} \{w \in L^2(\Omega);\ w \leq \psi \quad \text{on } \Omega\}, \end{aligned} \tag{16.15}$$

where $y = y(w) \in H_0^1(\Omega)$ is the weak solution of the elliptic state equation

$$Ey = w \quad \text{on } \Omega$$

with $Ey = -\sum_{i,j=1}^{n} \frac{\partial}{\partial x_i}\left(a_{ij}\frac{\partial y}{\partial x_j}\right)$.

Therefore, the aim is to compute a control w subject to the given constraints in such a way that the best possible matching of the corresponding state y with the "desired" state y_d is achieved. The second term in the objective function is a regularization with parameter $\alpha > 0$ to avoid ill-posedness.

We assume $a_{ij} \in L^\infty(\Omega)$ and that E is strongly elliptic, i.e., for some $C_E > 0$ holds

$$\sum_{i,j=1}^{n} a_{ij}\xi_i\xi_j \geq C_E\|\xi\|_2^2 \quad \forall \xi \in \mathbb{R}^n \quad \text{on } \Omega. \tag{16.16}$$

Furthermore, let $y_d \in L^2(\Omega)$ and $w_d, \psi \in L^\infty(\Omega)$.

The strong ellipticity (16.16) implies [25] that there exists $c_E > 0$ such that

$$\langle v, Ev\rangle_{H_0^1,H^{-1}} \geq c_E\|v\|_{H_0^1(\Omega)}^2 \quad \forall v \in H_0^1(\Omega).$$

The Lax-Milgram Lemma yields that $E \in \mathcal{L}(H_0^1(\Omega), H^{-1}(\Omega))$ is continuously invertible and that the same holds true for the adjoint E^*. The objective function

$$J(w) = \frac{1}{2}\|E^{-1}w - y_d\|_{L^2(\Omega)}^2 + \frac{\alpha}{2}\|w - w_d\|_{L^2(\Omega)}^2$$

is quadratic with gradient $\nabla J : L^2(\Omega) \to L^2(\Omega)$,

$$\nabla J(w) = (E^{-1})^*(E^{-1}w - y_d) + \alpha(w - w_d).$$

and strongly positive Hessian operator $D^2J : L^2(\Omega) \to \mathcal{L}(L^2(\Omega), L^2(\Omega))$,

$$D^2J(w) = (EE^*)^{-1} + \alpha I.$$

Since $\mathcal{F}$ is closed and convex, problem (16.15) possesses a unique solution $w \in L^2(\Omega)$ [8], which is characterized by the variational inequality

$$w \in \mathcal{F}, \quad (\nabla J(w), v - w)_{L^2(\Omega)} \geq 0 \quad \forall v \in \mathcal{F}. \tag{16.17}$$

Obviously, (16.17) holds if and only if

$$\nabla J(w) \leq 0, \quad w \leq \psi, \quad \nabla J(w)(w - \psi) = 0 \quad \text{on } \Omega. \tag{16.18}$$

From $E^{-1}, (E^{-1})^* \in \mathcal{L}(H^{-1}(\Omega), H_0^1(\Omega))$ it follows by standard embedding results [1] that $E^{-1}, (E^{-1})^* \in \mathcal{L}(L^2(\Omega), L^p(\Omega))$, where

$$p \begin{cases} \leq \infty & \text{if } n = 1, \\ < \infty & \text{if } n = 2, \\ \leq \frac{2n}{n-2} & \text{if } n \geq 3. \end{cases} \tag{16.19}$$

Now let $p > 2$ satisfy (16.19). For the case of Laplace's equation on a rectangular domain of dimension $n = 2$, which will be considered in our numerical experiments, it can be shown [15] that $E^{-1}, (E^{-1})^* \in \mathcal{L}(L^2(\Omega), H^2(\Omega))$, which allows to choose also $p = \infty$ in this case.

On $\Omega_I = \{x \in \Omega;\ w(x) \neq \psi(x)\}$ holds by (16.18) that $\nabla J(w)(x) = 0$ and thus

$$w(x) = \left(w_d - \alpha^{-1}(E^{-1})^*(E^{-1}w - y_d)\right)(x).$$

The function on the right is contained in $L^p(\Omega)$. Since on $\Omega \setminus \Omega_I$ holds $w(x) = \psi(x)$ and $\psi \in L^\infty(\Omega)$, we conclude that $w \in L^p(\Omega)$.

We set $q = p$, $r = 2$, and define $F : L^p(\Omega) \to L^q(\Omega)$,

$$F(u) \overset{\text{def}}{=} -\nabla J(\psi - u). \tag{16.20}$$

Then the unique solution of (16.15) satisfies $w \in L^p(\Omega)$ and, due to (16.18), is given by $w = \psi - \bar{u}$, where $\bar{u} \in L^p(\Omega)$ solves NCP with F as defined in (16.20). The operator

$$F : u \mapsto F(u) = ((EE^*)^{-1} + \alpha I)u + (E^{-1})^*(y_d - E^{-1}\psi) + \alpha(w_d - \psi)$$

is an affine linear endomorphism of $L^p(\Omega) = L^q(\Omega)$, which shows that Assumption (A) is satisfied with $DF(u) = D^2J(u) = (EE^*)^{-1} + \alpha I$. Since DF is constant and for all $v \in L^2(\Omega)$ holds

$$(v, DF(u)v)_{L^2(\Omega)} = \left\|E^{-1}v\right\|^2_{L^2(\Omega)} + \alpha\|v\|^2_{L^2(\Omega)} \geq \alpha\|v\|^2_{L^2(\Omega)},$$

we see that Assumption (R′) also holds. By Theorem 16.6, this implies Assumption (R). Also, F has the form $F(u) = G(u) + \alpha I$, where G is an affine linear bounded operator from $L^r(\Omega) = L^2(\Omega)$ to $L^p(\Omega)$. Hence, smoothing steps can be computed as described in section 3. Therefore, Algorithm 16.4 is applicable and converges locally q-superlinearly to the unique solution $\bar{u}$ according to Theorem 16.5.

6 NUMERICAL RESULTS

Finally, we present preliminary numerical results for the solution of the elliptic control problem described in section 5. Our test problem is taken from [3]:

$$\Omega = (0,1)^2 \subset \mathbb{R}^2, \quad E = -\Delta = -\left(\frac{\partial^2}{\partial x_1^2} + \frac{\partial^2}{\partial x_2^2}\right),$$

$$y_d(x_1, x_2) = \frac{1}{6}\sin(2\pi x_1)\sin(2\pi x_2)\exp(2x_1), \quad w_d \equiv 0 \equiv \psi.$$

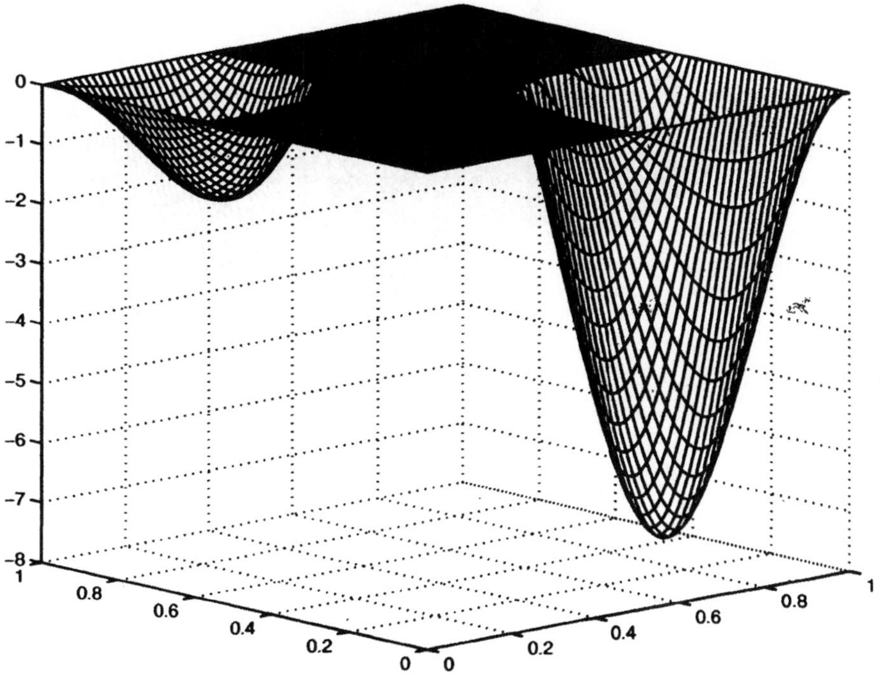

Figure 16.1 The resulting optimal control w.

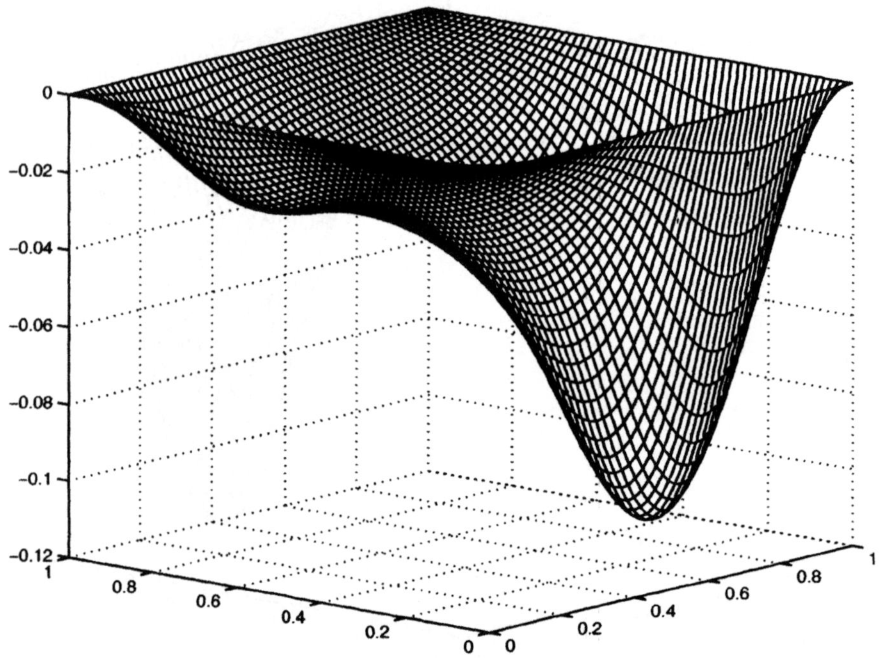

Figure 16.2 The resulting state y.

For the regularization, $\alpha = 10^{-3}$ was chosen. In accordance with the results in section 5, we worked with $p = q = \infty$ and $r = 2$. The initial point was $u_0 \equiv 1$, which corresponds to $w_0 \equiv -1$. For our numerical computations we worked with a trust-region globalization of Algorithm 16.4. It uses the merit function $\|\Phi(u)\|^2_{L^2(\Omega)}/2$, an affine-scaled fraction of Cauchy decrease, and an L^∞-trust-region, see [27, 29, 30]. Trial steps were obtained by computing projected Newton steps as described in Algorithm 16.4 with $\mathcal{B} = \mathcal{B}_+$. These steps were projected onto the trust-region. In our numerical tests these trial steps always satisfied the fraction of Cauchy decrease condition. The decision whether to perform a smoothing step or not was based on the L^∞/L^2-norm ratio of s_k and the smoothing step was implemented as described in section 3.

Table 16.1 Numerical results for three different mesh sizes.

	k	$\|\Phi(u_k)\|_{L^\infty}$	$\|u_k - \bar{u}\|_{L^\infty}$	$\dfrac{\|u_k^n - \bar{u}\|_{L^\infty}}{\|u_k^n - \bar{u}\|_{L^2}}$
	0	1.32e−02	6.42e+00	3.94e+00
	1	6.24e−03	5.42e+00	4.36e+00
N=20	2	3.87e−03	3.42e+00	5.89e+00
	3	1.18e−05	8.18e−03	4.18e+00
	4	4.87e−10	4.58e−07	1.22e+01
	5	2.17e−18	1.78e−15	
	0	1.31e−02	6.47e+00	4.07e+00
	1	6.28e−03	5.47e+00	4.55e+00
N=40	2	3.91e−03	3.47e+00	6.24e+00
	3	1.25e−05	7.93e−03	3.81e+00
	4	1.65e−09	1.62e−06	2.71e+01
	5	1.73e−18	1.78e−15	
	0	1.31e−02	6.49e+00	4.12e+00
	1	6.29e−03	5.49e+00	4.64e+00
N=80	2	3.92e−03	3.49e+00	6.37e+00
	3	1.30e−05	8.78e−03	4.03e+00
	4	3.92e−09	2.59e−06	4.13e+00
	5	8.29e−14	8.28e−11	

The discretization of the problem was done on an equidistant $(N \times N)$-grid. The Laplace operator was approximated by a 5-point finite difference operator. Table 16.1 shows the results for the three different mesh sizes $N = 20$, $N = 40$, and $N = 80$. Reported are the (discretized) residual $\|\Phi(u_k)\|_{L^\infty(\Omega)}$, the distance $\|u_k - \bar{u}\|_{L^\infty(\Omega)}$ from the solution, and the L^∞/L^2-norm ratio of $u_k^n - \bar{u}$, which indicates if a smoothing step is needed. The results in Table 16.1 confirm that the algorithm is very efficient with mesh independent performance. Only five iterations are needed for convergence to machine accuracy. On all meshes we observe a locally superlinear, even quadratic rate of convergence. The moderate norm ratios of $u_k^n - \bar{u}$ show that for this particular problem no smoothing steps are needed. The norm ratio of s_k proved to be a reliable indicator. In Figure 16.1 and 16.2 the optimal control and the corresponding state are displayed. Figure 16.1 shows nicely on which part of Ω the upper bound is active.

Acknowledgments

The author wishes to thank the two referees for their helpful suggestions.

References

[1] R. A. Adams. *Sobolev Spaces.* Academic Press, New York - San Francisco - London, 1975.

[2] J. Appell. The superposition operator in function spaces - A survey. *Expo. Math.*, 6:209–270, 1988.

[3] M. Bergounioux, M. Haddou, M. Hintermüller, and K. Kunisch. A comparison of interior point methods and a Moreau-Yosida based active set strategy for constrained optimal control problems. Preprint, 1998.

[4] F. H. Clarke. *Optimization and Nonsmooth Analysis.* John Wiley, New York, 1983.

[5] T. De Luca, F. Facchinei, and C. Kanzow. A semismooth equation approach to the solution of nonlinear complementarity problems. *Math. Program.*, 75:407–439, 1996.

[6] J. C. Dunn. On L^2 sufficient conditions and the gradient projection method for optimal control problems. *SIAM J. Control Optimization*, 34:1270–1290, 1996.

[7] G. Duvaut and J. L. Lions. *Inequalities in Mechanics and Physics.* Springer-Verlag, Berlin - Heidelberg - New York, 1976.

[8] I. Ekeland and R. Temam. *Convex Analysis and Variational Problems.* North Holland, Amsterdam, 1976.

[9] F. Facchinei and C. Kanzow. A nonsmooth inexact Newton method for the solution of large-scale nonlinear complementarity problems. *Math. Program.*, 76:493–512, 1997.

[10] F. Facchinei and J. Soares. A new merit function for nonlinear complementarity problems and a related algorithm. *SIAM J. Optim.*, 7:225–247, 1997.

[11] M. C. Ferris and J.-S. Pang. Engineering and economic applications of complementarity problems. *SIAM Rev.*, 39:669–713, 1997.

[12] A. Fischer. A special Newton-type optimization method. *Optimization*, 24:269–284, 1992.

[13] A. Fischer. Solution of monotone complementarity problems with locally Lipschitzian functions. *Math. Program.*, 76B:513–532, 1997.

[14] R. Glowinski, J. L. Lions, and R. Trémolières. *Analyse Numérique des Inéquations Variationnelles, Tome 1: Théorie Générale Premières Applications, Tome 2: Applications aux Phénomènes Stationnaires et d'Évolution.* Dunod, Paris, 1976.

[15] P. Grisvard. *Elliptic Problems in Nonsmooth Domains.* Pitman Publishing Inc., Boston - London - Melbourne, 1985.

[16] S. Ito, C. T. Kelley, and E. W. Sachs. Inexact primal-dual interior point iteration for linear programs in function spaces. *Comput. Optim. Appl.*, 4:189–201, 1995.

[17] H. Jiang and L. Qi. A new nonsmooth equations approach to nonlinear complementarity problems. *SIAM J. Control Optimization*, 35:178–193, 1997.

[18] C. T. Kelley and E. W. Sachs. Multilevel algorithms for constrained compact fixed point problems. *SIAM J. Sci. Comput.*, 15:645–667, 1994.

[19] B. Kummer. Newton's method for nondifferentiable functions. In J. Guddat et al. (eds.), *Advances in Mathematical Optimization*, 114–125, Akademie-Verlag, Berlin, 1988.

[20] P. D. Panagiotopoulos. *Inequality Problems in Mechanics and Applications - Convex and Nonconvex Energy Functions.* Birkhäuser, Boston - Basel - Stuttgart, 1985.

[21] J.-S. Pang and L. Qi. Nonsmooth equations: Motivation and algorithms. *SIAM J. Optim.*, 3:443–465, 1993.

[22] L. Qi. Convergence analysis of some algorithms for solving nonsmooth equations. *Math. Oper. Res.*, 18:227–244, 1993.

[23] L. Qi. C-differentiability, C-differential operators and generalized Newton methods. Tech. Report, School of Mathematics, The University of New South Wales, Sydney, Australia, 1996.

[24] L. Qi and J. Sun. A nonsmooth version of Newton's method. *Math. Program.*, 58:353–367, 1993.

[25] M. Renardy and R. C. Rogers. *An Introduction to Partial Differential Equations*. Springer-Verlag, 1993.

[26] T. Tian and J. C. Dunn. On the gradient projection method for optimal control problems with nonnegative L^2 inputs. *SIAM J. Control Optimization*, 32:516–537, 1994.

[27] M. Ulbrich. Non-monotone trust-region methods for bound constrained semismooth equations with applications to nonlinear mixed complementarity problems. Tech. Report TUM M9906, Fakultät für Mathematik, Technische Universität München, 1999. Accepted for publication (in revised form) in SIAM J. Optimization.

[28] M. Ulbrich. Semismooth Newton methods for operator equations in function spaces. Technical Report TR00-11, Department of Computational and Applied Mathematics, Rice University, Houston, Texas 77005-1892, 2000.

[29] M. Ulbrich and S. Ulbrich. Superlinear convergence of affine-scaling interior-point Newton methods for infinite-dimensional nonlinear problems with pointwise bounds. To appear in SIAM J. Control Optimization, 2000.

[30] M. Ulbrich, S. Ulbrich, and M. Heinkenschloss. Global convergence of trust-region interior-point algorithms for infinite-dimensional nonconvex minimization subject to pointwise bounds. *SIAM J. Control Optimization*, 37:731–764, 1999.

[31] M. M. Vainberg. *Variational Methods for the Study of Nonlinear Operators*. Holden-Day, San Francisco - London - Amsterdam, 1964.

Chapter 17

THE PROXIMAL POINT ALGORITHM FOR THE P_0 COMPLEMENTARITY PROBLEM

Nobuo Yamashita
Department of Applied Mathematics and Physics, Graduate School of Informatics, Kyoto University, Kyoto 606-8501, Japan
nobuo@amp.i.kyoto-u.ac.jp

Junji Imai
Toyota Central R&D Labs., Inc., 41-1 Yokomichi, Nagakute, Nagakute-cho, Aichi-gun, Aichi 480-1192, Japan
imai@mcl.tytlabs.co.jp

Masao Fukushima
Department of Applied Mathematics and Physics, Graduate School of Informatics, Kyoto University, Kyoto 606-8501, Japan
fuku@amp.i.kyoto-u.ac.jp

Abstract In this paper we consider a proximal point algorithm (PPA) for solving the nonlinear complementarity problem (NCP) with a P_0 function. PPA was originally proposed by Martinet and further developed by Rockafellar for monotone variational inequalities and monotone operator problems. PPA is known to have nice convergence properties under mild conditions. However, until now, it has been applied mainly to monotone problems. In this paper, we propose a PPA for the NCP involving a P_0 function and establish its global convergence under appropriate conditions by using the Mountain Pass Theorem. Moreover, we give conditions under which it has a superlinear rate of convergence.

Keywords: nonlinear complementarity problem, proximal point algorithm, P_0 function

M.C. Ferris et al. (eds.), Complementarity: Applications, Algorithms and Extensions, 361–379.

1 INTRODUCTION

In this paper we consider the following nonlinear complementarity problem NCP(F) [8, 12]: Find an $x \in \boldsymbol{R}^n$ such that

$$x \geq 0, \; F(x) \geq 0, \; x^T F(x) = 0, \tag{17.1}$$

where F is a mapping from $\boldsymbol{R}^n$ into $\boldsymbol{R}^n$. We assume that F is a continuously differentiable function throughout this paper. Applications of NCP can be found in many important fields such as mathematical programming, economics, engineering and mechanics [1, 8].

A variety of methods for solving the NCP have been proposed so far. One of the solution methods for solving monotone NCP is the proximal point algorithm (PPA) proposed by Martinet [10] and further studied by Rockafellar [15]. PPA is a procedure for finding a vector z satisfying $0 \in T(z)$, where T is a maximal monotone operator. So PPA can be applied to a broad class of problems such as convex programming problems, monotone variational inequality problems, and monotone NCP. Furthermore, it is known that it has nice convergence properties under mild conditions [15, 18]. However, it has been applied mainly to monotone problems, and, to the authors' knowledge, has not been discussed for nonmonotone NCP. For NCP(F), given the current point x^k, PPA produces the next iterate by approximately solving the subproblem

$$x \geq 0, \; F^k(x) \equiv F(x) + c_k(x - x^k) \geq 0, \; x^T F^k(x) = 0, \tag{17.2}$$

where $c_k > 0$. If F is a P_0 function, then F^k is a P function. Moreover, the subproblem NCP(F^k) always has a unique solution, so that PPA is well defined. The main purpose of this paper is to apply PPA to NCP(F) with F being a P_0 function. Unfortunately, to show global convergence for NCP with a P_0 function, we cannot apply the techniques of [15] that rely on the properties of monotone operators. Here we adopt the technique used by Facchinei and Kanzow [4] for a regularization method that solves a sequence of complementarity problems

$$x \geq 0, \; \hat{F}^k(x) \equiv F(x) + \hat{c}_k x \geq 0, \; x^T \hat{F}^k(x) = 0, \tag{17.3}$$

where $\hat{c}_k$ is a positive parameter converging to 0 as $k \to \infty$. (Note that, in PPA, it is not necessary to assume that $c_k \to 0$ as $k \to \infty$. See Remark 17.12 below.) By using the Mountain Pass Theorem, they showed that the regularization method converges globally if F is a P_0 function and the solution set of NCP(F) is nonempty and bounded. Using a similar technique, we will show that PPA also has a global convergence property under the same assumptions. We also discuss superlinear convergence of PPA.

This paper is organized as follows. In Section 2, we review some definitions and preliminary results that will be used in the subsequent analysis. In Section 3, we propose a prototype of PPA for solving NCP(F) with P_0 function, and establish global and superlinear convergence theorems. In Section 4, we discuss how the algorithm can be implemented so as to attain superlinear convergence in a genuine sense.

2 PRELIMINARIES

In this section we review some mathematical concepts used in this paper.

2.1 BASIC CONCEPTS

We first show some definitions related to the mapping F.

Definition 17.1 *The function $F : \boldsymbol{R}^n \to \boldsymbol{R}^n$ is called a*
(a) P_0 function if, for all $x, y \in \boldsymbol{R}^n$ with $x \neq y$, there is an index i such that

$$x_i \neq y_i \text{ and } (x_i - y_i)[F_i(x) - F_i(y)] \geq 0;$$

(b) P function if, for all $x, y \in \boldsymbol{R}^n$ with $x \neq y$, there is an index i such that

$$(x_i - y_i)[F_i(x) - F_i(y)] > 0.$$

Note that if F is monotone, then F is a P_0 function.

For a given point $x^k \in \boldsymbol{R}^n$ and a given constant $c_k > 0$, we define the function $F^k : \boldsymbol{R}^n \to \boldsymbol{R}^n$ by

$$F^k(x) \equiv F(x) + c_k(x - x^k). \tag{17.4}$$

The following lemma shows that F^k is a regularization of F.

Lemma 17.2 *[4, Lemma 3.2] If F is a P_0 function, then the function F^k defined by (17.4) is a P function.*

Next we recall some concepts of generalized differentials. Let $G : \boldsymbol{R}^n \to \boldsymbol{R}^m$ be locally Lipschitzian. Then G is differentiable almost everywhere. Let $D_G \subseteq \boldsymbol{R}^n$ be the set where G is differentiable. Then the B subdifferential of G at x is defined as

$$\partial_B G(x) \equiv \{V \in \boldsymbol{R}^{m \times n} \mid \exists \{x^k\} \subseteq D_G \text{ such that } \lim_{x^k \to x} G'(x^k) = V\},$$

and the Clarke subdifferential of G at x is defined by

$$\partial G(x) \equiv \operatorname{co} \partial_B G(x),$$

where co denotes the convex hull of a set.

We recall the notion of semismoothness that plays an important role in the analysis of the generalized Newton method.

Definition 17.3 *Let $G : \boldsymbol{R}^n \to \boldsymbol{R}^m$ be locally Lipschitzian. Then G is said to be semismooth at x if*

$$\lim_{\substack{V \in \partial G(x+td') \\ d' \to d, t \downarrow 0}} Vd'$$

exists for all $d \in \boldsymbol{R}^n$. Moreover, G is said to be strongly semismooth at x if for any $V \in \partial G(x+d)$ and any $d \to 0$,

$$Vd - G'(x; d) = O(\|d\|^2),$$

where $G'(x;d)$ denotes the directional derivative of G at x along direction d.

Many useful functions such as min and max functions are not only B-differentiable but also semismooth. However, there are no exact calculus rules for B-derivatives of general functions. Recently, Qi [14] proposed the following concept of differentials that have exact calculus rules.

Definition 17.4 *Suppose that $G : \boldsymbol{R}^n \to \boldsymbol{R}^m$ is a continuous vector function and $T : \boldsymbol{R}^n \to \boldsymbol{R}^m \times \boldsymbol{R}^n$ is a set-valued operator. Then G is said to be C-differentiable at $x \in \boldsymbol{R}^n$ with T, and T is called a C-differential operator of G if*

i) *$T(y)$ is nonempty and compact for any y in a neighborhood of x;*

ii) *T is upper semicontinuous at x;*

iii) *for any $V \in T(x+d)$,*

$$F(x+d) = F(x) + Vd + o(\|d\|).$$

If furthermore

iv) *for any $V \in T(x+d)$,*

$$G(x+d) = G(x) + Vd + O(\|d\|^2),$$

then F is said to be strongly C-differentiable at x with T and T is called a strong C-differential operator of G.

Note that the above definition does not assume F to be locally Lipschitz. For C-differential operators, Qi [14] showed the following property, which will be used in the subsequent analysis. For other interesting properties of C-differential operators, see [14].

Proposition 17.5 *Let $F : \mathbf{R}^n \to \mathbf{R}^l$ and $G : \mathbf{R}^l \to \mathbf{R}^m$ be Clarke subdifferentiable and strongly semismooth. Then $H \equiv G \circ F : \mathbf{R}^n \to \mathbf{R}^m$ is strongly C-differentiable with the operator T defined by*

$$T(x) \equiv \{AB \in \mathbf{R}^{m \times n} \mid A \in \partial G(F(x)), B \in \partial F(x)\}.$$

That is, there exists a positive constant κ such that, for all $d \in \mathbf{R}^n$ sufficiently small and any $V \in T(x+d)$,

$$\|H(x+d) - H(x) - Vd\| \leq \kappa \|d\|^2.$$

Proof Directly follows from Proposition 2.2 and Theorem 2.1 in [14]. □

The following definition is concerned with the regularity of B subdifferentials of a mapping.

Definition 17.6 *Let $G : \mathbf{R}^n \to \mathbf{R}^n$ be a locally Lipschitzian function. Then a point $x \in \mathbf{R}^n$ is BD-regular with G if all the elements in $\partial_B G(x)$ are nonsingular.*

It is well known that x^* is an isolated solution of the equation $G(x) = 0$ if x^* is BD-regular with G and $G(x^*) = 0$.

2.2 EQUIVALENT REFORMULATION

We consider the reformulation of NCP(F) using the Fischer-Burmeister function

$$\psi(a, b) \equiv \sqrt{a^2 + b^2} - a - b.$$

The most fundamental property of this function is that

$$\psi(a, b) = 0 \iff a \geq 0, \ b \geq 0, \ ab = 0.$$

Note that ψ is continuously differentiable everywhere except at the origin. This function was introduced by Fischer [6] and has played a central role in designing several nonsmooth Newton-type methods for the solution of NCP(F) [2, 5]. By using this function, we obtain the following system of equations equivalent to NCP(F):

$$H(x) \equiv \begin{pmatrix} \psi(x_1, F_1(x)) \\ \vdots \\ \psi(x_n, F_n(x)) \end{pmatrix} = 0.$$

Note that H is not differentiable at x if $x_i = F_i(x) = 0$ for some i. With this function, we define a merit function $\Phi : \boldsymbol{R}^n \to \boldsymbol{R}$ for NCP(F) by

$$\Phi(x) \equiv \frac{1}{2}\|H(x)\|^2.$$

This function can be used to determine a step size in an iterative method for solving the equation $H(x) = 0$ and turns out to be important in establishing global convergence of the method. For subproblems NCP(F^k) in PPA, we also define the corresponding functions by

$$H^k(x) \equiv \begin{pmatrix} \psi(x_1, F_1^k(x)) \\ \vdots \\ \psi(x_n, F_n^k(x)) \end{pmatrix}$$

and

$$\Phi^k(x) \equiv \frac{1}{2}\|H^k(x)\|^2.$$

We summarize some useful properties of these functions.

Proposition 17.7 *[2, 5, 9, 17] The following statements hold:*

(a) H is semismooth everywhere. Furthermore, if ∇F is Lipschitz continuous, then H is strongly semismooth everywhere.

(b) $x^* \in \boldsymbol{R}^n$ solves NCP(F) (resp. NCP(F^k)) if and only if x^* solves the nonlinear equation $H(x) = 0$ (resp. $H^k(x) = 0$).

(c) The merit functions Φ and Φ^k are continuously differentiable on the whole space $\boldsymbol{R}^n$.

(d) If F is a P_0 function, then every stationary point of Φ (resp. Φ^k) is a solution of NCP(F) (resp. NCP(F^k)).

Proposition 17.8 *The merit function Φ^k is coercive, i.e.,*

$$\lim_{\|x\| \to \infty} \Phi^k(x) = +\infty.$$

Proof This can be shown in a way similar to [4, Proposition 3.4]. □

Proposition 17.8 guarantees that the sequence generated by a descent method to minimize Φ^k is bounded.

The following lemma will be useful in analyzing the rate of convergence of the PPA proposed in this paper.

Lemma 17.9 *For any a, b and c, we have*

$$|\psi(a, b+c) - \psi(a, b)| \leq 2|c|. \tag{17.5}$$

Moreover, we have

$$\|H^k(x) - H(x)\| \leq 2c_k \|x - x^k\|. \tag{17.6}$$

Proof Since

$$\psi(a, b+c) - \psi(a, b) = \sqrt{a^2 + (b+c)^2} - \sqrt{a^2 + b^2} - c,$$

it suffices to show

$$|\sqrt{a^2 + (b+c)^2} - \sqrt{a^2 + b^2}| \leq |c|.$$

In fact, this inequality follows from the fact that $\|(a, b)^T\| = \sqrt{a^2 + b^2}$, $\|(a, b+c)^T\| = \sqrt{a^2 + (b+c)^2}$, $\|(0, c)^T\| = |c|$, and the triangle inequality. Moreover, the inequality (17.6) directly follows from (17.5) □

3 PROTOTYPE ALGORITHM AND ITS CONVERGENCE

In this section, we propose a prototype of PPA for solving NCP(F) with P_0 function and show its global convergence.

Algorithm PPA-0 (Prototype)

Step 0: Choose parameters $c_0 > 0, \delta_0 \in (0, 1)$ and $x^0 \in \boldsymbol{R}^n$. Set $k := 0$.

Step 1: If x^k satisfies a termination condition, then stop.

Step 2: Let $F^k : \boldsymbol{R}^n \to \boldsymbol{R}^n$ be defined by (17.4) and obtain an approximate solution x^{k+1} of NCP(F^k) such that

$$\Phi^k(x^{k+1})^{\frac{1}{2}} \leq \delta_k \min\{1, \|x^{k+1} - x^k\|\}. \tag{17.7}$$

Step 3: Choose $c_{k+1} \in (0, c_k)$ and $\delta_{k+1} \in (0, \delta_k)$. Set $k := k + 1$ and go to Step 1.

Remark 17.10 *Condition (17.7) is reminiscent of the criterion*

$$\|x^{k+1} - \bar{x}^k\| \leq \delta_k \min\{1, \|x^{k+1} - x^k\|\},$$

where $\bar{x}^k$ is a solution of NCP(F^k). This criterion with $\delta_k \to 0$ was used by Rockafellar [15] to show that PPA for monotone operators has global and superlinear convergence.

Now, we give conditions under which the algorithm globally converges to a solution of NCP(F). To this end, we need the following assumption on parameters c_k.

Assumption 17.11 *The sequence $\{c_k\}$ satisfies the following conditions:*

(a) $c_k(x^{k+1} - x^k) \to 0$ *if $\{x^k\}$ is bounded.*

(b) $c_k x^k \to 0$ *if $\{x^k\}$ is unbounded.*

In the next section, we will show how the algorithm is implemented so that the conditions in Assumption 17.11 can be realized.

Remark 17.12 *When $\{x^k\}$ is bounded, we do not need to let $c_k \to 0$ for (a) to hold. If $\{c_k\}$ is bounded away from 0, then subproblems NCP(F^k) will remain well-conditioned numerically. On the other hand, the regularization method of [4] requires $\hat{c}_k \to 0$, which may deteriorate the tractability of subproblem (17.3) eventually.*

Remark 17.13 *When $\{x^k\}$ is unbounded, we have eventually*

$$F^k(x) = F(x) + c_k(x - x^k) \approx F(x) + c_k x.$$

In this case, the algorithm essentially reduces to the regularization method of [4]. Therefore, we can use the same technique as [4] to show global convergence of PPA-0.

Under the above assumption, we have the following lemma.

Lemma 17.14 *Suppose that Assumption 17.11 holds. Let $S \subset \boldsymbol{R}^n$ be an arbitrary compact set. If $\{x^k\}$ is unbounded, then, for any $\delta > 0$, there exists a sufficiently large $\bar{k}$ such that for all $k \geq \bar{k}$*

$$|\Phi^k(x) - \Phi(x)| \leq \delta \quad \textit{for all} \quad x \in S.$$

Proof By the definition of F^k, we have

$$\|F^k(x) - F(x)\| = \|c_k(x - x^k)\|.$$

Suppose $\{x^k\}$ is unbounded. It then follows from the assumption on $\{c_k\}$ that $c_k \to 0$ and $\|F^k(x) - F(x)\| \to 0$ uniformly as $k \to \infty$ on any compact set S. Moreover, using Lemma 17.9, it is not difficult to see that $\{\Phi^k\}$ also converges to Φ uniformly on S. This completes the proof.
□

The next lemma, which is derived from the Mountain Pass Theorem [11, Theorem 9.2.7], will be used to establish a global convergence theorem for the proposed algorithm.

Lemma 17.15 *Let $f : \mathbf{R}^n \to \mathbf{R}$ be continuously differentiable and coercive, and S be a nonempty compact set in $\mathbf{R}^n$. Let*

$$m := \min_{x \in \partial S} f(x).$$

where ∂S denotes the boundary of S. Suppose that there are two points $a \in S$ and $b \notin S$ such that $f(a) < m$ and $f(b) < m$. Then there exists a point c such that $\nabla f(c) = 0$ and $f(c) \geq m$.

By using these lemmas, global convergence of PPA-0 for NCP(F) with P_0 function can be shown in a way similar to [4, Theorem 5.4]. For completeness, we give a proof.

Theorem 17.16 *Suppose that F is a P_0 function and the solution set of NCP(F) is nonempty and bounded. Suppose also that Assumption 17.11 holds. If $\delta_k \to 0$, then $\{x^k\}$ is bounded and any accumulation point of $\{x^k\}$ is a solution of NCP(F).*

Proof First we show that $\{x^k\}$ is a bounded sequence. Suppose the sequence $\{x^k\}$ is not bounded. Then there exists a subsequence $\{x^k\}_K$ such that $\|x^k\| \to \infty$ as $k \to \infty$ with $k \in K$. Since the solution set of NCP(F) is bounded, there exists a compact set $S \subset \mathbf{R}^n$ such that its interior contains the solution set and

$$x^k \notin S \tag{17.8}$$

for all $k \in K$ sufficiently large. If x^* is a solution of NCP(F), then we have $\Phi(x^*) = 0$. Let

$$\bar{\nu} := \min_{x \in \partial S} \Phi(x) > 0.$$

Applying Lemma 17.14 with $\delta := \bar{\nu}/4$, there exists a $\bar{k}$ such that for all $k \geq \bar{k}$

$$\Phi^k(x^*) \leq \frac{1}{4}\bar{\nu}$$

and

$$\min_{x \in \partial S} \Phi^k(x) \geq \frac{3}{4}\bar{\nu}.$$

Since $\Phi^k(x^{k+1}) \leq \delta_k^2$ by Step 2 of PPA-0 and $\delta_k \to 0$, there exists a $\tilde{k}$ such that for all $k \geq \tilde{k}$

$$\Phi^k(x^{k+1}) \leq \frac{1}{4}\bar{\nu}. \tag{17.9}$$

Now consider a fixed index $k \geq \max\{\bar{k}, \tilde{k}\}$ with $k \in K$ such that (17.8) and (17.9) are satisfied. Letting $a := x^*$ and $b := x^k$, we have from the Mountain Pass Theorem (Lemma 17.15) that there exists a vector $c \in \boldsymbol{R}^n$ such that

$$\nabla \Phi^k(c) = 0 \text{ and } \Phi^k(c) \geq \frac{3}{4}\bar{\nu} > 0.$$

Therefore c is a stationary point of Φ^k, which does not minimize Φ^k globally. However this contradicts Proposition 17.7(d) saying that any stationary point of Φ^k must be a global minimizer of Φ^k. Hence $\{x^k\}$ is bounded.

Next, we show that any accumulation point of $\{x^k\}$ is a solution of NCP(F). Since $\{x^k\}$ is bounded, we have $\|F^k(x^{k+1}) - F(x^{k+1})\| \to 0$ by Assumption 17.11, and hence $|\Phi^k(x^{k+1}) - \Phi(x^{k+1})| \to 0$. By condition (17.7) in Step 3 and the assumption on $\{\delta_k\}$, we have $\Phi^k(x^{k+1}) \to 0$. Consequently we have $\Phi(x^{k+1}) \to 0$, which means that every accumulation point of the sequence $\{x^k\}$ is a solution of NCP(F). □

When F is a monotone function, we can prove global convergence of PPA-0 without the boundedness assumption on the solution set of NCP(F).

Theorem 17.17 *Let F be a Lipschitz continuous monotone function and $\{c_k\}$ be bounded. Assume NCP(F) has at least one solution. Then the sequence $\{x^k\}$ generated by PPA-0 converges to a solution of NCP(F).*

Proof The theorem can be proven in a similar way to [18]. □

Remark 17.18 *The above theorem assumes the Lipschitz continuity of F, which is somewhat restrictive. We can remove this assumption by introducing an additional criterion for approximate solution of NCP(F^k) as in the PPA proposed in [18].*

Remark 17.19 *Theorem 17.17 does not require Assumption 17.11. This is because the generated sequence converges to a solution when F is monotone and the solution set is nonempty.*

Next we proceed to establishing the superlinear rate of convergence of PPA-0. To this end, we need the following assumption.

Assumption 17.20 *The sequence $\{x^k\}$ converges to a solution x^* of NCP(F) that is BD-regular with H.*

Since F is assumed to be a P_0 function, the solution set of NCP(F) is connected whenever it is bounded [3, 7]. In this case, Assumption 17.20

ensures that x^* is the unique solution of NCP(F). Moreover, Assumption 17.20 yields the regularity of $\partial_B H^k(x^k)$ as shown in the following lemma, which is useful in analyzing convergence of the generalized Newton method.

Lemma 17.21 *Under Assumption 17.20, there exists a $C > 0$ such that for all k sufficiently large*

$$\|V_k^{-1}\| \leq C,$$

where $V_k \in \partial_B H^k(x^k)$.

Proof First we show that there exists a sequence $\{N^k\}$ of neighborhoods of $\{x^k\}$ such that if H^k is differentiable at $x \in N^k$, then H is differentiable at x. Let index set $J(x)$ be defined by

$$J(x) \equiv \{j \mid (x_j, F_j(x)) \neq (0,0)\}.$$

Now let N^k be an open ball centered at x^k such that $J(x^k) \subseteq J(x)$ for all $x \in N^k$. Since F is continuous, N^k is nonempty. If H is not differentiable at $x \in N^k$, then there exists $j \notin J(x)$. By the definition of N^k, $j \notin J(x^k)$. Suppose that H_j^k is differentiable x, that is, $(x_j, F_j(x) + c_k(x_j - x_j^k)) \neq (0,0)$. If $x_j \neq 0$, then $j \in J(x)$, which is a contradiction. Therefore we have $F_j(x) + c_k(x_j - x_j^k) \neq 0$. Since $x_j = 0$ and $x_j^k = 0$ by the fact that $j \notin J(x^k)$, we have $F_j(x) \neq 0$, which contradicts that $j \notin J(x)$. Hence, H^k is not differentiable at x. Consequently, for the sequence $\{N_k\}$ of neighborhoods of $\{x^k\}$, if H^k is differentiable at $x \in N^k$, then H is also differentiable at x.

Now suppose that $\|V_k^{-1}\| \to \infty$. By the definition of $\partial_B H^k$, there exists a sequence $\{\hat{x}^k\}$ such that $\hat{x}^k \in N^k$, $\hat{x}^k \to x^*$, H^k is differentiable at $\hat{x}^k$, and $\|\nabla H^k(\hat{x}^k)^T - V_k\| \to 0$. Since $\hat{x}^k \in N^k$ and H^k is differentiable at $\hat{x}^k$, H is differentiable at $\hat{x}^k$. Furthermore, since $\hat{x}^k \to x^*$ and $\partial_B H(x^*)$ is nonempty and bounded, $\{\nabla H(\hat{x}^k)^T\}$ is bounded and its accumulation point V^* belongs to $\partial_B H(x^*)$. By the BD-regularity of x^* with H, V^* is nonsingular. Hence, $\{V_k\}$ is also bounded and its accumulation point is a nonsingular matrix. This is a contradiction. □

Another consequence of Assumption 17.20 is the following lemma.

Lemma 17.22 *Let Assumption 17.20 hold. Then there exist positive constants κ and ε such that*

$$\|x - x^*\| \leq \kappa \|H(x)\| \quad \text{for all } x \text{ such that } \|H(x)\| < \varepsilon.$$

Proof Directly follows from [13, Proposition 3]. □

Now, we have the following superlinear convergence result for PPA-0.

Theorem 17.23 *Suppose that Assumption 17.20 holds and the solution set of NCP(F) is bounded. Suppose also that $\delta_k \to 0$ and $c_k \to 0$. Then the sequence $\{x^k\}$ generated by PPA-0 converges superlinearly to the solution x^* of NCP(F).*

Proof As mentioned after Assumption 17.20, under the given assumptions, x^* is the unique solution of NCP(F), and hence $\{x^k\}$ converges to x^* by Theorem 17.16. Moreover, by (17.6) in Lemma 17.9, we have

$$\|H^k(x^{k+1}) - H(x^{k+1})\| \leq 2c_k\|x^{k+1} - x^k\|.$$

It follows from the triangle inequality that

$$\|H(x^{k+1})\| \leq \|H^k(x^{k+1})\| + 2c_k\|x^{k+1} - x^k\|,$$

that is,

$$\Phi(x^{k+1})^{\frac{1}{2}} \leq \Phi^k(x^{k+1})^{\frac{1}{2}} + \sqrt{2}c_k\|x^{k+1} - x^k\|. \tag{17.10}$$

By Lemma 17.22, there exists a positive constant κ such that, for all k sufficiently large

$$\|x^{k+1} - x^*\| \leq \kappa\Phi(x^{k+1})^{\frac{1}{2}}. \tag{17.11}$$

Moreover, by criterion (17.7) in Step 2, we have

$$\Phi^k(x^{k+1})^{\frac{1}{2}} \leq \delta_k\|x^{k+1} - x^k\|. \tag{17.12}$$

Consequently, by (17.10)-(17.12), we have

$$\begin{aligned}
\|x^{k+1} - x^*\| &\leq \kappa\Phi^k(x^{k+1})^{\frac{1}{2}} + \sqrt{2}\kappa c_k\|x^k - x^{k+1}\| \\
&\leq \kappa(\delta_k + \sqrt{2}c_k)\|x^k - x^{k+1}\| \\
&\leq \kappa(\delta_k + \sqrt{2}c_k)(\|x^k - x^*\| + \|x^{k+1} - x^*\|).
\end{aligned}$$

Rearranging terms in this inequality, we have

$$\|x^{k+1} - x^*\| \leq \frac{\kappa(\delta_k + \sqrt{2}c_k)}{1 - \kappa(\delta_k + \sqrt{2}c_k)}\|x^k - x^*\|.$$

Since $\delta_k \to 0$ and $c_k \to 0$, we have the desired property. □

4 IMPLEMENTATION ISSUES

In the prototype of PPA presented in the previous section, no rules are specified for updating parameters and no procedure is given for solving subproblems. In this section we discuss implementation issues for the algorithm. In particular we consider how to update parameters c_k and δ_k and how to solve subproblems NCP(F^k).

4.1 HOW TO CHOOSE THE PARAMETERS

First, we consider the updating rule of $\{c_k\}$. Note that Theorem 17.16, which shows global convergence of PPA-0, relies on Assumption 17.11, that is,

(a) $c_k(x^{k+1} - x^k) \to 0$ if $\{x^k\}$ is bounded;

(b) $c_k x^k \to 0$ if $\{x^k\}$ is unbounded.

Taking into account these conditions, we determine c_k by

$$c_k = \min\{1, \frac{1}{\|x^k\|^2}\} \min\{(\beta)^k, \Phi(x^k)\}, \tag{17.13}$$

where $\beta \in (0,1)$ is a given constant. The term $\min\{1, \frac{1}{\|x^k\|^2}\}$ forces $\{c_k\}$ to satisfy (b), while the term $\min\{(\beta)^k, \Phi(x^k)\}$ let $\{c_k\}$ converge to 0. In particular, the presence of $\Phi(x^k)$ in the latter term plays a crucial role in establishing superlinear convergence in a genuine sense, as will be seen later.

As to parameter δ_k, it suffice to let δ_k converge to 0. Thus we simply determine δ_k by

$$\delta_k = (\beta)^k.$$

4.2 HOW TO SOLVE SUBPROBLEMS

For solving subproblem NCP(F^k), we adopt the generalized Newton method proposed by De Luca, Facchinei and Kanzow [2].

Generalized Newton method for NCP(F^k).

Step 0: Let $z^0 := x^k$, and choose $\rho \in (0, \frac{1}{2})$. Set $j := 0$.

Step 1: Select an element $V_j \in \partial_B H^k(z^j)$. Find a solution d^j of the linear equation

$$V_j d = -H^k(z^j). \tag{17.14}$$

Step 2: If criterion (17.7) is satisfied with $z^j + d^j$ substituted for x^{k+1}, let $x^{k+1} := z^j + d^j$ and terminate the iteration. Otherwise, find the smallest nonnegative integer l such that

$$\Phi^k(z^j + 2^{-l}d^j) \leq \Phi^k(z^j) + \rho 2^{-l} \nabla \Phi^k(z^j)^T d^j.$$

Set $z^{j+1} := z^j + 2^{-l}d^j$ and go to Step 3.

Step 3: If criterion (17.7) is satisfied with z^{j+1} substituted for x^{k+1}, let $x^{k+1} := z^{j+1}$ and terminate the iteration. Otherwise, set $j := j+1$, and go to Step 1.

Since F^k is a P function, the Jacobian matrices of F^k are P matrices. Hence all elements of $\partial H^k(x)$ are nonsingular [2]. Therefore the equation (17.14) in Step 1 is solvable for all j. Moreover, we have the following theorem.

Theorem 17.24 *For each k, the generalized Newton method finds an approximate solution of NCP(F^k) that satisfies (17.7) in a finite number of iterations.*

Proof It can be proven by using Proposition 17.8 and [2, Theorem 11]. □

4.3 SPECIFIC PPA

By incorporating the above-mentioned ingredients in the algorithm, we propose a specific version of PPA-0 for NCP(F) with P_0 function.

Algorithm PPA-1 (Specific PPA)

Step 1: Choose a constant $\beta \in (0,1)$ and $x^0 \in \boldsymbol{R}^n$. Set

$$c_0 := \min\{1, \frac{1}{\|x^0\|^2}\} \min\{1, \Phi(x^0)\}$$

and $k := 0$.

Step 2: If x^k satisfies a termination condition, then stop.

Step 3: Let $F^k : \boldsymbol{R}^n \to \boldsymbol{R}^n$ be defined by (17.4). By using the generalized Newton method, obtain an approximate solution x^{k+1} of NCP(F^k) such that

$$\Phi^k(x^{k+1})^{\frac{1}{2}} \leq (\beta)^k \min\{1, \|x^{k+1} - x^k\|\}.$$

Step 4: Set $c_{k+1} := \min\{1, \frac{1}{\|x^{k+1}\|^2}\} \min\{(\beta)^{k+1}, \Phi(x^{k+1})\}$, $k := k+1$ and go to Step 2.

The convergence results established in Theorems 17.16, 17.17 and 17.23 for PPA-0 remain valid as summarized in the following theorem.

Theorem 17.25 *Suppose that either of the following conditions holds:*

(a) *F is a P_0 function and the solution set of NCP(F) is bounded.*

(b) *F is monotone and Lipschitz continuous.*

Then the sequence $\{x^k\}$ generated by PPA-1 is bounded and its accumulation points are solutions of NCP(F). Moreover, if Assumption 17.20 holds, then the algorithm has a superlinear rate of convergence.

Theorem 17.25 says that the sequence $\{x^k\}$ generated by PPA-1 converges to a BD-regular solution x^* superlinearly. However, this does not necessarily mean that the algorithm is practically efficient, because it says nothing about computational costs to solve a subproblem at each iteration. So it is important to estimate the number of iterations the generalized Newton method spends at each iteration of the algorithm. Moreover, it is particularly interesting to see under what conditions the generalized Newton method spends just a single iteration. In the next subsection, we answer this question.

4.4 GENUINE SUPERLINEAR CONVERGENCE

In this section we give conditions under which a single Newton step of the generalized Newton method for NCP(F^k) attains criterion (17.7) in Step 3 of PPA-1, thereby genuine superlinear convergence of the algorithm is ensured. We note that such genuine superlinear convergence of PPA was first proposed by Solodov and Svaiter [16].

Suppose that $\{x^k\}$ converges to a solution x^* of NCP(F) that is BD-regular with H. Furthermore, let x_N^k be a point obtained by a single iteration of the generalized Newton method for NCP(F^k), that is,

$$x_N^k \equiv x^k - V_k^{-1} H^k(x^k), \tag{17.15}$$

where $V_k \in \partial_B H^k(x^k)$. The goal of this subsection is to show that the inequality

$$\Phi^k(x_N^k)^{\frac{1}{2}} \leq (\beta)^k \min\{1, \|x_N^k - x^k\|\} \tag{17.16}$$

holds for all k sufficiently large. Now observe that

$$\|x_N^k - x^k\| = \|V_k^{-1} H^k(x^k)\| \leq C\|H^k(x^k)\|,$$

where the inequality follows from Lemma 17.21. Since $\|H^k(x^k)\| \to 0$, we have $\|x_N^k - x^k\| < 1$ for k sufficiently large. Therefore, in view of the definition of $\Phi^k(x)$, it suffices to verify the following inequality instead of (17.16):

$$\|H^k(x_N^k)\| \leq \sqrt{2}(\beta)^k \|x_N^k - x^k\|. \tag{17.17}$$

The next lemma shows that the generalized Jacobian of H^k may serve as an approximation to the generalized Jacobian of H when k is sufficiently large.

Lemma 17.26 *Suppose that ∇F is Lipschitz continuous. Let $x^k \to x^*$ and $V_k \in \partial_B H^k(x^k)$ for each k. Then there exist a sequence $\{\bar{V}_k\} \subset \mathbf{R}^{n\times n}$ and positive constants κ_1 and κ_2 such that*

$$\|\bar{V}_k(x^k - x^*) - H(x^k) + H(x^*)\| \le \kappa_1 \|x^k - x^*\|^2$$

and

$$\|\bar{V}_k - V_k\| \le \kappa_2 c_k$$

for all k sufficiently large.

Proof Note that V_k is represented as

$$V_k = \mathrm{diag}(\gamma_i^k) + \mathrm{diag}(\mu_i^k)(\nabla F(x^k) + c_k I)^T,$$

where the pair (γ_i^k, μ_i^k) satisfies $(\gamma_i^k + 1)^2 + (\mu_i^k + 1)^2 = 1$ [9]. Now consider the sequence $\{\bar{V}_k\}$ given by

$$\bar{V}_k = \mathrm{diag}(\gamma_i^k) + \mathrm{diag}(\mu_i^k)\nabla F(x^k)^T,$$

where the pair (γ_i^k, μ_i^k) is the same as in V_k. Then, since $\{(\gamma_i^k, \mu_i^k)\}$ is bounded, there exists $\kappa_2 > 0$ such that

$$\|\bar{V}_k - V_k\| \le \kappa_2 c_k.$$

Furthermore, by the definition of $\bar{V}_k$, we have

$$\bar{V}_k \in \bar{T}(x^k) \equiv \left\{ AB \in \mathbf{R}^{n\times n} \;\middle|\; \begin{array}{l} A = (\mathrm{diag}(a_i^k), \mathrm{diag}(b_i^k)), \\ B = (I, \nabla F(x^k)^T)^T, \\ (a_i^k, b_i^k) \in \partial\psi(x_i^k, F_i(x^k)) \end{array} \right\}.$$

Since ψ and F are strongly C-differentiable, it then follows from Proposition 17.5 that the mapping $\bar{T}$ is a strong C-differential operator of H. Hence, there exists a positive constant κ_1 satisfying

$$\|\bar{V}_k(x^k - x^*) - H(x^k) + H(x^*)\| \le \kappa_1 \|x^k - x^*\|^2$$

for all k sufficiently large. The proof is complete. □

Now we have the main result in this section.

Theorem 17.27 *Suppose that ∇F is Lipschitz continuous and Assumption 17.20 holds. Then x_N^k given by (17.15) satisfies (17.17), and hence $x^{k+1} = x_N^k$, for all k large enough.*

Proof Since H is locally Lipschitz continuous, there exist $C_1 > 0$ and $\varepsilon > 0$ such that

$$\|H(x)\| = \|H(x) - H(x^*)\| \leq C_1\|x - x^*\| \tag{17.18}$$

for all x satisfying $\|x - x^*\| \leq \varepsilon$. It follows from (17.13) that

$$c_k \leq C_1^2\|x^k - x^*\|^2 \tag{17.19}$$

for all k sufficiently large. Furthermore, by Lemmas 17.21 and 17.26, we have

$$\begin{aligned}
\|x_N^k - x^*\| &= \|x^k - x^* - V_k^{-1}(H^k(x^k) - H(x^*))\| \\
&\leq \|V_k^{-1}\|\|V_k(x^k - x^*) - H(x^k) + H(x^*)\| \\
&\leq C(\|\bar{V}_k - V_k\|\|x^k - x^*\| \\
&\quad +\|\bar{V}_k(x^k - x^*) - H(x^k) + H(x^*)\|) \\
&\leq C(\kappa_2 c_k\|x^k - x^*\| + \kappa_1\|x^k - x^*\|^2),
\end{aligned}$$

where $\bar{V}_k$ is as indicated in Lemma 17.26. It then follows from (17.19) that there exists a constant $\kappa_3 > 0$ such that

$$\|x_N^k - x^*\| \leq \kappa_3\|x^k - x^*\|^2. \tag{17.20}$$

From (17.6) in Lemma 17.9, (17.18) and (17.19), we have

$$\begin{aligned}
\|H^k(x_N^k)\| &\leq \|H(x_N^k)\| + 2c_k\|x_N^k - x^k\| \\
&\leq C_1\|x_N^k - x^*\| + 2C_1^2\|x_N^k - x^k\|\|x^k - x^*\|^2.
\end{aligned}$$

It then follows from (17.20) that there exists $\kappa_4 > 0$ such that

$$\|H^k(x_N^k)\| \leq \kappa_4\|x^k - x^*\|^2. \tag{17.21}$$

On the other hand, by the triangle inequality and (17.20), we have

$$\begin{aligned}
\|x^k - x_N^k\| &\geq \|x^k - x^*\| - \|x_N^k - x^*\| \\
&\geq (1 - \kappa_3\|x^k - x^*\|)\|x^k - x^*\|,
\end{aligned}$$

which together with the fact that $x^k \to x^*$ implies

$$\|x^k - x_N^k\| \geq \frac{1}{2}\|x^k - x^*\| \tag{17.22}$$

for all k large enough. It follows from (17.21) and (17.22) that

$$\|H^k(x_N^k)\| \leq 2\kappa_4\|x^k - x^*\|\|x_N^k - x^k\|.$$

Since x^k converges to x^* superlinearly, we have for all k large enough

$$2\kappa_4\|x^k - x^*\| \leq \sqrt{2}(\beta)^k.$$

Consequently we have the desired inequality (17.17). □

5 CONCLUDING REMARKS

In this paper, we have proposed the PPA for solving NCP(F) with P_0 function. We have proved that the proposed algorithm has a global convergence property under the condition that the solution set of NCP(F) is bounded. Furthermore we have showed that it has a superlinear convergence property under some regularity conditions. These conditions have also been assumed in [4] and seem to be the weakest assumption for convergence of a method for NCP(F) with P_0 function.

As mentioned in Remark 17.12, one of the nice features of the PPA is that it does not require $c_k \to 0$ for global convergence. However, the condition $c_k \to 0$ is actually needed to ensure convergence of the specific algorithm PPA-1 presented in Section 4. It remains as an open problem to develop a specific PPA that converges globally even if $\{c_k\}$ is bounded away from 0.

Acknowledgments

We would like to thank two anonymous referees for their helpful suggestions on the earlier version of this paper. The work of the first and the third authors was supported in part by the Scientific Research Grant-in-Aid from the Ministry of Education, Science, Sports and Culture, Japan.

References

[1] R.W. COTTLE, J.S. PANG AND R.E. STONE, *The Linear Complementarity Problem*, Academic Press, New York, 1992.

[2] T. DE LUCA, F. FACCHINEI AND C. KANZOW, A semismooth equation approach to the solution of nonlinear complementarity problems, *Mathematical Programming*, 75 (1996), 407–439.

[3] F. FACCHINEI, Structural and stability properties of P_0 nonlinear complementarity problems, *Mathematics of Operations Research*, 23 (1998), 735–745.

[4] F. FACCHINEI AND C. KANZOW, Beyond monotonicity in regularization methods for nonlinear complementarity problems, *SIAM Journal on Control and Optimization*, 37 (1999), 1150–1161.

[5] F. FACCHINEI AND J. SOARES, A new merit function for nonlinear complementarity problems and a related algorithm, *SIAM Journal on Optimization*, 7 (1997), 225–247.

[6] A. FISCHER, A special Newton-type optimization method, *Optimization*, 24 (1992), 269–284.

[7] M.S. GOWDA AND R. SZNAJDER, Weak univalence and connectedness of inverse images of continuous functions, *Mathematics of Operations Research*, 24 (1999), 255–261.

[8] P.T. HARKER AND J.S. PANG, Finite dimensional variational inequality and nonlinear complementarity problems: A survey of theory, algorithms and applications, *Mathematical Programming*, 48 (1990), 161–220.

[9] H. JIANG AND L. QI, A new nonsmooth equations approach to nonlinear complementarity problems, *SIAM Journal on Control and Optimization*, 35 (1997), 178–193.

[10] B. MARTINET, Regularisation d'inequations variationelles par approximations successives, *Revue Française d'Informatique et de Recherche Opérationelle*, 4 (1970), 154–159.

[11] R.S. PALAIS AND C.-L. TERNG, *Critical Point Theory and Submanifold Geometry, Lecture Note in Mathematics, 1353*, Springer Verlag, Berlin, 1988.

[12] J.-S. PANG, Complementarity problems, *Handbook of Global Optimization*, R. Horst and P. Pardalos (eds.), Kluwer Academic Publishers, Boston, Massachusetts, 1994, pp. 271–338.

[13] J.-S. PANG AND L. QI, Nonsmooth equations: Motivation and algorithms, *SIAM Journal on Optimization*, 3 (1993), 443–465.

[14] L. QI, C-differentiability, C-differential operators and generalized Newton methods, Technical Report, School of Mathematics, University of New South Wales, Sydney, Australia, 1996.

[15] R.T. ROCKAFELLAR, Monotone operators and the proximal point algorithm, *SIAM Journal on Control and Optimization*, 14 (1976), 877–898.

[16] M.V. SOLODOV AND B.F. SVAITER A globally convergent inexact Newton method for systems of monotone equations, *Reformulation: Nonsmooth, Piecewise Smooth, Semismooth and Smoothing Methods*, M. Fukushima and L. Qi (eds.), Kluwer Academic Publishers, Boston, Massachusetts, 1998, pp. 355–369.

[17] N. YAMASHITA AND M. FUKUSHIMA, Modified Newton methods for solving a semismooth reformulation of monotone complementarity problems, *Mathematical Programming*, 76 (1997), 469–491.

[18] N. YAMASHITA AND M. FUKUSHIMA, The proximal point algorithm with genuine superlinear convergence for the monotone complementarity problem, *SIAM Journal on Optimization*, to appear.

Chapter 18

FREE BOUNDARY PROBLEMS IN ASSET PRICING WITH TRANSACTION COSTS

Thaleia Zariphopoulou
Department of Mathematics and School of Business, University of Wisconsin, Madison, Wisconsin
zariphop@math.wisc.edu

Abstract This presentation provides an overview of a class of free boundary problems that arise in valuation models in markets with transaction costs. Transaction costs are a realistic feature in numerous financial transactions and their presence affects considerably the theoretical asset and derivative prices.

In the area of optimal portfolio management, the valuation models give rise to singular stochastic control problems and the goal is to characterize the value function (maximal utility) and to specify the optimal control policies.

In the area of derivative pricing, the classical Black and Scholes valuation theory, based on exact replication, breaks down completely when transaction costs are present. Various approaches have been developed which lead to free boundary problems for the derivative prices. These methods include among others, the method of super-replicating strategies and the utility maximization theory.

1 INTRODUCTION

In his seminal papers, Merton ([28], [29]) introduced an optimal portfolio management model of a single agent in a stochastic setting. Trading takes place between a riskless security (e.g. a bond) and one or more stocks whose prices are modeled as diffusion processes. A crucial simplification in Merton's work is the absence of *transaction costs* on the various trades. The first to incorporate proportional transaction costs in Merton's model were Magill and Constantinides [27] in an effort to understand how these costs affect the trading policies and the equilibrium asset prices. They were the first to bring out the important insight about

M.C. Ferris et al. (eds.), Complementarity: Applications, Algorithms and Extensions, 381–400.

the different nature of the optimal investment policies, the one of *singular trading policies.* Under these policies, lump-sum transactions take place which amount to instantaneously altering the portfolio holdings in the bond and the stock account(s).

Taksar, Klass and Assaf [38] were the first to formulate a transaction cost model as a singular stochastic control problem in the context of maximizing the long term expected rate of wealth. Subsequently, Davis and Norman [12] provided a rigorous mathematical formulation and extensive analysis of the Merton problem in the presence of proportional costs for CRRA utilities. Their paper is considered a landmark in the literature on transaction costs and contains useful insights and fundamental results, both theoretical and numerical, for the value function and the optimal investment policies. Their work was later revisited by Shreve and Soner [34].

Departing from the special class of CRRA utilities, Zariphopoulou ([44], [45]) was the first to study optimal portfolio management models with proportional transaction costs for general individual preferences. In [44], Zariphopoulou introduced a simple investment model with two securities, a riskless bond and a risky security whose rate of return is modeled as a continuous-time Markov chain, and she provided characterization results for the maximal utilities.

As it will be apparent from the discussion in the next sections, the stochastic optimization problems with transaction costs do not have in general closed form solutions. Thus, it is highly desirable — mainly for the practical applications — to provide numerical results for their value function and the optimal investment policies and consumption plans. Such results were first provided by Davis and Norman [12] and later by Tourin and Zariphopoulou [40] for general utility functions. Other numerical schemes have been proposed by Akian, Menaldi and Sulem [1] for a model of portfolio selection with more than one risky asset and by Sulem [37] for a mixed portfolio problem with transaction costs. Pichler [32] developed a different class of schemes for the Davis Norman (DN) model and he also studied the probability distributions of the relevant expected gains.

In the entire volume of the aforementioned work, the mathematical analysis is carried out through the so-called *Hamiltonian-Jacobi-Bellman* (HJB) *equation.* This equation, as it will be discussed subsequently in detail, is the offspring of the optimality principle and stochastic analysis. Generally speaking, the value function (maximal utility) is expected to solve the HJB equation and the first order conditions in HJB usually give the candidate optimal policies in a feedback form. The verification that the solution of the HJB equation and the optimality of the candidate policies are established via classical verification results, provided that the value function has the appropriate regularity properties (for a general overview of the theory of controlled Markov processes, we refer the reader to the book of Fleming and Soner [18]). In the majority of stochastic control problems arising in asset pricing models, the required regularity properties might not hold. This is the result of various characteristics of the particular nature of the involved financial models, like trading constraints, transaction costs, certain assumptions on the price processes as well as an imperfect information structure. Therefore, it is imperative to relax the notion of solutions of the associated HJB equation in order to develop a viable method to study this class of problems. It turns out that a rich class of weak solutions of the HJB equation are the so-called *viscosity solutions.* These solutions were introduced by Crandall and Lions [10] for first-order non-linear partial differential equations and by Lions [26] for the second-order case. The strength of this theory lies in the fact that it provides rigorous characterization of the value function as the unique viscosity solution of the HJB equation. Moreover, the strong stability properties of viscosity solutions provide excellent convergence results for a large class of numerical schemes for the value function and the optimal policies.

In stochastic optimization problems arising in optimal investment and consumption models, viscosity solutions were first employed by Zariphopoulou [44] for the Merton model with trading constraints (see also Zariphopoulou [46]) and for the aforementioned transaction cost model with the Markov chain parameters (see Zariphopoulou [45]) Later on, this class of solutions was used by Fleming and Zariphopoulou [19], Duffie and Zariphopoulou [15], and in the context of transaction costs by Shreve and Soner [34]. Shreve and Soner [34] provided a viscosity characterization of the value function of the DN model for a broad range of market parameters. For general utilities, this characterization was provided by Tourin and Zariphopoulou [40] and it was subsequently used to establish the convergence of the numerical schemes they developed (see also Tourin and Zariphopoulou [41] and [42]).

Besides the area of optimal portfolio management, the area of *derivative securities* has been one of the fastest growing areas of Finance as well as one of the most active areas of research on stochastic analysis, stochastic control and computations. Derivatives are financial instruments whose values depend on the price levels of the so-called primitive securities, like stocks. The fundamental problem of derivative valuation amounts in determining the derivative's fair value and in specifying the hedging policy which eliminates the risk inherent to the contract. Derivative contracts had always existed in financial environments but it was after the seminal work of Fisher Black and Myron Scholes [4] (in collaboration with Robert Merton) that this area blossomed and started expanding rapidly. The Black and Scholes valuation approach brought to the modern finance the powerful methodologies of martingale theory and stochastic calculus. Today, numerous different kinds of derivative instruments are traded all around the world and various new contracts are being created every day. The valuation of these contracts gives rise to a number of challenging problems in the areas of stochastic analysis, martingale theory, stochastic control and partial differential equations.

Despite the ever growing activity in derivatives' markets, very few questions have been successfully addressed to date when derivatives are created, traded and hedged in *markets with frictions*. The most important kind of frictions comes from the stochastic nature of the volatility of the primitive stock security. Other market frictions come from trading constraints, incomplete information and finally from transaction costs.

The fundamental difficulty for pricing derivatives in the presence of transaction costs lies in the fact that the Black and Scholes approach breaks down completely. In fact, in a frictionless market, Black and Scholes [4] and Merton [30] relied on an ingenious no-arbitrage argument to price an option on a stock when the interest rate is constant and the stock price follows a geometric Brownian motion. They presented a self-financing, dynamic trading policy between the bond and stock accounts which replicates the payoff of the option.

The Achilles' heel of the argument is that the frictionless market assumption must be taken literally. The dynamic replication policy incurs an *infinite volume* of transactions over any finite trading interval, given the fact that the Brownian motion which drives the stock price has infinite variation.

Merton ([31], Chapter 14) maintained the goal of a dynamic trading policy as that of replicating the option payoff and modeled the path of the stock price as a two-period binomial process. The initial cost of the replication policy is finite and serves as an upper bound to the write price of a call which is arbitrage-free. Shen [33] and Boyle and Vorst [5] extended Merton's model to a multiperiod binomial process for the stock price and provided numerical solutions to the initial cost of the replicating portfolio. As the number of periods increases within the given lifetime of a call option, the initial cost of the replicating portfolio tends to infinity.

Bensaid et al. [3] and Edirisinghe, Naik and Uppal [16] noted that a tighter upper bound on the write price of a call option is obtained by replacing the goal of replicating the payoff of the option with the goal of *dominating the payoff*. Davis and Clark [11] conjectured and

Soner, Shreve and Cvitanic [36] proved, that the cost of initially buying one share of stock is indeed the cost of the cheapest dominating policy in the presence of finite proportional transaction costs. Their result on feasible *super-replicating strategies* was subsequently generalized by Levental and Skorohod [25].

Leland [24] initiated a novel approach by introducing a class of *imperfectly replicating policies* in the presence of proportional transaction costs. Imperfectly replicating policies were further studied by Figlewski [17], Grannan and Swindle [20], Hoggard, Whalley and Wilmott [22] and Toft [39]. Avellaneda and Parás [2] extended the notion of imperfectly replicating policies to that of imperfectly dominating policies.

An alternative approach, initiated by Hodges and Neuberger [21] and developed further by Davis, Panas and Zariphopoulou [13], is the so-called *utility maximization method.* In this approach, the price of the derivative is determined by comparing the value functions of an investor with and without the opportunity to trade the available derivative. By considering the utility functionals (with and without the derivative), this methodology incorporates the individual's attitude towards the risk which cannot be eliminated, in contradistinction to the case of no transaction costs. The above results were considerably generalized by Constantinides and Zariphopoulou [7] who applied utility methods to establish price bounds for all types of European claims and for general preferences.

Besides the claims of European-type, the valuation of American options was examined by Davis and Zariphopoulou [14] for the class of exponential utilities. More recently, Constantinides and Zariphopoulou [8] extended their results to the cases of American-type and path-dependent claims, written on many stocks and for CRRA utilities.

2 FREE BOUNDARY PROBLEMS IN OPTIMAL INVESTMENT-CONSUMPTION MODELS

We start with the description of the benchmark optimal investment-consumption model of Davis and Norman incorporating general utilities in the payoff functional. This is a model of a single agent, or a small investor, as it is otherwise known, in the sense that his/her actions cannot influence the prices of the underline securities.

We consider an economy with two securities, a *bond* with price B_t and a *stock* with price S_t at date $t \geq 0$. Prices are denominated in units of a consumption good, say dollars.

The bond pays no coupons, is default free and has price dynamics

$$B_t = e^{rt}B_0, \qquad t \geq 0 \tag{18.1}$$

where r is the *constant rate of interest.*

We denote by W_t a one-dimensional standard Brownian motion which generates the filtration $\mathcal{F}_t$ on a fixed, complete probability space $(\Omega, \mathcal{F}, P)$. The stock price is the diffusion process

$$S_t = S_0 \exp\left\{\left(\mu - \frac{\sigma^2}{2}\right)t + \sigma W_t\right\} \tag{18.2}$$

where μ is the *mean rate of return* and σ is the *volatility*; μ and σ are constants such that $\mu > r$ and $\sigma \neq 0$.

The investor holds x_t dollars of the bond and y_t dollars of the stock at date t. We consider a pair of right-continuous with left limits (CADLAG), non-decreasing processes (L_t, M_t) such that L_t represents the cumulative dollar amount transferred into the stock account and M_t the cumulative dollar amount transferred out of the stock account. By convention, L_0=M_0=0. The stock account process is

$$y_t = y + \int_0^t \mu y_\tau d\tau + \int_0^t \sigma y_\tau dW_\tau + L_t - M_t. \tag{18.3}$$

with $y_0 = y$.

Transfers between the stock and the bond accounts incur *proportional transaction costs.* In particular, the cumulative transfer L_t into the stock account reduces the bond account by βL_t and the cumulative transfer M_t out of the stock account increases the bond account by αM_t, where $0 < \alpha < 1 < \beta$.

The investor consumes at the rate c_t dollars out of the bond account. There are no transaction costs in transfers from the bond account into the consumption good.

The bond account process is

$$x_t = x + \int_0^t \{r x_\tau - c_\tau\} d\tau - \beta L_t + \alpha M_t. \tag{18.4}$$

with $x_0 = x$. The integral represents the accumulation of interest and the drain due to consumption. The last two terms represent the cumulative transfers between the stock and bond accounts, net of transaction costs.

A policy is a $\mathcal{F}_t$-progressively measurable triple (c_t, L_t, M_t). We restrict our attention to the set of admissible policies $\mathcal{A}$ such that

$$c_t \geq 0 \text{ and } E\int_0^t c_\tau d\tau < \infty \text{ a.s. for } t \geq 0 \tag{18.5}$$

and

$$w_t = x_t + \binom{\alpha}{\beta} y_t \geq 0 \text{ a.s. for } t \geq 0 \tag{18.6}$$

where we adopt the notation

$$\binom{\alpha}{\beta} z = \begin{cases} \alpha z & \text{if } z \geq 0 \\ \beta z & \text{if } z < 0. \end{cases} \tag{18.7}$$

We refer to w_t as the *net worth*. It represents the investor's bond holdings, if the investor were to transfer the holdings from the stock account into the bond account, incurring in the process the transaction costs.

The investor has von Neumann-Morgenstern preferences

$$E\left[\int_0^{+\infty} e^{-\rho t} U(c_t) dt\right]$$

over the consumption stream $\{c_t, t \geq 0\}$, where ρ is the *subjective discount rate* and the *utility function* $U : R_0^+ \to R_0^+$ is assumed to have the following properties:

i) $U \in C([0, +\infty)) \cap C^1((0, +\infty))$ is increasing and concave.

ii) $U(c) \leq K(1 + c)^\gamma$, $\forall c \geq 0$, for some positive constants K and γ with $0 < \gamma < 1$.

Given the initial endowment (x, y) in

$$D = \left\{(x, y) \in \mathcal{R} \times \mathcal{R} : x + \binom{\alpha}{\beta} y \geq 0\right\},$$

we define the *value function* V as

$$V(x, y) = \sup_{\mathcal{A}} E\left[\int_0^{+\infty} e^{-\rho t} U(c_t) dt \mid x_0 = x, y_0 = y\right]. \tag{18.8}$$

3 THE HJB EQUATION AND VISCOSITY SOLUTIONS

In this section we discuss the Hamilton Jacobi Bellman (HJB) equation that the value function is expected to solve. It turns out that the

HJB equation is a *Variational Inequality with gradient constraints.* Because such free boundary problems do not have in general smooth solutions, one needs to work with a weak class of solutions, namely the constrained viscosity solutions. The characterization of V as a constrained solution is natural because of the presence of the state constraint given by (18.6).

The notion of *viscosity solutions* was introduced by Crandall and Lions [10] for first-order, and by Lions [26] for second-order equations. For a general overview of the theory we refer to the "User's Guide" by Crandall, Ishii and Lions [9] and Fleming and Soner [18].

Next, we recall the notion of *constrained viscosity solutions*, which was introduced by Soner [35] and Capuzzo-Dolcetta and Lions [6] for first-order equations and by Lions [26] for second-order equations (see also Ishii and Lions [23]). To this end, we consider a non-linear second order partial differential equation of the form

$$F(X, v, Dv, D^2 v) = 0 \text{ in } D, \tag{18.9}$$

with the open domain $D \subset \mathcal{R}^N$, and Dv and $D^2 v$ standing respectively for the gradient vector and the second derivative matrix of v; F is continuous in all its arguments and *degenerate elliptic*, meaning that

$$F(X, p, q, A + B) \leq F(X, p, q, A) \text{ if } B \geq 0.$$

Definition 18.1 *A continuous function $v : \mathcal{R} \to \mathcal{R}$ is a constrained viscosity solution of (18.9) on $\overline{D}$ if*

i) v is a viscosity subsolution of (18.9) on $\overline{D}$, that is for any $\phi \in C^2(\overline{D})$ and any local maximum point $X_0 \in \overline{D}$ of $v - \phi$

$$F(x_0, v(x_0), D\phi(x_0), D^2\phi(x_0)) \leq 0$$

and

ii) v is a viscosity supersolution of (18.9) in D, that is for any $\phi \in C^2(\overline{D})$ and any local minimum point $X_0 \in D$ of $v - \phi$

$$F(x_0, v(x_0), D\phi(x_0), D^2\phi(x_0)) \geq 0.$$

The following result was proved by Tourin and Zariphopoulou [40] and by Shreve and Soner [34] for the case of CRRA utilities.

Theorem 18.2 *The value function V is uniformly continuous on $\overline{D}$, jointly concave and increasing in x and y. Moreover it is a constrained viscosity solution on $\overline{D}$ of the Hamilton–Jacobi–Bellman equation*

$$\min\Big[\rho V -\frac{1}{2}\sigma^2 y^2 V_{yy} - \mu y V_y - rxV_x - \max_{c\geq 0}(-cV_x + U(c)), \qquad \beta V_x - V_y, \; -\alpha V_x + V_y\Big] = 0 \tag{18.10}$$

and it is unique in the class of concave functions that are bounded from below and they grow like $O(x^\gamma + y^\gamma)$.

4 THE CASE OF CRRA UTILITIES

A special class of utility functions are the so-called Constant Relative Risk Aversion utilities. These functions $U : \mathcal{R} \to \mathcal{R}$ are of the form

$$\begin{cases} U(c) = \frac{1}{\gamma}c^\gamma & \text{for } \gamma < 1, \gamma \neq 0 \\ U(c) = \log c & \text{for } \gamma = 0 \end{cases} \tag{18.11}$$

where the variable c typically plays the role of the consumption stream or the state wealth. The main characteristic of these utilities is that their *Relative Risk Aversion Coefficient*, defined as $-\dfrac{cU''(c)}{U'(c)}$, is constant across consumption (or wealth) levels.

CRRA utilities result in homogeneous value functions which are in turn used to reduce the dimensionality of the underlying optimization problem. This is the central feature in the benchmark work of Davis and Norman who reduced the dimensionality of (18.10) via the transformation

$$V(x, y) = y^\gamma F(\frac{x}{y}). \tag{18.12}$$

The function F solves the one-dimensional Variational Inequality

$$\min\left\{\tilde{\rho}F - \frac{1}{2}\sigma^2 z^2 F'' - \tilde{\mu} zF' - \frac{1-\gamma}{\gamma}(F')^{\frac{\gamma}{\gamma-1}}, \qquad (\beta\gamma + z)F' - \gamma F, \; -(\alpha\gamma + z)F' + \gamma F\right\} = 0 \tag{18.13}$$

for $z = \dfrac{x}{y}$ and $\tilde{\rho} = \rho - \mu\gamma + \frac{1}{2}\sigma^2\gamma(1-\gamma)$ and $\tilde{\mu} = \rho - r - \sigma^2(1-\gamma)$; the non-linear term $\dfrac{1-\gamma}{\gamma}(F')^{\frac{\gamma}{\gamma-1}}$ comes from the reduced form of $\max_{c\geq 0}\left\{-cV_x + \dfrac{c^\gamma}{\gamma}\right\}$ using that $V_x(x,y) = y^{\gamma-1}F'\left(\dfrac{x}{y}\right)$.

Davis and Norman analyzed the above equation and under certain assumptions on the market coefficients, they constructed a solution ψ satisfying, for some positive constants A and B, and points z_1 and z_2

$$\begin{cases} \psi(z) = A(\alpha\gamma + z)^\gamma & z \leq z_1 \\ \tilde{\rho}\psi = \frac{1}{2}\sigma^2 z^2\psi'' + \tilde{\mu}z\psi' + \frac{1-\gamma}{\gamma}(\psi')^{\frac{\gamma}{\gamma-1}} & z_1 < z < z_2 \\ \psi(z) = B(\beta\gamma + z)^\gamma & z \geq z_2. \end{cases} \tag{18.14}$$

The function ψ was constructed as the solution of a two point boundary problem of second order with endpoints z_1 and z_2. These endpoints were specified by the so-called "principle of smooth fit" which is used to produce a smooth solution of (18.14).

The set of equations (18.14) indicates that when the ratio of account holdings $\dfrac{x}{y}$ is between the threshold levels z_1 and z_2, then it is optimal not to rebalance the portfolio but only to consume. In other words, the individual must refrain from trading in the region $\mathcal{NT} = \left\{(x,y) \in D : z_1 \leq \dfrac{y}{x} \leq z_2\right\}$.

If the holdings ratio, say $\dfrac{y_0}{x_0}$, is below z_1 then it is optimal to instantaneously rebalance the portfolio components by moving from the original point to the point $(\overline{y}, \overline{x})$ with $\overline{y} = z_1\overline{x}$ with $\overline{x} = \dfrac{x_0 + \beta y_0}{1 + \beta z_1}$. This corresponds to a transaction of *buying shares of stock* and this is the optimal policy that one should apply to all points $(x,y) \in \overline{D}$ with $\dfrac{y}{x} < z_1$. Similarly, if the holdings ratio $\dfrac{y_0}{x_0}$ is above z_2, then it is optimal to instantaneously rebalance the portfolio components by moving to the point $\tilde{y} = z_2\tilde{x}$ with $\tilde{x} = \dfrac{x_0 + \alpha y_0}{1 + \alpha z_2}$. This corresponds to a transaction of *selling stock shares* and this is the optimal policy for all point $(x,y) \in \overline{D}$ such that $\dfrac{y}{x} > z_2$.

5 FREE BOUNDARY PROBLEMS IN DERIVATIVE PRICING

In their seminal paper, Black and Scholes [4] developed a theory for the valuation of derivative securities in frictionless markets. They considered the problem of determining the value of a *European call* which is written on an underlying stock whose price S_s follows the diffusion process described in (18.2). The market is also endowed with a riskless security whose price is given by (18.1). The European claim is written at the time $t > 0$ and expires at maturity time T. Its payoff, at expiration, is given by $(S_T - K)^+$ where K is the (prespecified) *exercise price.*

The valuation problem amounts to specifying the *fair value* of the security at its birth time t.

Black and Scholes had the novel idea of constructing a dynamic portfolio whose value coincides with the terminal payoff, $(S_T - K)^+$, of the call. Then they argued that the amount needed to set up this hedging portfolio, at time t, yields the correct price of the European call. Moreover, the components of this portfolio, across time, give the *perfectly replicating (hedging) strategies* which reproduce the current value of the security.

The strategies of Black and Scholes yield that both components of the replicating portfolio, are diffusion processes which is a direct consequence of the diffusion nature of the underlying stock price. Clearly in markets with transaction costs, these hedging strategies will immediately produce an infinite voluem of transactions no matter how small the transaction costs are. Therefore, a *perfectly replicating portfolio no longer exists!*

Abandoning the idea of exact replication, one might look for a portfolio strategy which results, at expiration time T, in portfolio value at least as great as the value of the European call. Such strategies are known as *super-replicating* strategies.

Bensaid, Lesne, Pagès and Scheinkman [3] uncovered the intriguing idea that super-replication may be feasible, in the sense that the cost of the super-replicating portfolio is actually finite. This cost then may provide a sensible bound on the price of the option. (See also, Edirisinghe, Naik and Uppal [16].)

Unfortunately, in the (limiting) case of continuous time, the super-replication approach cannot form the basis of a viable valuation theory. In fact, Davis and Clark [11] conjectured that the *minimal* cost of the super-replication of a European call is the *value of one share* of the underlying stock. Therefore, even though super-replication techniques might provide finite values, their minimal value yields a trivial bound, the value of one share of stock, which is of little economic interest.

Using convex analysis arguments, Soner, Shreve and Cvitanic [36] established the conjecture of Davis and Clark. Below we state their result by adopting the notation used in (18.7). The bond and stock account processes, x_s and y_s, are given by the state equations (18.4) and (18.3). The European call has exercise time T, strike price K and it is written on the underlying stock whose price is given in (18.2).

Theorem 18.3 *Consider the payoff $(S_T - K)^+$ of a European call written of a stock with price S_s, $t \leq s \leq T$, as in (18.2). Then in order to have*

$$x_T + \binom{\alpha}{\beta} y_T \geq (S_T - K)^+ \quad a.e. \tag{18.15}$$

the following constraint must hold for <u>*all*</u> $t \leq s < T$

$$x_s + \binom{\alpha}{\beta}\left(y_s - \frac{S_s}{\alpha}\right) \geq 0 \quad a.e. \tag{18.16}$$

The above result on *trivial super-replicating strategies* was later established by Levental and Skorohod [25] in a general framework as it is stated below.

Proposition 18.4 *Consider a European claim with payoff $g(S_T)$ at expiration time T where g is increasing, convex and $g(0) = 0$ with*

$$\lim_{S \to \infty} \frac{g(S)}{S} = \ell > 0.$$

The stock price S_s, $t \leq s \leq T$ is the diffusion process described in (18.2). Then in order to have

$$x_T + \binom{\alpha}{\beta} y_T \geq g(S_T) \quad a.e. \tag{18.17}$$

the following constraint must hold for <u>*all*</u> $t \leq s < T$

$$x_S + \binom{\alpha}{\beta}\left(y_s - \ell\frac{S_s}{\alpha}\right) \geq 0 \quad a.e. \tag{18.18}$$

Even though super-replicating strategies produce price bounds of little economic interest, these results are of fundamental importance in the utility maximization theory. In fact, the *state constraints* (18.16) and (18.18) essentially characterize the set of *feasible strategies* that the writer, or the buyer of the derivative may use in their valuation problems. As it will be demonstrated in the next section, even under the stringent constraints (18.16), (18.18), the presence of risk aversion — through the utility functionals — allows us to derive non-trivial bounds for the so-called reservation derivative prices.

Finally, from the previous results an interesting question arises, namely, how can the constraints (18.16) and (18.18) be relaxed, if at all, when the super-replication requirements (18.15) or (18.17) are allowed to hold with probability $1 - \epsilon$, instead of almost surely. This problem is interesting especially from the practical point of view where some "slippage" might be tolerated. Numerical results for the case of European calls can be found in Tourin and Zariphopoulou [43].

6 THE UTILITY MAXIMIZATION THEORY

The Black and Scholes valuation method produces derivative prices which are *independent* of the individual portfolio holdings as well as the individual attitude towards risk. Clearly, these universal properties stem from the ability to exactly replicate the payoff of the security, in the absence of market frictions. As it was discussed earlier, this possibility disappears in the presence of transaction costs and thus, these universal features might not be preserved. Indeed, the utility maximization approach brings in the individual attitude towards the derivative-inherent risk which cannot be any more eliminated.

Hodges and Neuberger [21] were the first to apply the utility approach to price European calls when the agents are endowed with exponential utilities. Their results were further developed by Davis, Panas and Zariphopoulou [13] for the same class of options, and by Davis and Zariphopoulou [14] for American options.

The utility maximization approach looks at the value functions of the investor with and without the opportunity to trade (write or buy) the derivative security.

If the investor chooses not to trade the available claim, his value function is given by $V(x, y)$, as it is defined in (18.8). Suppose now that a third asset is introduced, a cash settled European-style contingent claim with expiration at date T and payoff $g(S_T)$ at expiration. If the investor writes the claim at date t with $0 \leq t \leq T$, the bond account is credited with an amount, say C dollars, which represents the price of the claim, and is debited $g(S_T)$ dollars at the expiration date T. To keep the problem tractable, we assume that the investor may not trade the claim in the open interval (t, T).

Let x_t and y_t be the initial endowment at time t *after* the bond account has been credited with the proceeds from writing the claim. Once the claim is written, the writer's objective is to maximize his expected utility from consumption with the extra obligation to surrender to the buyer $g(S_T)$ dollars at time T.

The *value function of the writer* is

$$J(x,y,S,t) = \sup_{\mathcal{A}_1} E\Bigg[\int_t^T e^{-\rho(s-t)} U(c_s) ds \\ + e^{-\rho(T-t)} V(x_T - g(S_T), y_T) \mid x_t = x, y_t = y, S_t = S \Bigg] \tag{18.19}$$

where V is given in (18.8) and $\mathcal{A}_1$ is the set of admissible policies defined subsequently.

It is assumed that the payoff function g satisfies the following assumptions

$$\begin{cases} g : [0,+\infty) \to [0,+\infty) \qquad \text{is convex} \\ g(0) = 0, \quad \lim_{S\to\infty} \dfrac{g(S)}{S} = 1. \end{cases} \tag{18.20}$$

The previous exposition on feasible super-replication strategies suggests that the set of the writer's admissible policies must be determined as follows. From (18.19) we have that the writer's terminal (liquidated) wealth must be nonnegative; in other words, the terminal constraint

$$x_T + \left(\frac{\alpha}{\beta}\right) y_T \geq g(S_T) \quad \text{a.e.} \tag{18.21}$$

must be fulfilled. The payoff function g satisfies the assumptions of Levental and Skorohod [25]. Therefore, Proposition 18.4 yields that at *all* previous times $t \leq s < T$, the state wealth of the writer must satisfy the stringent constraint

$$x_s + \left(\frac{\alpha}{\beta}\right)\left(y_s - \frac{S_s}{\alpha}\right) \geq 0 \quad \text{a.e.} \tag{18.22}$$

So, we define the set $\mathcal{A}_1$ of admissible policies of the investor who has written a contingent claim, as the set of $\mathcal{F}_t$-progressively measurable processes (c_t, L_t, M_t), with L_t and M_t being CADLAG which also satisfy the conditions

$$c_s \geq 0 \text{ and } E \int_t^s c_\tau d\tau < \infty \text{ a.s. for } t \leq s \leq T \tag{18.23}$$

and

$$w_s = x_s + \binom{\alpha}{\beta}\left(y_s - \frac{S_s}{\alpha}\right) \geq 0 \text{ a.s. for } t \leq s < T \tag{18.24}$$

and, we define the set of admissible policies $\{c_t, L_t, M_t; T < t\}$ of the investor who has written a claim by $\mathcal{A}$, as given in (18.5) and (18.6).

The value function $J(x, y, S, t)$ is given by (18.19) and is defined for $(x, y, S) \in \overline{D}_1$ where

$$\overline{D}_1 = \left\{(x, y, S) : x + \binom{\alpha}{\beta}\left(y - \frac{S}{\alpha}\right) \geq 0,\ S \geq 0\right\}.$$

Definition 18.5 *The reservation write price $C(x, y, S, t)$, for initial endowment (x, y), is defined as the minimum value at which the investor is willing to write the claim. Therefore, C satisfies for $(x + C(x, y, S, t), y, S) \in \overline{D}_1$*

$$V(x, y) = J(x + C(x, y, S, t), y, S, t) \tag{18.25}$$

Definition 18.6 *The write price $\overline{C}(S, t)$ is defined as the maximum of reservation write prices across all admissible states (x, y, S). Therefore, $\overline{C}$ satisfies for all $(x + \overline{C}(S, t), y, S) \in \overline{D}_1$,*

$$V(x, y) \leq J(x + \overline{C}(S, t), y, S, t) \tag{18.26}$$

The above inequality guarantees that the *writer will be willing to write the option at any price higher than $\overline{C}(S, t)$*, independently of his current portfolio position.

The case of exponential utilities was first examined by Hodges and Neuberger [21] and subsequently by Davis, Panas and Zariphopoulou [13]. Constantinides and Zariphopoulou [7] generalized all previous results on the subject for general individual preferences and they derived an upper bound $h = h(S, t)$ for the write price which satisfies (18.26) on $\overline{D}_1$. The main steps for the construction and characterization of the upper bound are presented below. The proof of their main result can be found in Constantinides and Zariphopoulou [7].

Theorem 18.7 *The value function is a constrained viscosity solution on $\overline{D}_1 \times [0, T)$ of the Variational Inequality*

$$\min\left[\mathcal{L}J - \overline{\mathcal{L}}J, \beta\frac{\partial J}{\partial x} - \frac{\partial J}{\partial y}, -\alpha\frac{\partial J}{\partial x} + \frac{\partial J}{\partial y}\right] = 0 \tag{18.27}$$

with

$$J(x,y,S,T) = V(x - g(S), y) \tag{18.28}$$

where the operators $\mathcal{L}$ *and* $\overline{\mathcal{L}}$ *are given by*

$$\begin{cases} \mathcal{L}J = \rho J - \frac{1}{2}\sigma^2 y^2 \frac{\partial^2 J}{\partial y^2} - \mu y \frac{\partial J}{\partial y} - rx\frac{\partial J}{\partial x} - \max_{c\geq 0}\Big(- c\frac{\partial J}{\partial x} + U(c)\Big) \\ \overline{\mathcal{L}}J = \frac{\partial J}{\partial t} + \frac{1}{2}\sigma^2 S^2 \frac{\partial^2 J}{\partial S^2} + \sigma^2 yS\frac{\partial^2 J}{\partial y \partial S} + \mu S\frac{\partial J}{\partial S}. \end{cases} \tag{18.29}$$

Moreover, J *is the unique constrained viscosity solution of (18.27) in the class of uniformly continuous and concave functions, with respect to the state variables* (x, y, S).

The underlying idea for the derivation of *analytic bounds* for the write price of a European-type claim, is to construct suitable subsolutions of the free boundary problems (18.10) and (18.27) in order to use a comparison result to establish (18.25). The main difficulty stems from the fact that the value functions V and J are defined on *different* domains and that there are no explicit or closed-form solutions for the two associated free-boundary problems (18.8) and (18.19).

The basic idea of Constantinides and Zariphopoulou [7] for the choice of the candidate bound is first to find the write price in the case that $(x, y, S) \in \partial D_1$, i.e. when the writer holds the *minimal allowed position* which amounts to the value of one stock share, taking into account the transaction costs. Then one needs to show that this price works for all wealth levels greater than the minimal one.

To this end, we start with the following lemma which describes the value function on

$$\partial D_1 = \left\{(x,y,S) : x + \binom{\alpha}{\beta}\left(y - \frac{S}{\alpha}\right) = 0\right\}.$$

Lemma 18.8 *For* $(x, y, S) \in \partial D_1$*, the value function* J *is given by*

$$J(x,y,S,t) = E\left[e^{-\rho(T-t)}V\left(-g(S_T), \frac{S_T}{\alpha}\right) \Big| S_t = S\right] \tag{18.30}$$

Proof. The proof follows directly from the fact that the only admissible policy for the boundary points (x, y, S) is to move instantaneously at time t, to the point $\left(0, \frac{S}{\alpha}, S\right)$ and remain there until time T.

The next result gives the main ingredient for the construction of the candidate solution. Its proof can be found in Constantinides and Zariphopoulou [7].

Lemma 18.9 *If $h_{\hat\rho} = h_{\hat\rho}(S,t)$ is such that*

$$V\left(\frac{\beta}{\alpha}S, -\frac{h_{\hat\rho}}{\beta}\right) = E\left[e^{-\hat\rho(T-t)}V(S_T - g(S_T), 0) \mid S_t = S\right] \qquad (18.31)$$

with $0 \le h_{\hat\rho} \le \frac{\beta}{\alpha}S$ and $\hat\rho \ge \rho$, then (18.26) holds for $(x,y,S) \in \partial D_1$.

Theorem 18.10 *Assume that the $\mathcal{NT}$ region for the utility maximization problem (18.8) satisfies $\mathcal{NT} \subseteq \{(x,y) : Ax \le y \le Bx, x \ge 0$ with $B > A > 0\}$. Also, assume that the utility function U satisfies $\lambda_1 \frac{c^\gamma}{\gamma} \le U(c) \le \lambda_2 \frac{c^\gamma}{\gamma}$ for some positive constants λ_1 and λ_2. Let*

$$m = \left[\left(\frac{\lambda_2\left(\beta + \frac{1}{A}\right)}{\lambda_1\left(\alpha + \frac{1}{B}\right)^\gamma}\right)^{\frac{1}{1-\gamma}} \frac{\rho - r\gamma}{\alpha\gamma} + \left(\mu + \frac{r}{\alpha A}\right) - \frac{\rho}{2}\left(1 + \frac{1}{\beta B}\right)\right]\sigma^{-2}$$

and consider the discount factor $\hat\rho$ in (18.31) given by

$$\hat\rho = \max\left[\rho, \mu + \frac{m\sigma^2(1-\beta A)}{2\beta A}\right].$$

Then, the derivative price bound $h(S,t)$ is given by (18.31) with $\hat\rho$ as above.

References

[1] Akian, M., J. L. Menaldi, and A. Sulem. (1992). "Multi-asset portfolio selection problem with transaction costs. Probabilités numériques," Mathematics and Computers in Simulation, 38, 163–172.

[2] Avellaneda, M., and A. Parás. (1994). "Optimal hedging portfolios for derivative securities in the presence of large transaction costs," Applied Mathematical Finance, 1, 165–193.

[3] Bensaid, B., J. Lesne, H. Pages and J. Scheinkman. (1992). "Derivative asset pricing with transaction costs," Mathematical Finance, 2, 63–86.

[4] Black, F., and M. Scholes. (1973). "The pricing of options and corporate liabilities," Journal of Political Economy, 81, 637–654.

[5] Boyle, P., and T. Vorst. (1992). "Option replication in discrete time with transaction costs," Journal of Finance, 47, 271–293.

[6] Capuzzo-Dolcetta, I., and P.-L. Lions. (1990). "Hamilton-Jacobi equations with state constraints," Transactions of the American Mathematical Society, 318, 543–583.

[7] Constantinides, G. M., and T. Zariphopoulou. (1999). "Bounds on prices of contingent claims in an intertemporal economy with proportional transaction costs and general preferences," Finance and Stochastics, 3(3), 345–369.

[8] Constantinides, G. M. and T. Zariphopoulou. (2000). "Price bounds on derivative prices in an intertemporal setting with proportional costs and multiple securities," preprint.

[9] Crandall, M. G., H. Ishii, and P.-L. Lions. (1992). "User's guide to viscosity solutions of second order partial differential equations," Bulletin of the American Mathematical Society, 27, 1–67.

[10] Crandall, M. G., and P.-L. Lions. (1983). "Viscosity solutions of Hamilton-Jacobi equations," Transactions of the American Mathematical Society, 277, 1–42.

[11] Davis, M. H. A., and J. M. C. Clark. (1994). "A note on superreplicating strategies," Philosophical Transactions of the Royal Society of London A, 485–494.

[12] Davis, M. H. A., and A. R. Norman. (1990). "Portfolio selection with transaction costs," Mathematics of Operations Research. 15, 676–713.

[13] Davis, M. H. A., V. Panas, and T. Zariphopoulou. (1993). "European option pricing with transaction costs," SIAM Journal on Control and Optimization, 31, 470–493.

[14] Davis, M. H. A., and T. Zariphopoulou. (1995). "American options and transaction fees," *Mathematical Finance*, IMA Volumes in Mathematics and its Applications, New York, NY: Springer-Verlag.

[15] Duffie, D., and T. Zariphopoulou. (1993). "Optimal investment with undiversifiable income risk," Mathematical Finance, 3, 135–148.

[16] Edirisinghe, C., V. Naik and R. Uppal. (1993). "Optimal replication of options with transaction costs and trading restrictions," Journal of Finance, 28, 117–138.

[17] Figlewski, S. (1989). "Options arbitrage in imperfect markets," Journal of Finance, 44, 1289–1311.

[18] Fleming, W. H., and H. M. Soner. (1993). *Controlled Markov Processes and Viscosity Solutions.* New York, NY: Springer Verlag.

[19] Fleming, W. H., and T. Zariphopoulou. (1991). "An optimal investment/consumption model with borrowing," Mathematics of Operations Research, 16, 802–822.

[20] Grannan, E. R., and G. H. Swindle. "Minimizing transaction costs of option hedging strategies," Mathematical Finance, to appear.

[21] Hodges, S. D., and A. Neuberger. (1989). "Optimal replication of contingent claims under transactions costs," The Review of Futures Markets, 8(2), 222-239.

[22] Hoggard, T., E. Whalley and P. Wilmott. (1994). "Hedging option portfolios in the presence of transaction costs," Advances in Futures and Options Research, 7, 21–35.

[23] Ishii, H., and P.-L. Lions. (1990). "Viscosity solutions of fully nonlinear second-order elliptic partial differential equations," Journal of Differential Equations, 83, 26–78.

[24] Leland, H. E. (1985). "Option pricing and replication with transaction costs," Journal of Finance, 40, 1283–1301.

[25] Levental, S., and A. Skorohod. (1997). "On the possibility of hedging options in the presence of transaction costs," Annals of Applied Probability, 7, 410–443.

[26] Lions, P.-L. (1983). "Optimal control of diffusion processes and Hamilton-Jacobi-Bellman equations 1: The dynamic programming principle and applications; 2: Viscosity solutions and uniqueness," Communications in Partial Differential Equations, 8, 1101–1174; 1229–1276.

[27] Magill, M. J. P., and G. Constantinides. (1976). "Portfolio selection with transaction costs," Journal of Economic Theory, 13, 245–263.

[28] Merton, R. C. (1969). "Lifetime portfolio selection under uncertainty: the continuous-time case," Journal of Economic Theory, 3, 247–257.

[29] Merton, R. C. (1971). "Optimum consumption and portfolio rules in a continuous-time model," Journal of Economic Theory, 3, 373–413.

[30] Merton, R. C. (1973). "Theory of rational option pricing," Bell Journal of Economics and Management Science, 4, 141–183.

[31] Merton, R. C. (1990). *Continuous Time Finance*, Oxford, UK: Basil Blackwell.

[32] Pichler, A. (1996). "On transaction costs and HJB equations," preprint.

[33] Shen, Q. (1990). "Bid-ask prices for call options with transaction costs," University of Pennsylvania Working Paper.

[34] Shreve, S. E., and H. M. Soner. (1994). "Optimal investment and consumption with transaction costs," Annals of Applied Probability, 4(3), 609–692.

[35] Soner, H. M. (1986). "Optimal control with state space constraints," SIAM Journal on Control and Optimization, 24, 552–562, 1110–1122.

[36] Soner, H. M., S. Shreve and J. Cvitanic. (1995). "There is no non-trivial hedging portfolio for option pricing with transaction costs," Annals of Applied Probability, 5(2), 327–355.

[37] Sulem, A. (1997). "Dynamic optimization for a mixed portfolio with transaction costs," *Numerical Methods in Finance*, Newton Institute, Cambridge, UK: Cambridge University Press.

[38] Taksar, M., M. J. Klass and D. Assaf. (1988). "A diffusion model for optimal portfolio selection in the presence of brokerage fees," Mathematics of Operations Research, 13, 277–294.

[39] Toft, K. B. (1996). "On the mean-variance tradeoff in option replication with transactions costs," Journal of Financial and Quantitative Analysis, 31, 233–263.

[40] Tourin, A., and T. Zariphopoulou. (1994). "Numerical schemes for investment models with singular transactions," Computational Economics, 7, 287–307.

[41] Tourin, A., and T. Zariphopoulou. (1995). "Portfolio selection with transactions costs," Progress in Probability, 36, 385–391.

[42] Tourin, A., and T. Zariphopoulou. (1997). "Viscosity solutions and numerical schemes for investment/ consumption models with transaction costs, *Numerical Methods in Finance*, Newton Institute, Cambridge, England: Cambridge University Press, 245–269

[43] Tourin, A., and T. Zariphopoulou. (1998). "Super-replicating strategies with probability less than one in the presence of transaction costs," preprint.

[44] Zariphopoulou, T. (1989). "Investment-consumption models with constraints," Providence, RI: Ph.D. Thesis, Brown University.

[45] Zariphopoulou, T. (1992). "Investment/consumption model with transaction costs and markov-chains parameters," SIAM Journal on Control and Optimization, 30, 613–636.

[46] Zariphopoulou, T. (1994). "Investment and consumption models with constraints," SIAM Journal on Control and Optimization, 32, 59–84.

Applied Optimization

1. D.-Z. Du and D.F. Hsu (eds.): *Combinatorial Network Theory.* 1996
ISBN 0-7923-3777-8
2. M.J. Panik: *Linear Programming: Mathematics, Theory and Algorithms.* 1996
ISBN 0-7923-3782-4
3. R.B. Kearfott and V. Kreinovich (eds.): *Applications of Interval Computations.* 1996 ISBN 0-7923-3847-2
4. N. Hritonenko and Y. Yatsenko: *Modeling and Optimimization of the Lifetime of Technology.* 1996 ISBN 0-7923-4014-0
5. T. Terlaky (ed.): *Interior Point Methods of Mathematical Programming.* 1996
ISBN 0-7923-4201-1
6. B. Jansen: *Interior Point Techniques in Optimization.* Complementarity, Sensitivity and Algorithms. 1997 ISBN 0-7923-4430-8
7. A. Migdalas, P.M. Pardalos and S. Storøy (eds.): *Parallel Computing in Optimization.* 1997 ISBN 0-7923-4583-5
8. F.A. Lootsma: *Fuzzy Logic for Planning and Decision Making.* 1997
ISBN 0-7923-4681-5
9. J.A. dos Santos Gromicho: *Quasiconvex Optimization and Location Theory.* 1998
ISBN 0-7923-4694-7
10. V. Kreinovich, A. Lakeyev, J. Rohn and P. Kahl: *Computational Complexity and Feasibility of Data Processing and Interval Computations.* 1998
ISBN 0-7923-4865-6
11. J. Gil-Aluja: *The Interactive Management of Human Resources in Uncertainty.* 1998
ISBN 0-7923-4886-9
12. C. Zopounidis and A.I. Dimitras: *Multicriteria Decision Aid Methods for the Prediction of Business Failure.* 1998 ISBN 0-7923-4900-8
13. F. Giannessi, S. Komlósi and T. Rapcsák (eds.): *New Trends in Mathematical Programming.* Homage to Steven Vajda. 1998 ISBN 0-7923-5036-7
14. Ya-xiang Yuan (ed.): *Advances in Nonlinear Programming.* Proceedings of the '96 International Conference on Nonlinear Programming. 1998 ISBN 0-7923-5053-7
15. W.W. Hager and P.M. Pardalos: *Optimal Control.* Theory, Algorithms, and Applications. 1998 ISBN 0-7923-5067-7
16. Gang Yu (ed.): *Industrial Applications of Combinatorial Optimization.* 1998
ISBN 0-7923-5073-1
17. D. Braha and O. Maimon (eds.): *A Mathematical Theory of Design: Foundations, Algorithms and Applications.* 1998 ISBN 0-7923-5079-0

Applied Optimization

18. O. Maimon, E. Khmelnitsky and K. Kogan: *Optimal Flow Control in Manufacturing.* Production Planning and Scheduling. 1998 ISBN 0-7923-5106-1
19. C. Zopounidis and P.M. Pardalos (eds.): *Managing in Uncertainty: Theory and Practice.* 1998 ISBN 0-7923-5110-X
20. A.S. Belenky: *Operations Research in Transportation Systems:* Ideas and Schemes of Optimization Methods for Strategic Planning and Operations Management. 1998 ISBN 0-7923-5157-6
21. J. Gil-Aluja: *Investment in Uncertainty.* 1999 ISBN 0-7923-5296-3
22. M. Fukushima and L. Qi (eds.): *Reformulation: Nonsmooth, Piecewise Smooth, Semismooth and Smooting Methods.* 1999 ISBN 0-7923-5320-X
23. M. Patriksson: *Nonlinear Programming and Variational Inequality Problems.* A Unified Approach. 1999 ISBN 0-7923-5455-9
24. R. De Leone, A. Murli, P.M. Pardalos and G. Toraldo (eds.): *High Performance Algorithms and Software in Nonlinear Optimization.* 1999 ISBN 0-7923-5483-4
25. A. Schöbel: *Locating Lines and Hyperplanes.* Theory and Algorithms. 1999 ISBN 0-7923-5559-8
26. R.B. Statnikov: *Multicriteria Design.* Optimization and Identification. 1999 ISBN 0-7923-5560-1
27. V. Tsurkov and A. Mironov: *Minimax under Transportation Constrains.* 1999 ISBN 0-7923-5609-8
28. V.I. Ivanov: *Model Development and Optimization.* 1999 ISBN 0-7923-5610-1
29. F.A. Lootsma: *Multi-Criteria Decision Analysis via Ratio and Difference Judgement.* 1999 ISBN 0-7923-5669-1
30. A. Eberhard, R. Hill, D. Ralph and B.M. Glover (eds.): *Progress in Optimization.* Contributions from Australasia. 1999 ISBN 0-7923-5733-7
31. T. Hürlimann: *Mathematical Modeling and Optimization.* An Essay for the Design of Computer-Based Modeling Tools. 1999 ISBN 0-7923-5927-5
32. J. Gil-Aluja: *Elements for a Theory of Decision in Uncertainty.* 1999 ISBN 0-7923-5987-9
33. H. Frenk, K. Roos, T. Terlaky and S. Zhang (eds.): *High Performance Optimization.* 1999 ISBN 0-7923-6013-3
34. N. Hritonenko and Y. Yatsenko: *Mathematical Modeling in Economics, Ecology and the Environment.* 1999 ISBN 0-7923-6015-X
35. J. Virant: *Design Considerations of Time in Fuzzy Systems.* 2000 ISBN 0-7923-6100-8

Applied Optimization

36. G. Di Pillo and F. Giannessi (eds.): *Nonlinear Optimization and Related Topics*. 2000 ISBN 0-7923-6109-1
37. V. Tsurkov: *Hierarchical Optimization and Mathematical Physics*. 2000 ISBN 0-7923-6175-X
38. C. Zopounidis and M. Doumpos: *Intelligent Decision Aiding Systems Based on Multiple Criteria for Financial Engineering*. 2000 ISBN 0-7923-6273-X
39. X. Yang, A.I. Mees, M. Fisher and L.Jennings (eds.): *Progress in Optimization*. Contributions from Australasia. 2000 ISBN 0-7923-6175-X
40. D. Butnariu and A.N. Iusem: *Totally Convex Functions for Fixed Points Computation and Infinite Dimensional Optimization*. 2000 ISBN 0-7923-6287-X
41. J. Mockus: *A Set of Examples of Global and Discrete Optimization*. Applications of Bayesian Heuristic Approach. 2000 ISBN 0-7923-6359-0
42. H. Neunzert and A.H. Siddiqi: *Topics in Industrial Mathematics*. Case Studies and Related Mathematical Methods. 2000 ISBN 0-7923-6417-1
43. K. Kogan and E. Khmelnitsky: *Scheduling: Control-Based Theory and Polynomial-Time Algorithms*. 2000 ISBN 0-7923-6486-4
44. E. Triantaphyllou: *Multi-Criteria Decision Making Methods*. A Comparative Study. 2000 ISBN 0-7923-6607-7
45. S.H. Zanakis, G. Doukidis and C. Zopounidis (eds.): *Decision Making: Recent Developments and Worldwide Applications*. 2000 ISBN 0-7923-6621-2
46. G.E. Stavroulakis: *Inverse and Crack Identification Problems in Engineering Mechanics*. 2000 ISBN 0-7923-6690-5
47. A. Rubinov and B. Glover (eds.): *Optimization and Related Topics*. 2001 ISBN 0-7923-6732-4
48. M. Pursula and J. Niittymäki (eds.): *Mathematical Methods on Optimization in Transportation Systems*. 2000 ISBN 0-7923-6774-X
49. V. Tsurkov: *Large-scale Optimization – Problems and Methods*. 2001 ISBN 0-7923-6817-7
50. M.C. Ferris, O.L. Mangasarian and J.-S. Pang (eds.): *Complementarity: Applications, Algorithms and Extensions*. 2001 ISBN 0-7923-6816-9

KLUWER ACADEMIC PUBLISHERS – DORDRECHT / BOSTON / LONDON

GPSR Compliance

The European Union's (EU) General Product Safety Regulation (GPSR) is a set of rules that requires consumer products to be safe and our obligations to ensure this.

If you have any concerns about our products, you can contact us on ProductSafety@springernature.com

In case Publisher is established outside the EU, the EU authorized representative is:

Springer Nature Customer Service Center GmbH
Europaplatz 3
69115 Heidelberg, Germany

Batch number: 10755284

Printed by Printforce, the Netherlands